Antimicrobials in Pharmaceutical and Medicinal Research

The need for state-of-the-art antimicrobial agents is greater than ever because of the development of multidrug resistance in communal pathogens, the rapid rise of new infections, and the potential for use of multidrug-resistant agents in biological protection. Although the need for novel antimicrobials is increasing, the development of such agents faces significant obstacles. Pharmaceutical research and development costs are estimated to be $400–$800 million per approved agent. The most important natural antimicrobial compounds derived from various plant sources contain a wide variety of secondary metabolites. With collected contributions from international subject experts, this volume focuses primarily on antimicrobials.

This book deliberates recent developments in microbial science in combating infectious diseases and explores advances in antimicrobial constituents and their applications in the fight against bacteria. In addition, it also provides a variety of photographs, diagrams, and tables to help illustrate the material. The novel strategies to combat antimicrobial resistance are also described, emphasizing collaborative measures of control. We describe the concerted efforts undertaken by global communities to combat antimicrobial resistance in detail. The most efficient strategy could be a behavioral change toward indiscriminate consumption, usage, and prescription of antibiotics.

Students, research scientists, academicians and policy makers can benefit from *Antimicrobials in Pharmaceutical and Medicinal Research* as a resource that addresses biotechnology, applied microbiology, healthcare/pharmaceutical products, medicinal plant products, and all disciplines related to antimicrobial research.

Features of the book:

- Covers development in plant-based antimicrobials for sepsis management and progress;
- Describes modern approaches for phyto-nanoconjugates in combating multidrug resistance in biomedicine;
- Details methods to improve antimicrobial properties to have a longer service life in combating infection;
- Describe bacteriocins and plant metabolites as biotechnological tools in food, pharmaceuticals and therapeutics applications;
- Highlights natural antimicrobial therapeutic peptides;
- Offers current and future applications of emerging antimicrobial technologies.

Current Trends in Antimicrobial Research
Series Editor: Arti Gupta

Antimicrobials in Pharmaceutical and Medicinal Research
Edited by Arti Gupta and Ram Prasad

Antimicrobials in Pharmaceutical and Medicinal Research

Edited by
Arti Gupta and Ram Prasad

CRC Press
Taylor & Francis Group
Boca Raton London New York

CRC Press is an imprint of the
Taylor & Francis Group, an informa business

First edition published 2023
by CRC Press
6000 Broken Sound Parkway NW, Suite 300, Boca Raton, FL 33487-2742

and by CRC Press
4 Park Square, Milton Park, Abingdon, Oxon, OX14 4RN

CRC Press is an imprint of Taylor & Francis Group, LLC

Library of Congress Cataloging-in-Publication Data
Names: Gupta, Arti (Professor of biotechnology), editor. |
Prasad, Ram (Professor of microbiology), editor.
Title: Antimicrobials in pharmaceutical and medicinal research /
edited by Arti Gupta, Ram Prasad.
Description: Boca Raton : CRC Press, 2023. |
Series: Current trends in antimicrobial research |
Includes bibliographical references and index.
Identifiers: LCCN 2022058736 (print) | LCCN 2022058737 (ebook) |
Subjects: LCSH: Anti-infective agents. | Drugs–Research. | Medicine–Research.
Classification: LCC RM267 .A58 2023 (print) | LCC RM267 (ebook) |
DDC 615.7/92–dc23/eng/20230119
LC record available at https://lccn.loc.gov/2022058736
LC ebook record available at https://lccn.loc.gov/2022058737

ISBN: 9781032215532 (hbk)
ISBN: 9781032215549 (pbk)
ISBN: 9781003268932 (ebk)

DOI: 10.1201/9781003268932

Typeset in Times
by Newgen Publishing UK

Contents

Preface

Microbial infections are responsible for millions of deaths each year worldwide. More than 1.2 million people and potentially millions more died in 2019 as a direct result of antibiotic-resistant bacterial infections, according to the most comprehensive estimate to date of the global impact of antimicrobial resistance. The emergence and spread of drug-resistant pathogens that have acquired new resistance mechanisms, leading to antimicrobial resistance, continues to threaten our ability to treat common infections. The occurrence of the evolution of resistance has caused the existing anti-bacterial drugs to become ineffective. In current years, several approaches have been recommended to overcome the resistance of antibiotics, which apparently restores the desirable antibacterial action and that can create opportunities for innovative therapeutic approaches. So, antimicrobial resistance is a global health and development threat. It requires urgent multisectoral action in order to achieve the Sustainable Development Goals.

The first chapter by Perçin et al. reviews antimicrobial cryogels for biomedical use. Chapter 2 by Abd El-Baky and Al Fattah Amara highlights the roles of natural antimicrobial therapeutic peptides, especially on milk lactoferricin and Spirulina platensis peptides. In Chapter 3, Joon et al. describe plant-based antimicrobials for sepsis management and attempted to recapitulate the findings from the *in vitro*, *in vivo*, and clinical studies on the anti-sepsis activities of various plant extracts and bioactive isolated from them. Extensive studies on antimicrobials in pharmaceutical and medical research are discussed by Sharma et al. in Chapter 4 and in Chapter 5, Gupta and Prakash highlight the phyto-nanoconjugates in combating multidrug resistance in biomedicine research. In Chapter 6, Jadhav and Karuppayil emphasize natural plant products, that is, essential oil components and anti-viral properties. In Chapter 7, Joon et al. describe antimicrobials for sepsis management and in Chapters 8 and 9, Praveen et al. highlight bacteriocins and plant metabolites as biotechnological tools in food, pharmaceuticals and therapeutics applications. Finally, in Chapter 10, Palomo-Ligas et al. highlight strategies and applicability of treatment against intestinal parasite infection.

Arti Gupta and Ram Prasad

Acknowledgments

First and foremost, we would like to thank God for his never-ending grace, mercy and provision during what ended up being one of the toughest times of our life.

We are indebted to the many people who helped to bring this book to light. I wish to thank series editors Dr Gustavo Molina and Dr Vijai Kumar Gupta; Dr Marc Gutierrez and Editorial assistants, Dr Jyotsna Jangra and Dr Renu Upadhyaya; and Taylor & Francis/CRC Press for their generous assistance, constant support and patience in initializing the series volume. The editors are very grateful to Vice Chancellor Faizabad University, Ayoodhaya and Mahatma Gandhi Central University, Motihari, for their kind support and constant encouragement. Special thanks are due to my well-wishers, family members and friends.

Dr Arti Gupta gives special thanks to her wonderful husband, Saket Agrawal, and her son Reyansh who encouraged the work, helping to preserve both family and professional life. Dr Ram Prasad gives his special regards to his family for their constant invaluable support and their motivation to publish it. We both also give special thanks to all faculty colleagues and senior faculties of our team.

Shri Avadh Raj Singh Smarak Degree College, Gonda, UP, India — Arti Gupta
Mahatma Gandhi Central University, Motihari, Bihar, India — Ram Prasad

Acknowledgements

Editor Biographies

Arti Gupta obtained her Ph.D. from Mahatma Jyotiba Phule Rohilkhand University, Bareilly, India in 2010 in Animal Science. Dr Gupta's current research interests are Animal Biotechnology, Molecular Plant Biotechnology, Molecular Animal Biotechnology, Bioprocess Technology and Microbiology. Since July 2014 Dr Gupta has served for Sri Avadh Raj Singh Smarak Degree College, Gonda, Uttar Pradesh, India and engaged in some editorial work with Springer Nature. Dr Gupta has 10 years of teaching and research experience in Plant and Animal Biotechnology and Microbial Biotechnology.

She has been awarded the University Topper (Gold Medal), an M.Sc. (Biotech.) from Ch.C.S.University, Meerut, the Young Scientist Award (Gold Medal) awarded by the Zoological Society of India, Lucknow, Best Poster Presenter awarded by the *Asian Journal of Experimental Science*, Jaipur, Best Poster Presenter awarded by the International Consortium of Contemporary Biologists (ICCB) and Madhawai-Shyam Educational Trust, Ranchi, the Fellowship Award granted by the International Consortium of Contemporary Biologists (FICCB) and Madhawi Shayam Educational Trust (FMSET) and the Dr V.P. Agarwal (Gold Medal) awarded by D.A.V. (P.G.) College, Muzaffarnagar. To date she has published a monograph, contributed seven book chapters, published around 50 national and international research papers, edited seven books, and issued 28 abstracts, as well as attended symposia, seminars, conferences, and workshops. Dr Gupta has lifetime membership of the Indian Science Congress Association, the Biotech Research Society of India, the Zoological Society of India and the International Consortium of Contemporary Biologists. She has served as a member on several editorial boards. Previously, she served as Teaching Personnel, G.B. Pantnagar University, Pantnagar, Uttrakhand, India, and Visiting Academic Researcher, BAMRC, SRUC, University of Edinburgh, Scotland.

Ram Prasad, Ph.D. is associated with the Department of Botany, Mahatma Gandhi Central University, Motihari, Bihar, India. His research interests include Applied and Environmental Microbiology, plant-microbe interactions, Sustainable Agriculture and Nanobiotechnology. Dr Prasad has more than 250 publications to his credit, including research papers, review articles and book chapters and seven patents issued or pending. He has also edited or authored several books. He has 12 years of teaching experience and has been awarded the Young Scientist Award and the Prof. J.S. Datta Munshi Gold Medal by the International Society for Ecological Communications; FSAB fellowship by the Society for Applied Biotechnology; the American Cancer Society UICC International Fellowship for Beginning Investigators, USA; Outstanding Scientist Award in the field of Microbiology; BRICPL Science Investigator Award and Research Excellence Award, to list a few. He has served on several editorial boards. Dr Prasad has lifetime membership of the Association of Microbiologists of India, the Biotech Research Society of India, Mycological Society of India, Indian Science Congress Association, Previously, he served as Assistant Professor at Amity University, Uttar Pradesh, India; Visiting Assistant Professor, Whiting School of Engineering, Department of Mechanical Engineering at Johns Hopkins University, Baltimore, United States; and Research Associate Professor at the School of Environmental Science and Engineering, Sun Yat-sen University, Guangzhou, China.

Contributors

Amro Abd Al Fattah Amara: Protein Research Department, Genetic Engineering and Biotechnology Research Institute (GEBRI), City of Scientific Research and Technological Applications (SRTA-City), New Borg El-Arab City, Alexandria, Egypt

Nawal Abd El-Baky: Protein Research Department, Genetic Engineering and Biotechnology Research Institute (GEBRI), City of Scientific Research and Technological Applications (SRTA-City), New Borg El-Arab City, Alexandria, Egypt

M.F. Abu Bakar: Faculty of Applied Sciences & Technology, Universiti Tun Hussein Onn Malaysia, Muar, Johor, Malaysia

Jameel Mohammed Al-Khayri: Department of Agricultural Biotechnology, College of Agriculture and Food Sciences, King Faisal University, Saudi Arabia

Gülşen Bayrak: Hacettepe University, Department of Chemistry, Ankara, Turkey

Rakesh Bhatnagar: Molecular Biology and Genetic Engineering Laboratory, School of Biotechnology, Jawaharlal Nehru University, New Delhi, India, and Amity University of Rajasthan, Jaipur, Rajasthan, India

Lizeth Guadalupe Campos-Muzquiz: Departamento de Investigación en Alimentos, Facultad de Ciencias Químicas, Universidad Autónoma de Coahuila, Blvd. V. Carranza s/n. Col. República Oriente Saltillo, Coahuila, México

Rocio Guadalupe Castillo-Godina: Departamento de Investigación en Alimentos, Facultad de Ciencias Químicas, Universidad Autónoma de Coahuila, Blvd. V. Carranza s/n. Col. República Oriente Saltillo, Coahuila, México

Adil Denizli: Hacettepe University, Department of Chemistry, Ankara, Turkey

Jyothis Devasia: Department of Chemistry, CHRIST (Deemed to be University), Hosur Road, Bangalore, India

Faculty of Engineering and Technology, North-Chiang Mai University, Thailand

Adriana Carolina Flores-Gallegos: Departamento de Investigación en Alimentos, Facultad de Ciencias Químicas, Universidad Autónoma de Coahuila, Blvd. V. Carranza s/n. Col. República Oriente Saltillo, Coahuila, México

Mariana Garza-Ontiveros: Departamento de Investigación en Alimentos, Facultad de Ciencias Químicas, Universidad Autónoma de Coahuila, Blvd. V. Carranza s/n. Col. República Oriente Saltillo, Coahuila, México

Charu Gupta: Amity Institute of Herbal Research & Studies, Amity University-UP, Sector-125, NOIDA (UP, India)

Filiberto Gutiérrez-Gutiérrez: Departamento de Química, Centro Universitario de Ciencias Exactas e Ingenierías, Universidad de Guadalajara, Blvd. Gral. Marcelino García Barragán, Olímpica, Guadalajara, Jalisco, México

BaraiyaDivyeksha Harishchandra: Institute for Stem Cell Science and Regenerative Medicine (inStem), GKVK Campus, Bangalore, India

Raúl Rodríguez Herrera: Departamento de Investigación en Alimentos, Facultad de Ciencias Químicas, Universidad Autónoma de Coahuila, Blvd. V. Carranza s/n. Col. República Oriente Saltillo, Coahuila, México

Ashwini Khanderao Jadhav: Department of Stem Cell and Regenerative Medicine and Medical Biotechnology, Centre for Interdisciplinary Research, DY Patil Education Society, Deemed University, (NAAC Accredited with 'A' grade), Kolhapur, Maharashtra, India

Arpita Jayan: Department of Life Sciences, CHRIST (Deemed to be University), Hosur Road, Bangalore, India

Shikha Joon: Institutes for Systems Genetics, Frontiers Science Center for Disease-related Molecular Network, West China Hospital, Sichuan University, Chengdu, Sichuan, China

Sankunny Mohan Karuppayil: Department of Stem Cell and Regenerative Medicine and Medical Biotechnology, Centre for Interdisciplinary Research, DY Patil Education Society, Deemed University, (NAAC Accredited with 'A' grade), Kolhapur, Maharashtra, India

Ayşe Feyza Keleş: Hacettepe University, Department of Biology, Ankara, Turkey

Worapong Kitdamrongtham: Pharmaceutical Chemistry & Division, Division of Cosmetic Technology,

Himani Meena: Department of Microbiology, School of Life Sciences, Pondicherry University, Puducherry, India

Mohit Mishra: Molecular Biology and Genetic Engineering Laboratory, School of Biotechnology, Jawaharlal Nehru University, New Delhi, India

Epsita Mukherjee: Amity Institute of Biotechnology, Amity University, Noida, India

Sendar Daniel Nery-Flores: Departamento de Investigación en Alimentos, Facultad de Ciencias Químicas, Universidad Autónoma de Coahuila, Blvd. V. Carranza s/n. Col. República Oriente Saltillo, Coahuila, México

Suresh Nithya: Department of Life Sciences, CHRIST (Deemed to be University), Hosur Road, Bangalore, India

Aatika Nizam: Department of Chemistry, CHRIST (Deemed to be University), Hosur Road, Bangalore, India

Vishakha Pagare: Department of Life Sciences, CHRIST (Deemed to be University), Hosur Road, Bangalore, India

Lissethe Palomo-Ligas: Departamento de Investigación en Alimentos, Facultad de Ciencias Químicas, Universidad Autónoma de Coahuila, Blvd. V. Carranza s/n. Col. República Oriente Saltillo, Coahuila, México

Lipsa Panda: Department of Biotechnology, School of Life Sciences, Pondicherry University, Puducherry, India

Ketan Parmar: Department of chemistry, Sir P. T. Sarvajanik College of Science, Veer Narmad South Gujarat University, Surat, India

Işık Perçin: Hacettepe University, Department of Biology, Ankara, Turkey

Dhan Prakash: Amity Institute of Herbal Research & Studies, Amity University-UP, Sector-125, NOIDA (UP, India)

Nagella Praveen: Department of Life Sciences, CHRIST (Deemed to be University), Hosur Road, Bangalore, India

Ravi Shankar Thenmozhi Preetha: Department of Life Sciences, CHRIST (Deemed to be University), Hosur Road, Bangalore, India

Ajit Ramesh Sawant: Department of Biotechnology, School of Life Sciences, Pondicherry University, Puducherry, India

Rahul Sharma: Department of Biotechnology, School of Life Sciences, Pondicherry University, Puducherry, India

Bairong Shen: Institutes for Systems Genetics, Frontiers Science Center for Disease-related Molecular Network, West China Hospital, Sichuan University, Chengdu, Sichuan, China

Udisha Singh: Department of Life Sciences, CHRIST (Deemed to be University), Hosur Road, Bangalore, India

Rajeev K. Singla: Institutes for Systems Genetics, Frontiers Science Center for Disease-related Molecular Network, West China Hospital, Sichuan University, Chengdu, Sichuan, China

Enty Tjoa: Department of Microbiology, School of Medicine and Sciences, Atma Jaya Catholic University of Indonesia, Jakarta, Indonesia

Akash Tripathi Satsangi: Department of Biotechnology, School of Engineering and Technology, Sharda University, Knowledge Park III, Greater Noida, Uttar Pradesh, India

José Roberto Vargas-Villanueva: Departamento de Investigación en Alimentos, Facultad de Ciencias Químicas, Universidad Autónoma de Coahuila, Blvd. V. Carranza s/n. Col. República Oriente Saltillo, Coahuila, México

Veerappa Lakshmaiah Vasantha: Department of Life Sciences, CHRIST (Deemed to be University), Hosur Road, Bangalore, India

1 Antimicrobial Cryogels for Biomedical Use

Işık Perçin,[1] Gülşen Bayrak,[1] Ayşe Feyza Keleş,[2] and Adil Denizli[2]*
[1]Hacettepe University, Department of Biology, Ankara, Turkey
[2]Hacettepe University, Department of Chemistry, Ankara, Turkey
*Correspondng author: denizli@hacettepe.edu.tr

CONTENTS

1.1 INTRODUCTION

Hydrogels are polymeric crosslinked three-dimensional network structures. Thanks to its three-dimensional networks, hydrogels are able to absorb a large amount of water. Hydrogels can be synthesized from natural or synthetic polymers or both together. Hydrogels have attracted great interest from scientists over the past few decades due to their outstanding success and potential in a wide variety of applications. The ability of hydrogels to absorb water is due to hydrophilic functional groups in their structure and cross-links between network chains. Also, the cross linked structure provides hydrogels' resistance to dissolution (Bashir et al. 2020). Besides natural hydrogels, interest in synthetic hydrogels with high water absorption capacity, long service life and high gel strength has increased in recent years (Madduma-Bandarage and Madihally 2021). Despite its popularity, some limitations of hydrogels, such as pore size, elasticity and mechanical strength, limited their use in some application areas.

Cryogels are structures containing various polymeric materials that have applications in multiple areas such as biology, biochemistry, medicine, food technology, environment, cell and tissue engineering (Razavi et al. 2019; Türkmen et al. 2022). Cryogels consist of polymeric structures with three-dimensional interconnected macropores and are synthesized by free radical polymerization and freeze drawing techniques under zero conditions (Hezarkhani et al. 2021). Many advantages of cryogels come from their three-dimensional supermacroporous interconnected structure, their flexibility and their chemical and physical stability (Razavi et al. 2019; Zhao et al. 2019). Similar to hydrogels, cryogels can be prepared with either natural or synthetic polymers, or both. Natural and synthetic polymers used for cryogel preparation and their application areas are given in Table 1.1. Cryogel synthesis basically depends on three stages: preparation of monomer/polymer solutions, polymerization and thawing. Gel precursors are crosslinked either physically, chemically or both

DOI: 10.1201/9781003268932-1

TABLE 1.1
Application Areas of Natural and Synthetic Polymers Used for Cryogel Preparations

POLYMER TYPES	APPLICATIONS	REFERENCES
	NATURAL POLYMERS	
Gelatin	Tissue engineering scaffold (cartilage, skin, skeletal, bone, cardiovascular), Bioreactor, Cell delivery, Drug delivery	(Rezaeeyazdi et al. 2018; Goodarzi et al. 2020; Y. He, Wang, Wang, Xiao and Lin, 2021)
Chitosan	Tissue engineering scaffold (cartilage, skin, bone, neural), Enzyme immobilization, Wound dressing, Drug delivery, Contact lens material, Affinity purification	(Kangkamano et al. 2017)
Alginate	Tissue engineering scaffold, Drug delivery, Affinity purification, Cell delivery	(Rodríguez-Dorado et al. 2019; Sahoo and Biswal 2021)
Cellulose	Tissue engineering scaffold (cartilage, skin, bone), Drug delivery, wound dressing	(Tyshkunova et al. 2022)
Silk Fibroin	Tissue engineering scaffold (bone, skin and cartilage)	(Zhang et al. 2022)
Collagen	Tissue engineering scaffold (cartilage, neural, bone), Wound dressing	(Y. He et al. 2021)
Hyaluronic Acid	Tissue engineering scaffold (cartilage, neural, bone, cardiovascular, skin)	(T. He et al. 2021)
Agarose	Tissue engineering scaffold (neural, cartilage), Protein purification, Microbial cell capture, Drug delivery	(Hixon et al. 2017; Razavi et al. 2019)
	SYNTHETIC POLYMERS	
PHEMA (2-hyroxyethyl methacrylate)	Tissue engineering scaffold (cartilage, bone), Protein purification, Protein depletion	(Şarkaya et al. 2022)
PVA (Polyvinyl alcohol)	Tissue engineering scaffold (cartilage, skin, cardiovascular), Enzyme and cell immobilization, Drug delivery, Bioreactor, Cell separation	(Berillo et al. 2019;Savina et al. 2021)
PVP (Polyvinylpyrrolidone)	Wound dressing	(Priya et al. 2016)
PEG (Polyethylene glycol)	Tissue engineering scaffold (cartilage, skin, bone, neural), Drug delivery systems	(Bruns et al. 2018; Pacelli et al. 2021)
PLG (Polylactic acid)	Tissue engineering scaffold (bone), Drug delivery systems	(Shirbin et al. 2016)

ways at sub-zero temperatures during polymerization (Shakya 2022). Most of the solvent starts to crystallize while monomer or polymer precursors initiate gelation surrounding ice crystals during polymerization. Thus, ice crystals act as porogens (Kumari, Karande and Kumar 2016; Bereli et al. 2020). Cryogel properties can be modified by changing synthesis conditions such as type of monomers, reaction time, temperature, and ratio of precursors (Milakin et al. 2018). The freezing rate has a significant effect on the physical characteristics of cryogels. Slower freezing speed results in larger pores and increased connectivity between pores. In contrast, faster freezing speed results in reduced interconnectivity and mechanically weaker cryogels (Hixon, Lu, and Sell 2017; Boonkanon et al. 2021). Another advantage of cryogels is that they can be prepared in various shapes and forms such as disc, rod, bead, column and powder, and in various sizes. Cryogels can also be synthesized as composites with various materials such as activated carbons, zeolites and nanoparticles in order to achieve many desired properties for different applications (Dinu et al. 2016; Busquets et al. 2016) Şarkaya, Bakhshpour and Denizli 2019). Additionally general properties of cryogels are summarized in Figure 1.1.

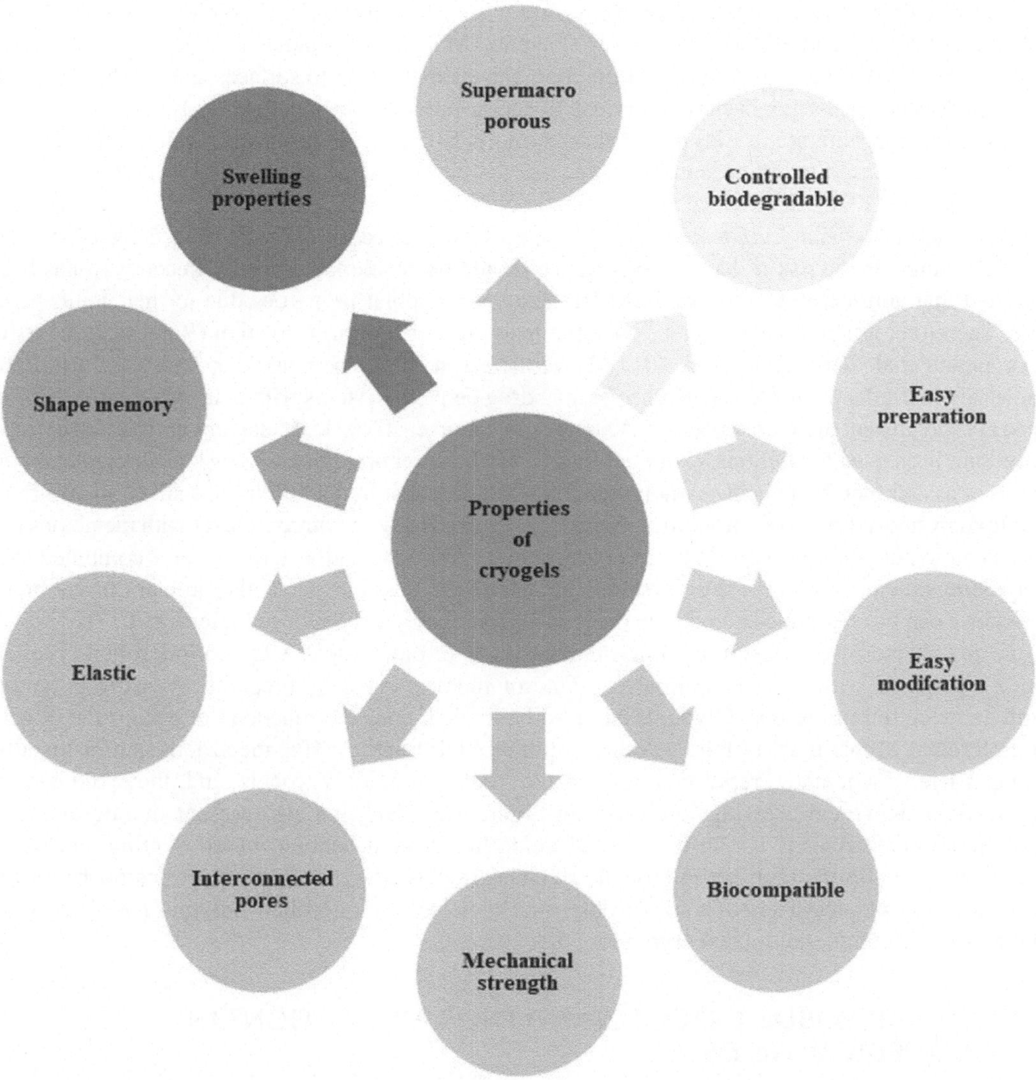

FIGURE 1.1 General properties of supermacroporous cryogels.

Human populations are exposed to microbial infections that endangered human life for centuries (Jamal et al. 2019). The discovery of penicillin is a breakthrough for antibiotic development in humanity's war against pathogens (Bhattacharjee et al. 2021). However, bacterial infections continue to be one of the serious complications that endangered human health, and some serious cases can result in death (Odonkor and Addo 2018). Therefore, intense efforts have been ongoing for decades to find new broad-spectrum antibiotics that do not pose any risk to the host cell, do not cause bacterial resistance, have low toxicity and are effective against various bacterial pathogens. Recently, healthcare-related infections (HAI) have caused serious problems for clinicians with the development of biological materials and medical equipment. For example, annual costs associated with HAIs in the USA are estimated for $33 billion. Medical devices can bring HAIs to hospital patients. Transplanted medical devices or biomaterials such as wound dressings, joint implants, pacemaker markers, catheters and contact lenses can cause infection (Yang et al. 2018). In addition, increased antibiotic resistance and an increasing number of drug-allergic patients complicate the selection of correct and effective antibacterial agents (Mazer-Amirshahi et al. 2017).

Biofilm can be defined as a colony of microorganisms that attach to surfaces and exhibit antibiotic resistance mechanisms for bacterial survival. Bacteria produces extracellular polymeric substances (EPS) and they proliferate in EPS matrix. Furthermore, EPS provide resistance against antibiotics by inhibiting the transportation of antibiotics through the biofilm (Rosli et al. 2021; Prasad et al. 2020; Sonawane et al. 2022). Traditional antimicrobial agents face increasing problems of drug resistance, overdose and cytotoxicity. As a solution to these problems, effective and safe drug delivery systems that can minimize the risk of bacterial resistance should be developed urgently. Recently, it has been reported that antimicrobial peptides (AMPs) have antimicrobial properties due to their ionic structure. Therefore, it is difficult to induce bacterial resistance or biofilm formation (Ramesh et al. 2016; Scorciapino et al. 2017; Kang et al. 2017). Nevertheless, it should be noted that AMPs are hemolytic and easily lose effectiveness. Therefore, effective drug delivery systems with delayed release performance need to prevent these drawbacks of AMPs (Boonkanon et al. 2021). Metal ions and natural extracts have long been used for fighting with microbes. It has been reported that silver, gold, copper and zinc are used as antimicrobial agents. Among these metals, the most frequently used is silver, thanks to its low toxicity and antibacterial properties (Yang et al. 2018). However, direct contact with metal ions can cause cytotoxicity for host cells in high concentrations. When the antibacterial agent is combined with polymeric structures, a lower dose systemic application is possible. Thus, the problem of antibiotic resistance can be overcome and other undesirable side effects can be reduced (Hemeg 2017).

In this chapter, we focus on antimicrobial cryogels in biomedical field. Antimicrobial cryogels can be prepared with various approaches. One of them is cryogels prepared by using polymers with antibacterial properties. Cryogels can also be loaded, coated, embedded or encapsulated with antimicrobial agents like antibiotics, other antimicrobial drugs, AMPs, metal ions and/or metallic nanoparticles. Schematic representation of the preparation of cryogels and their biomedical applications as wound dressings are given in Figure 1.2. Here, we also present literature related to antimicrobial cryogels use in biomedical fields like wound healing, blood clotting and tissue engineering. In this chapter, antimicrobial cryogels are classified into two categories for better the understanding of properties of antimicrobial mechanisms: antimicrobial polymer-based cryogels and antibacterial material loaded cryogels.

1.2 ANTIMICROBIAL CRYOGELS AND THEIR APPLICATIONS IN BIOMEDICAL FIELDS

Microbial infections are one of the biggest challenging medical concerns for human health around the world (González-Henríquez et al. 2017). For this reason, the importance of antimicrobial drugs, agents and materials is increasing, and the number of studies on the discovery of antimicrobial polymers is also increasing (Zare et al. 2020). Today, antibiotics are the most widely used drugs on living things. However, frequent use of antibiotics causes big problems like antibiotic resistance on bacteria (Bhattacharjee et al. 2021). Researchers investigate to find new antibiotics with low toxicity to host cells, a broad spectrum against bacterial pathogens and no bacterial resistance (González-Henríquez et al., 2017; Bhattacharjee et al. 2021). Hydrogels are preferred in many commercial applications with their healing-promoting, pain-reducing effects and superior properties that support new tissue formation. Biocompatible hydrogels and cryogels with chemically or physically incorporated, loaded, encapsulated antimicrobial agents in their polymeric structure are advantageous. Cryogels have interconnected macropores, high water holding capacity and are biocompatible, syringe injectable and mechanically stable (Abudula et al. 2021). Thus, cryogel systems are good candidates to enhance antimicrobial efficiency, decrease cytotoxicity, enable controlled release of antimicrobial agents and prevent loaded molecules from degradation (Bhattacharjee et al. 2021).

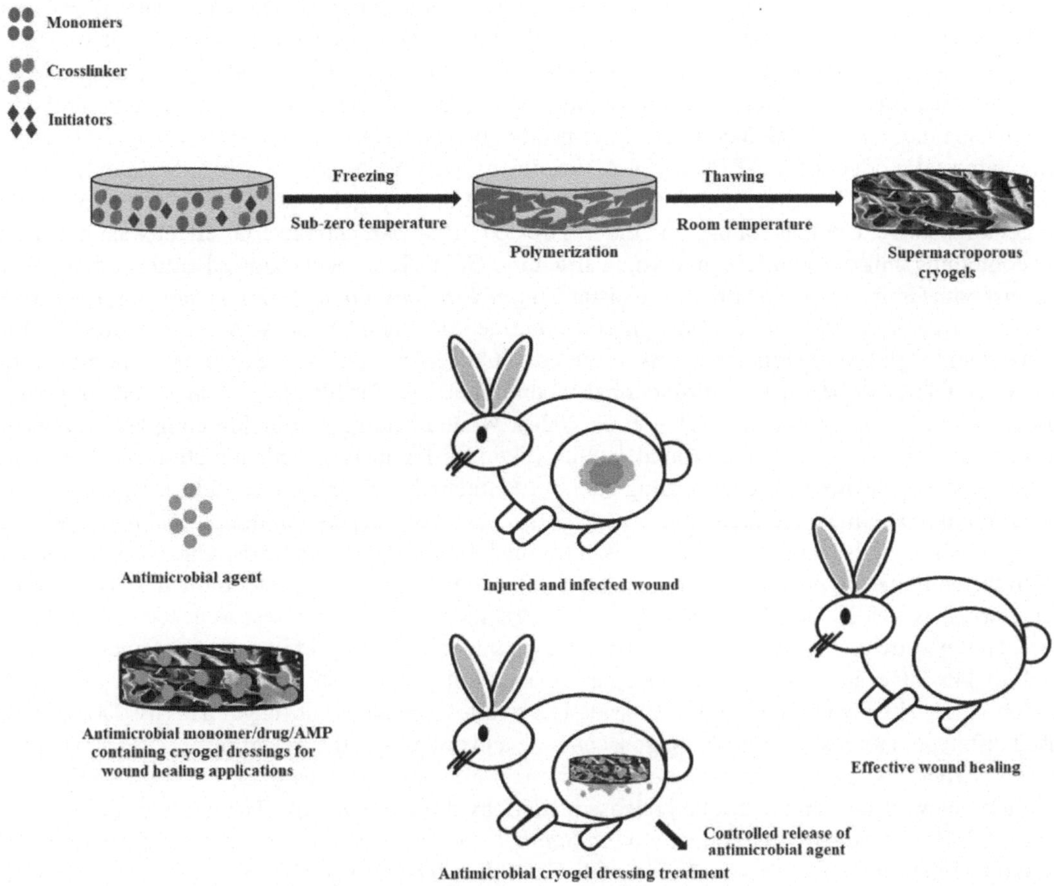

FIGURE 1.2 Schematic representation of preparation of antimicrobial supermacroporous cryogels and their biomedical application as a wound dressing.

1.2.1 Antimicrobial Polymer-based Cryogels

In this part of chapter, we will discuss cryogels containing polymers with antimicrobial properties and we will outline some literature studies. Antimicrobial polymers show their antimicrobial effect due to their positive charged functional groups. Specific interactions occur between positive charge of polymer and negative charge of cell surface of microorganisms. Thus, the cell membrane of microorganisms is disrupted and protein and nucleic acid synthesis is inhibited (Grigoras 2021).

Chitosan is a natural, biocompatible, biodegradable and antimicrobial material which shows potential for various biomedical applications (Xiao 2021; Grigoras 2021). Chitosan shows its antibacterial properties by disrupting the barrier properties of Gram-negative bacteria, thanks to the free amino groups on its surface (Thakar et al. 2019). Recently numerous wound dressings have been developed to protect injured wounds from infection, stop bleeding and to increase healing (Darpentigny et al. 2020). There is a clinical need for wound dressings that treat bacterial infections while also stopping bleeding and healing wounds. Multidrugresistant (MDR) bacterial infection has another major challenge for clinical wound care, because less effective antimicrobial agents are available for killing these bacteria. To meet the clinical need, Hou et al. developed antibacterial polysaccharide-peptide cryogel (GC-EPL) wound dressings with hemostatic property. GC-EPL cryogels were fabricated by copolymerization of glycol chitosan methacrylate (GC-MA) and different ratios of ε-polylysine acrylamide (EPL-A) concentration. Human dermal fibroblasts (HDF) and human bone marrow stromal cells (hBMSC) were used for cytotoxicity of GC-EPL cryogels and they showed good biocompatibility. GC-EPL cryogels showed high clotting rate as compared with controls which are Combat Gauze and gelatin sponge. Also, the hemostatic property of chitosan increased by the incorporation of positively charged EPL in GC-EPL cryogel structure. The hemostatic effect of GC-EPL cryogels was investigated with a rat liver-bleeding model for *in vivo* studies. Blood loss was significantly decreased in the cryogel treated group when compared to gelatin sponge, Combat Gauze, and the not treated group. An antibacterial effect was increased by combining chitosan and EPL in cryogel structure. GC-EPL cryogels showed antibacterial effect against both Gram positive methicillin-resistant *Staphylococcus aureus* (MRSA) and Gram-negative *Escherichia coli*. A full thickness MRS infected murine skin wound model was treated with GC-EPL cryogel and high healing efficiency was observed. GC-EPL cryogels with excellent properties were good candidates to become multifunctional wound dressings for bleeding control and supplying bacteria-infected wound healing (Hou et al. 2020). Wound healing is a highly complex biological mechanism. Initiation of acute wound healing involves hemostasis, inflammation, proliferation and remodeling. However, acute healing cannot be initiated in diabetic wound healing because of occurred hypoxia in wound area (Bai et al. 2020). Chen et al. prepared antimicrobial hydrogel and cryogel biomaterials by using glycol chitosan for the treatment of unhealed diabetic chronic wound. With the combined application of adipose stem cell-seeded cryogel/hydrogel biomaterials and acupuncture, they treated the diabetic skin wound in a rat model. As a result, researchers concluded that the combined treatment produced a synergistic immunomodulatory effect on the diabetic wound and 90.34% of wound closure was observed (Chen et al. 2021). Chitosan oligosaccharide (COS) which is a positively charged polysaccharide has a time-dependent antifungal activity. Dong et al. used different amounts of Chitosan oligosaccharide and poly(N-methacryl arginine) (PMarg) to prepare macroporous zwitterionic composite antifungal cryogel dressings (COS/PMarg). COS/Marg cryogels showed slow and controlled release of COS by diffusion *in vitro*. The antibacterial activity of COS/PMarg cryogels was evaluated with *Staphylococcus aureus* and *E. coli* and the antifungal activity of these cryogels was evaluated with *Candida albicans*. The COS/PMarg cryogel showed no inhibition zone for *S. aureus* and *E. coli* but showed for *C. albicans*. The COS/PMarg composite cryogel showed 76.4% antifungal activity for *C. albicans*. Macroporous zwitterionic COS/PMarg composite cryogel might be a potential antifungal dressing against body surface *C. albicans* infections (Dong et al. 2021). Some modifications of polymers, like increasing quaternary ammonium moieties, bring more effective antimicrobial properties (Grigoras 2021). Quaternized chitosan (QCS) is more water soluble and has greater antibacterial properties when compared with chitosan. Biocompatible QCS has excellent hemostatic properties. Zhao et al. developed antibacterial, injectable, and conductive QCSG/CNT cryogels. Cryogels were prepared based on carbon nanotube (CNT) and glycidyl methacrylate functionalized QCS for wound healing and noncompressible hemorrhage. QCSG/CNT cryogels showed excellent blood uptake capacity and blood-clotting ability. QCSG/CNT cryogel showed better hemostatic capability in *in vivo* studies (mouse liver injury model and mouse-tail amputation model) as compared to gauze and gelatin hemostatic sponge (Zhao et al. 2018).

Lignin is one of the most abundant natural polymers and has antimicrobial properties. Lignin is used as an antimicrobial material for variable biomedical applications (Zhong et al. 2020). Abudula et al. developed injectable, antibacterial, biocompatible, and antioxidant lignin-co-gelatin cryogels for biomedical applications. Lignin-co-gelatin cryogels were prepared by researchers with different lignin concentrations for improving the mechanical properties of cryogels. Lignin-co-gelatin cryogel with 0.2% lignin showed the highest compressive modulus. Also, all prepared lignin-co-gelatin cryogels showed over 95% shape recovery. Especially, lignin-co-gelatin cryogel with 0.2% lignin showed over 99% shape memory after 90% compression and injected with negligible shape loss in injectability tests. Researchers observed that concentration of polymer strongly affected cryogel injectability. Higher or lower polymer concentrations are able to cause a fracture in cryogel structure during the injection. Antibacterial properties of lignin-co-gelatin cryogels were evaluated with agar disc diffusion assay by using *S. aureus* and *E. coli*. Lignin containing cryogels showed an inhibition zone against both bacteria stains. Cytocompatibility studies of lignin-co-gelatin cryogels were evaluated with an alamar blue test by using 3T3 fibroblast cells. They found that lignin-co-gelatin cryogels have cytocompatibility and promote cell-matrix interactions. Biomaterials risk inducing immature dendritic cell maturation and activation which can cause inflammation in transplanted areas. However, lignin-co-gelatin cryogels minimally activated immature mouse bone marrow-derived dendritic cells. In conclusion, the designed bioresource-based multifunctional lignin-co-gelatin cryogel showed excellent promise to be used in multiple biomedical applications such as tissue engineering and wound healing (Abudula et al. 2021).

Polypyrrole is an antibacterial, electrically conductive and biocompatible polymer (Sanchez Ramirez et al. 2019). Milakin et al. developed electrically conductive polypyrrole-gelatin cryogels for potential biomedical applications. Researchers synthesized cryogels by oxidative cryopolymerization of pyrrole in the presence 2% to 8% concentraton of gelatin. Cytotoxicity of cryogels were tested with MTT assay by using mouse embryonic fibroblast cell line (NIH/3T3). Extraction of cryogels was prepared and extracts were diluted to 75, 50, 25, 10, and 1% concentrations for cytotoxicity experiments. Polypyrrole-gelatin cryogels prepared using 8% gelatin showed the lowest cytotoxic for all extracts except from the 100% extract concentration. The 100% extract concentration showed mild cytotoxicity. The antibacterial activity of polypyrrole-gelatin cryogels was studied using the colony counting method. All prepared cryogels showed antibacterial activity but cryogels with 6% and 7% gelatin showed weak antibacterial activity. Cryogels containing 8% gelatin showed the highest antibacterial effect. Polypyrrole-gelatin cryogels showed antibacterial properties without adding any other material (Milakin et al. 2020). More literature studies for antimicrobial polymer-based cryogels are given in Table 1.2.

1.2.2 Antimicrobial Material Loaded Cryogels

1.2.2.1 Metal Ion and Metallic Nanoparticle Loaded Cryogels

Metallic ion or metallic nanoparticle containing antimicrobial cryogels receive great attention in the biomedical field (Abdelgawad et al. 2020). Silver, copper and zinc are antibacterial metal ions and they are widely used to prevent bacterial growth on surfaces. Compared with antibiotics, metal ions and metallic nanoparticles have great advantages like having antibacterial properties against multi drug resistant microorganisms (Bai et al. 2020; Zhong et al., 2020). Metal ions and metallic nanoparticles electrostatically interact with the negatively charged cell surface of microorganisms. They show antibacterial effect by damaging cell membrane, binding to cytosolic proteins, binding DNA and causing reactive oxygen species (ROS) formation. ROS cause oxidative stress and damage of macromolecules including nucleic acids. This versatile oxidative damage can cause bacterial cell lysis (Gold et al. 2018). Nanoparticles can be loaded into cryogels in order to overcome aggregation tendency of nanoparticles. Nanoparticles spread inside large pores of cryogel structure. So, the antibacterial effect of nanoparticles and superior properties of cryogels such as mechanical strength and flexibility are combined in one polymeric structure. Thus, metallic nanoparticle loaded cryogels have effective antibacterial properties (Chaturvedi et al. 2016; Haleem et al. 2021). Recently, antimicrobial nanoparticles gained attention for treating infections. Nanoparticles are materials having size ranges from 1 to 100 nm size. Antimicrobial properties of NPs are originated from size, charge and surface morphology (Rosli et al. 2021). Silver, gold and copper nanoparticles are mostly used nanoparticles for antibacterial wound healing applications. Silver nanoparticles show anti-inflammatory and wound healing activities and have come to the forefront due to their antibacterial activity against MDR microorganisms. Silver shows its antimicrobial mechanism by changing the cell wall of microorganisms and blocking respiratory enzyme pathways (Bai et al. 2020). Silver

TABLE 1.2
Antimicrobial Polymer-based Cryogels and Their Biomedical Applications

Cryogel	Antimicrobial monomer/ polymer	Antimicrobial activity	*In vitro /In vivo* studies	Properties of cryogel	Reference
Poly(vinyl alcohol)–Gantrez® AN cryogel	Gantrez®	*S. aureus*	HDF cells and Porcine skin model	Antibacterial Wound dressing Biocompatible	(Caló et al. 2016)
QCSG/CNT Cryogel	Quaternized chitosan (QCS)	*E. coli S. aureus* and *P. aeruginosa*	L929 cells, New Zealand White Rabbit, Kunming Mice and Sprague Dawley Rats	Antibacterial Biocompatible Wound healing Hemostatic properties Injectable	(Zhao et al. 2018)
GC-EPL Cryogel	ε-polylysine (EPL)	*E. coli* and *S. aureus*	HDF cells, hBMSC, Sprague Dawley Rats and BALB/C Mice	Antibacterial Wound dressing Biocompatible Hemostatic property	(Hou et al. 2020)
Polypyrrole-Gelatin Cryogel	Polypyrrole	*E. coli* and *S. aureus*	NIH/3T3 cells	Antimicrobial Biocompatible Electrically conductive	(Milakin et al. 2020)
COS/PMarg Cryogels	Chitosan oligosaccharide (COS)	*C. albicans*	-	Antifungal Wound dressing	(Dong et al. 2021)
Lignin-*co*-Gelatin Cryogel	Lignin	*E. coli* and *S. aureus*	NIH/3T3 cells, BMDCs	Antibacterial Biocompatible İnjectable Antioxidant	(Abudula et al. 2021)

nanoparticles have antimicrobial activity against many pathogens but they have low toxicity against mammalian cells. Silver, which was used in the treatment of many diseases in the past, is now used to increase the antimicrobial properties of many biomedical materials such as wound dressings, bone cements and dental materials (Demir et al. 2020). Abdelgawad et al. prepared novel cryogels with high potential for use as an antibacterial wound dressing. Cryogels were prepared with carrageenan (CAR), cellulose nanocrystals (CNC) and different concentrations of silver nanoparticles. CNC were used for providing mechanical strength. Controlled release of silver nanoparticles from cryogels was exhibited. Silver nanoparticle loaded cryogels showed 100% antimicrobial activity for *E. coli* and *S. aureus* (Abdelgawad et al. 2020).

Tissues can be damaged and lose their function for various reasons and diseases. Tissue engineering in an interdisciplinary field aims to repair damaged tissue or to develop biocompatible artificial organs/tissues. Tissue scaffolds as three-dimensional microenvironments are used for therapeutic repair of injured tissues. These polymeric materials are intended to mimic the extracellular matrix (ECM) and allow interaction between cells. Scaffolds provide cells to survive, proliferate and differentiate (Bayrak et al. 2021). Infections can be observed and biofilm occurred between scaffold and host tissue after tissue scaffold transplantations. Biofilm resistance to the host immune system and the spreading of infections to the skeletal system through the blood stream are the main problems. The addition of antimicrobial agents into tissue scaffolds can prevent infections caused from transplantation. Thus, effective healing, regeneration and repair can be obtained. Demir et al. immobilized AgNPs into injectable chitosan cryogel microspheres for developed antimicrobial tissue scaffolds. Prepared all AgNPs-coated chitosan microspheres showed antimicrobial activity for *S. aureus* (ATCC 6538), *E. coli* (ATCC 10536), *Legionella pneumophila subsp. pneumophila* (ATCC 33152), *Enterococcus hirae* (ATCC 10541), *Pseudomonas aeruginosa* (ATCC 9027) *Bacillus cereus* and *C. albicans* (Demir et al. 2020). In another study, Zou et al. (2017) developed cryogel/silver nanocomposites from methacrylated carboxymethyl chitosan (mCMC) and poly(ethylene glycol) diacrylate (PEGDA). Silver nanoparticles were incorporated to cryogel structure by in situ chemical reduction. The obtained cryogel/AgNPs composite showed prominent and durable antibacterial activity against Gram-negative *E. coli* (*E. coli*, ATCC 25922) even at first 3h and could be utilized as potential antibacterial materials. Zou et al. (2017) also proved cryogel was synthesized through the cryogelation of methacrylated carboxymethyl chitosan (mCMC) and poly(ethylene glycol) diacrylate (PEGDA) precursors by photopolymerization and it is a further level for the potential use of silver and chitosan as antimicrobial materials in biomedical fields. Chopra et al. prepared a self-crosslinked polymer through the in situ reduction of graphene oxide (GO) in a gelatin cryogel (Gel-RGO). Different amounts of silver nanoparticle decorated nanohydroxyapatite (Ag@Hap) were incorporated into GelRGOAg@Hap cryogels. Continuous release of silver nanoparticles from GelRGOAg@Hap was observed over 14 days. GelRGOAg@Hap cryogels showed prolonged antimicrobial activity against *S. aureus*. By supporting biomineralization and differentiation of bone cells, biocompatible and hemocompatible GelRGOAg@Hap cryogels showed osteoconductive and osteoinductive potential. The results of study suggest that antibacterial GelRGOAg@Hap cryogels have a great potential for bone tissue engineering applications (Chopra et al. 2021).

Suner and Sahiner (2018) prepared humic acid particle embedded gum Arabic (GA/HA) cryogels for potential biomedical applications. Silver (Ag), copper (Cu) and iron (Fe) nanoparticles containing GA/HA-M cryogels were also prepared. Antimicrobial effectiveness of GA/HA and GA/HA-M cryogels were studied with *S. aureus, E. coli*, and *B. subtilis*. GA cryogel, GA/HA cryogel and GA/HA-Fe cryogels did not exhibit antibacterial effect. In contrast, GA/HA-Ag cryogel and GA/HA-Cu cryogel had an antibacterial effect against *S. aureus, E. coli*, and *B. subtilis*. Blood compatibility and blood-clotting tests were performed and GA/HA-M cryogels were found hemocompatible. Results of the study showed that silver, copper and iron containing nanoparticles combined with GA/HA cryogels are potential composites for biological applications (Suner and Sahiner 2018). Literature about metal ion and metallic nanoparticle containing antimicrobial cryogels are presented in Table 1.3.

1.2.2.2 Drug, AMPs and Antimicrobial Agent Loaded Cryogels

Drug loading efficiency is important for effective wound healing. Effective healing of the infected wound is provided with increased antimicrobial agent loading capacity. Drug loaded hydrogel system with controlled release is advantageous for wound healing applications. Drugs must be delivered even into deep layers of wounded skin by using drug loaded wound dressings. To increase the antimicrobial effectiveness and healing activities in the wound area, smart formulations that provide effective release of the antimicrobial agent need to be developed. The high water content of cryogels helps antimicrobial drugs maintain their activity and protects them from degradation. In addition, drugs are protected within the cryogel network, minimizing their enzymatic breakdown and providing continuous drug release (Huang et al., 2020; Vladimir I. Lozinsky et al. 2020; Bai et al., 2020). Another important parameter for wounds is hemorrhage control. Hemorrhage control is an important health issue around the world that uncontrolled hemorrhage causes death. The traumatized area must be applied with hemostatic agents for rapid control of hemorrhage. The ideal hemostatic agent must be biocompatible, antimicrobial economic, mechanically stable, easy to use and able to quickly control hemorrhage (Zhao et al. 2018). Huang et al. prepared antibacterial, biocompatible, biodegradable, and injectable interpenetrating polymer network (IPN) dry cryogel hemostats using gelatin and dopamine (GT/DA) in order to promote wound healing and stop deep noncompressible hemorrhage. The cytocompatibility and hemocompatibility properties of GT/DA

TABLE 1.3
Metal Ion or Metallic Nanoparticle Containing Antimicrobial Cryogels and Their Biomedical Applications

Polymer	Metal ion/ Metallic NPs	Antibacterial activity	*In vitro /In vivo* studies	Properties	Reference
PVA Cryogel-Zinc Oxide Nanocomposite	ZnO nanopaticles	*S. aureus, V. cholera, E. coli P. aeruginosa* and *Bacilli*	L929	Antibacterial Biocompatible Wound dressing Hemostatic property	(Chaturvedi et al. 2016)
Cryogel/AgNPs	AgNPs	*E. coli*	-	Antibacterial	(Zou et al. 2017)
GA/HA-M Cryogel Composite	Humic acid NPs AgNPs, CuNPs, And FeNPs	*E. coli, S. aureus* and *B. subtilis*	-	Antibacterial Hemostatic property	(Suner and Sahiner 2018)
PVA-CG/BC/ Hexahistidine-Tagged Organophosphorus Hydrolase (His6 - OPH) Composite	Meropene, Temporin and Indolicidin	*Pseudomonas sp*	-	Antibacterial	(Aslanli et al. 2019)
CSTFe Cryogels	Tannic acid/ ferric ion (TA/Fe^{3+})	*E. coli* and *S. aureus*	3T3 cells and SD male rats	Antibacterial Wound dressing Biocompatible Hemostatic property Photothermal activity	(Yu et al. 2019)
CAR/CNC AgNPs	AgNPs	*E. coli* and *S. aureus*	-	Antimicrobial Wound dressing	(Abdelgawad et al. 2020)
AgNPs-Coated Chitosan Cryogel Microspheres	AgNPs	*E. coli, S. aureus, E. hirae, B. cereus, P. aeruginosa, L. pneumophila subsp. pneumophila,* and *C. albicans*	-	Antibacterial Antifungal Tissue engineering scaffolds DNA cleavage activity	(Demir et al. 2020)
GelRGOAg@Hap Composite Cryogel	AgNPs	*S. aureus*	L929 cells and MG-63 cells	Antimicrobial Biocompatible Hemostatic property Controlled release Osteoconductive and Osteoinductive Bone scaffold	(Chopra et al. 2021)

cryogels were investigated and they showed good cytocompatibility as compared to tissue culture plates. All cryogels in this study showed less than 5% hemolysis. Furthermore, researchers found that GT/DA cryogels showed better blood-clotting capacity as compared to control groups which used gauze and gelatin hemostatic sponges. The effect of GT/DA cryogels was investigated for preventing blood loss and providing hemostasis in rat liver incision model, in the mouse liver trauma model and in the rabbit liver cross-incision model. GT/DA cryogels exhibited less blood loss and shorter hemostasis time as compared to gauze and gelatin hemostatic sponges. Antimicrobial activity of GT/DA cryogels was examined with *S. aureus* and *E. coli*, GT/DA cryogels showed 100% antibacterial properties against *S. aureus* and *E. coli*. Dopamine-free but gelatin-containing cryogels did not have antibacterial properties, so antibacterial mechanism of cryogels originated from dopamine. GT/DA cryogels had a great advantage as an antibacterial wound dressing with wound healing effect and rapid hemostasis (Huang et al. 2020). Yılmaz et al. prepared CS-g-PNIPAAm/PVA cryogels for smart delivery of antifungal drug voriconazole. Researchers used chitosan (CS)-graft-poly(N-isopropyl acrylamide) (PNIPAAm) and polyvinyl alcohol (PVA) as polymers. Voriconazole release from CS-g-PNIPAAm/PVA cryogels was decreased with the increase of PVA amount in the cryogel structure and voriconazole was released from this cryogel up to 8 h. Cytotoxicity studies of CS-g-PNIPAAm/PVA cryogels was performed with a human kidney proximal tubular epithelial cell line (HK-2) and NIH-3T3. Voriconazole-loaded CS-g-PNIPAAm/PVA cryogels significantly increased proliferation of HK-2 and NIH/3T3 cells. Unless antifungal experiments were missing for Voriconazole-loaded CS-g-PNIPAAm/PVA cryogels, by performing these experiments, biocompatible CS-g-PNIPAAm/PVA cryogels have a great potential for topical applications (Cheaburu-yilmaz et al. 2019).

Rosselle et al. fabricated reduced graphene oxide (rGO) incorporated near infrared (NIR) light activable cryogels. They prepared cryogels by using butyl methacrylate (BuMA) and poly(ethylene glycol) methyl ether methacrylate (PEGMEMA) for wound healing applications. rGO cryogels were prepared and antibiotic cefepime was loaded to cryogels. Cefepime release from rGO cryogels were evaluated by HPLC. Cytotoxicity experiments were performed with HeLa cells up to 48h and rGO cryogels were found to be biocompatible with no cell viability loss. Researchers also studied with *S. aureus* infected *ex vivo* human skin samples and inflammatory cytokines of wounded skin, infected skin, cefepime treated skin, photothermal activated rGO cryogel treated skins, photothermal activated cefepime loaded cryogel treated and nonactivated cefepime loaded cryogel treated skins. They found that activated cefepime loaded rGO cryogel treated infected wounds showed a decrease in proinflammatory responses. Photothermal activated cefepime loaded rGO cryogels have a potential to be used as a bandage for wound healing (Rosselle et al. 2020). Lozinsky et al. prepared serum-albumin-based cryogels by chemical modification (succinylation) of the cryogenic process. They investigated how chemical modifications affect antibacterial properties, biodegradability, swelling properties, drug (dioxidine) loading and drug release efficiency of cryogels. Succinylation of cryogels provided an increase in swelling degree, increase in resistance to enzymatic degradation, increase in loading capacity of antibacterial drug dioxidine and increase in antimicrobial activity. Both non-succinylated and succinylated serum-albumin-based cryogel sponges were found capable of biomedical applications like biodegradable drug-carriers (Lozinsky et al. 2020).

The need for potential antimicrobial materials has increased with the emergence of antibiotic drug resistance problem. Recently, AMPs found great attention in biomedical areas like wound healing applications because of the antimicrobial properties to drug resistant bacteria. AMPs are natural peptides produced by organisms and they help wound healing by improving immunity (Lin et al. 2019). AMPs are first-line defense peptides against pathogens and they are naturally produced from organisms. Positively charged AMPs interact with negatively charged bacteria cell membrane (Baixe et al. 2021). Instead of direct application of AMPs to the wound area, loading or encapsulating of AMPs on a moist polymeric material such as hydrogels are beneficial for providing continuous release and protecting AMPs from degradation (Atefyekta et al. 2021). Formulations of AMPs should be well designed to increase the stability, to extend the release time and to optimize the effect in the wound area. Chambre et al. developed AMP loaded cryogels to obtain better mechanical properties. Maleimide modified AMPs were covalently conjugated onto furan-based cryogels. Furfuryl methacrylate (FuMA) and poly(ethylene glycol) methyl ether methacrylate (PEGMEMA) were used as monomers. The furan group offers a handle for specific loading of the peptides, thus minimizing passive and burst drug release. *E. coli* (DSM 30083) and methicillin-resistant *S. aureus* (ATCC 43300) were used for antimicrobial studies and AMP loaded cryogels showed antibacterial activity against *S. aureus*, but not against *E. coli*. Researchers generated wounds on ready-to-use *ex vivo* human skin samples and *S. aureus* was used to infect the wounds. Heat-activated and

TABLE 1.4
Drug, Antimicrobial Agent, or AMP Containing Cryogels and Their Biomedical Applications

Polymer	Drug	Antimicrobial activity	*In vitro /In vivo* studies	Properties	Reference
CS-g-PNIPAAm/PVA cryogel	Voriconazole	-	HK-2 cells and NIH-3T3 cells	Antifungal and Drug carrier for topical applications	(Cheaburu-yilmaz et al. 2019)
rGO-CG	Cefepime	*S. aureus*	HeLa cells and *Ex vivo* human skin samples	Antibacterial Wound dressing Biocompatible	(Rosselle et al. 2020)
Serum-Albumin-Based Spongy Cryogel	Dioxidine	*E. coli* and *S. aureus*	-	Antimicrobial Biodegradable Drug carrier	(Lozinsky et al. 2020)
GT/DA (Gelatin-Dopamin) Cryogel	Dopamine doxycycline and vancomycin	*E. coli* and *S. aureus*	L929 cells, Kunming mice, SD rats and New Zealand white rabbit	Antibacterial Antioxidant Biocompatible Wound healing Hemostatic property Biodegradable	(Huang et al. 2020)
Furanyl-Based CG	RWRWRWC–NH2 (AMP)	*E. coli* and *S. aureus*	HeLa cells *ex vivo* skin	Antibacterial Biocompatible Wound bandage	(Chambre et al. 2020)
Hypericum perforatum Incorporated Chitosan Cryogel	*Hypericum perforatum*	*S. aureus, E. coli, E. hirae, P. aeruginosa B. cereus, L. pneumophila subsp. pneumophila,* and *C. albicans*	-	Antibacterial Antifungal Wound dressings Antioxidant	(Bölgen et al. 2020)
Methacrylated Hyaluronic Acid (HAGM) Cryogels	Calcium peroxide (CP)	*S. aureus* and *P. aeruginosa*	NIH/3T3 cells and BMDCs	Antimicrobial Biocompatible Tissue engineering scaffold Injectable	(Joshi Navare et al. 2020)
P(HEMA-MAH) Lyz-MIP	Lysosome	*S. aureus* and *E. coli*	L929 cells	Antibacterial Biocompatible	(Diken Gür et al. 2021)
Gelatin microcryogels	Lysozyme	*S. aureus, E. coli* and *B. subtilis*	L929	Antibacterial Biocompatible	(Çetin et al. 2021)

peptide-loaded cryogels were applied on infected wounds and they showed dry aspects in *ex vivo* skin samples after three days (Chambre et al. 2020).

Another antimicrobial agent, lysozyme, is an antimicrobial monomeric protein used in pharmacology and other applications. However, application of free lysozyme is limited due to its instability and easy inactivation properties. Loaded lysozyme in hydrogel-based system makes the use of lysozyme more advantageous, providing protection and increasing antimicrobial effectiveness (Wu et al. 2018). Diken et al. prepared novel antimicrobial, biocompatible and cost effective cryogel membranes to be used in biomedical applications such as tissue scaffolds and wound dressing.

Nmethacryloyl-(L)-histidine methyl ester (MAH) was used as the pseudo specific ligand and complexed with copper to provide metal ion coordination between MAH and template molecule which is lysozyme. Thus, lysozyme imprinted P(HEMA-MAH) cryogel membranes were obtained. The antibacterial activity of lysozyme imprinted P(HEMA-MAH) Lyz-MIP cryogel membranes was examined with *S. aureus* and *E. coli.* 100% bacterial inhibition was detected against *S. aureus* and *E. coli* by increasing lysozyme concentration in the structure of P(HEMA-MAH) Lyz-MIP cryogel (Diken Gür et al. 2021). Other studies in the literature about drug, antimicrobial agent or AMP containing cryogels are presented in Table 1.4.

1.3 CONCLUSIONS AND FUTURE PROSPECTS

Infectious diseases are an important issue for biomedical fields. The rapid emergence of antibiotic resistance is becoming a serious health problem worldwide. In addition, infections from medical devices are increasing noticeably. For solving these problems, antimicrobial polymeric materials like antimicrobial cryogels are great candidates for long-term antimicrobial properties. In addition, the prepared antimicrobial cryogels must be highly biocompatible since they will be used in biomedical fields. However, the fact that the polymers with antimicrobial properties are not high in number limits their application areas and limits the studies in literature. Antimicrobial agents like AMPs, metal ions and metallic nanoparticles are also great candidates for replacement with the antibiotic treatment. Nevertheless, disadvantages like rapid loss of effectiveness and host toxicity with higher doses are the main drawbacks for applications of these antimicrobials. In this case, the need and concern for systems like cryogels that will maintain the effectiveness of the antimicrobial agent and will not show a toxic effect by ensuring the release with appropriate doses is increasing for biomedical applications. The interconnected three-dimensional pore structure of antimicrobial cryogels will provide this need. Enhanced biocompatibility controlled and continuous release, possibility of stimulus-induced release, increased mechanical strength, and flexibility are unique advantages of antimicrobial cryogels. With the systematic release property for these antimicrobial agents, antimicrobial cryogels provide effective treatment of infections. In this chapter, antimicrobial cryogels and their biomedical applications like tissue engineering, drug delivery and wound healing are discussed and recent studies were summarized. Thanks to many features such as being biocompatible, biodegradable, prepared with antimicrobial polymers or containing antimicrobial agents and releasing these agents in a controlled manner, antimicrobial cryogels are a promising and emerging field for clinical use in biomedical fields. Nowadays, *in vitro* and *in vivo* studies on antimicrobial cryogels have been increasing in the biomedical field. For the potential future use of antimicrobial cryogels especially for clinical trials, prepared antimicrobial cryogels must be developed and it is important to test antimicrobial cryogels against particularly drug resistant strains and evaluate first *in vitro* and then *in vivo* the biocompatibility of antimicrobial cryogels. For further studies, in order for cryogels to show maximum antibacterial activity, cryogels can be prepared from antimicrobial polymers containing antimicrobial agents. In addition, it is expected that with the different dose trials, the toxicity of antimicrobial cryogels should be minimized for the host cells.

REFERENCES

Abdelgawad AM, El-Naggar ME, Elsherbiny DA, Ali S, Abdel-Aziz MS, Abd-Elmoneam YK. 2020. Antibacterial carrageenan/cellulose nanocrystal system loaded with silver nanoparticles, prepared via solid-state technique. J Environ Chem Eng [Internet]. 8(5):104276. https://doi.org/10.1016/j.jece.2020.104276

Abudula T, Colombani T, Alade T, Bencherif SA, Memić A. 2021. Injectable lignin-co-gelatin cryogels with antioxidant and antibacterial properties for biomedical applications. Biomacromolecules. 22(10):4110–4121. https://doi.org/10.1021/acs.biomac.1c00575

Aslanli A, Stepanov N, Razheva T, Podorozhko EA, Lyagin I, Lozinsky VI, Efremenko E. 2019. Enzymatically functionalized composite materials based on nanocellulose and poly(Vinyl Alcohol) cryogel and possessing antimicrobial activity. Materials (Basel). 12(21). https://doi.org/10.3390/ma12213619

Atefyekta S, Blomstrand E, Rajasekharan AK, Svensson S, Trobos M, Hong J, Webster TJ, Thomsen P, Andersson M. 2021. Antimicrobial Peptide-Functionalized Mesoporous Hydrogels. ACS Biomater Sci Eng. 7(4):1693–1702. https://doi.org/10.1021/acsbiomaterials.1c00029

Bai Q, Han K, Dong K, Zheng C, Zhang Y, Long Q, Lu T. 2020. Potential applications of nanomaterials and technology for diabetic wound healing. Int J Nanomedicine. 15:9717–9743. https://doi.org/10.2147/IJN.S276001

Baixe S, Ball V, Jierry L, Cianf S, Strub J, Haikel Y, Etienne O. 2021. applied sciences Strongly Adhesive and Antimicrobial Peptide-Loaded, Alginate – Catechol-Based Gels for Application against Periimplantitis.:1–15.

Bashir S, Hina M, Iqbal J, Rajpar AH, Mujtaba MA, Alghamdi NA, Wageh S, Ramesh K, Ramesh S. 2020. Fundamental concepts of hydrogels: Synthesis, properties, and their applications. Polymers (Basel). 12(11):1–60. https://doi.org/10.3390/polym12112702

Bayrak G, Perçin I, Kılıç Süloğlu A, Denizli A. 2021. Amino acid functionalized macroporous gelatin cryogels: Characterization and effects on cell proliferation. Process Biochem. 110(August):100–109. https://doi.org/10.1016/j.procbio.2021.08.002

Bereli N, Yavuz H, Denizli A. 2020. Protein chromatography by molecular imprinted cryogels. J Liq Chromatogr Relat Technol [Internet]. 43(15–16):657–670. https://doi.org/10.1080/10826076.2020.1780606

Berillo DA, Caplin JL, Cundy AB, Savina IN. 2019. A cryogel-based bioreactor for water treatment applications. Water Res [Internet]. 153:324–334. https://doi.org/10.1016/j.watres.2019.01.028

Bhattacharjee B, Ghosh S, Patra D, Haldar J. 2021. Advancements in release-active antimicrobial biomaterials: A journey from release to relief. Wiley Interdiscip Rev Nanomedicine Nanobiotechnology. (July):1–38. https://doi.org/10.1002/wnan.1745

Bölgen N, Demir D, Yalçın MS, Özdemir S. 2020. Development of Hypericum perforatum oil incorporated antimicrobial and antioxidant chitosan cryogel as a wound dressing material. Int J Biol Macromol. 161:1581–1590. https://doi.org/10.1016/j.ijbiomac.2020.08.056

Boonkanon C, Phatthanawiwat K, Chuenchom L, Lamthornkit N, Taweekarn T, Wongniramaikul W, Choodum A. 2021. Preparation and characterization of calcium cross-linked starch monolithic cryogels and their application as cost-effective green filters. Polymers (Basel). 13(22). https://doi.org/10.3390/polym13223975

Bruns J, McBride-Gagyi S, Zustiak SP. 2018. Injectable and cell-adhesive polyethylene glycol cryogel scaffolds: independent control of cryogel microstructure and composition. Macromol Mater Eng. https://doi.org/10.1002/mame.201800298

Busquets R, Ivanov AE, Mbundi L, Hörberg S, Kozynchenko OP, Cragg PJ, Savina IN, Whitby RLD, Mikhalovsky S V., Tennison SR et al. 2016. Carbon-cryogel hierarchical composites as effective and scalable filters for removal of trace organic pollutants from water. J Environ Manage. 182:141–148. https://doi.org/10.1016/j.jenvman.2016.07.061

Caló E, Barros J, Ballamy L, Khutoryanskiy V V. 2016. Poly(vinyl alcohol)-Gantrez® AN cryogels for wound care applications. RSC Adv. 6(107):105487–105494. https://doi.org/10.1039/c6ra24573k

Çetin K, Aslıyüce S, Idil N, Denizli A. 2021. Preparation of lysozyme loaded gelatin microcryogels and investigation of their antibacterial properties. J Biomater Sci Polym Ed [Internet]. 32(2):189–204. https://doi.org/10.1080/09205063.2020.1825303

Chambre L, Rosselle L, Barras A, Aydin D, Loczechin A, Gunbay S, Sanyal R, Skandrani N, Metzler-Nolte N, Bandow JE, et al. 2020. Photothermally Active Cryogel Devices for Effective Release of Antimicrobial Peptides: On-Demand Treatment of Infections. ACS Appl Mater Interfaces. 12(51):56805–56814. https://doi.org/10.1021/acsami.0c17633

Chaturvedi A, Bajpai AK, Bajpai J, Singh SK. 2016. Evaluation of poly (vinyl alcohol) based cryogel-zinc oxide nanocomposites for possible applications as wound dressing materials. Mater Sci Eng C [Internet]. 65:408–418. https://doi.org/10.1016/j.msec.2016.04.054

Cheaburu-Yilmaz CN, Yilmaz O, Kose FA, Bibire N. 2019. Chitosan-Graft-Poly(N-Isopropylacrylamide)/PVA cryogels as carriers for mucosal delivery of voriconazole. Polymers. 11(9):1432. https://doi.org/10.3390/polym11091432

Chen TY, Wen TK, Dai NT, Hsu SH. 2021. Cryogel/hydrogel biomaterials and acupuncture combined to promote diabetic skin wound healing through immunomodulation. Biomaterials [Internet]. 269(161):120608. https://doi.org/10.1016/j.biomaterials.2020.120608

Chopra V, Thomas J, Sharma A, Panwar V, Kaushik S, Ghosh D. 2021. A bioinspired, ice-templated multifunctional 3D cryogel composite crosslinked through in situ reduction of GO displayed improved mechanical, osteogenic and antimicrobial properties. Mater Sci Eng C [Internet]. 119(September 2020):111584. https://doi.org/10.1016/j.msec.2020.111584

Darpentigny C, Marcoux PR, Menneteau M, Michel B, Ricoul F, Jean B, Bras J, Nonglaton G. 2020. Antimicrobial Cellulose Nanofibril Porous Materials Obtained by Supercritical Impregnation of Thymol. ACS Appl Bio Mater. 3(5):2965–2975. https://doi.org/10.1021/acsabm.0c00033

Demir D, Özdemir S, Yalçın MS, Bölgen N. 2020. Chitosan cryogel microspheres decorated with silver nanoparticles as injectable and antimicrobial scaffolds. Int J Polym Mater Polym Biomater [Internet]. 69(14):919–927. https://doi.org/10.1080/00914037.2019.1631823

Diken Gür S, Bakhshpour M, Bereli N, Denizli A. 2021. Antibacterial effect against both Gram-positive and Gram-negative bacteria via lysozyme imprinted cryogel membranes. J Biomater Sci Polym Ed [Internet]. 32(8):1024–1039. https://doi.org/10.1080/09205063.2021.1892472

Dinu MV, Cocarta AI, Dragan ES. 2016. Synthesis, characterization and drug release properties of 3D chitosan/clinoptilolite biocomposite cryogels. Carbohydr Polym [Internet]. 153:203–211. https://doi.org/10.1016/j.carbpol.2016.07.111

Dong P, Shu X, Peng R, Lu S, Xie X, Shi Q. 2021. Macroporous zwitterionic composite cryogel based on chitosan oligosaccharide for antifungal application. Mater Sci Eng C [Internet]. 128(July):112327. https://doi.org/10.1016/j.msec.2021.112327

Gold K, Slay B, Knackstedt M, Gaharwar AK. 2018. Antimicrobial activity of metal and metal-oxide based nanoparticles. Adv Ther. 1(3):1–15. https://doi.org/10.1002/adtp.201700033

González-Henríquez CM, Sarabia-Vallejos MA, Rodriguez-Hernandez J. 2017. Advances in the fabrication of antimicrobial hydrogels for biomedical applications. Materials (Basel). 10(3):1–23. https://doi.org/10.3390/ma10030232

Goodarzi K, Jonidi Shariatzadeh F, Solouk A, Akbari S, Mirzadeh H. 2020. Injectable drug loaded gelatin based scaffolds as minimally invasive approach for drug delivery system: CNC/PAMAM nanoparticles. Eur Polym J [Internet]. 139(September):109992. https://doi.org/10.1016/j.eurpolymj.2020.109992

Grigoras AG. 2021. Natural and synthetic polymeric antimicrobials with quaternary ammonium moieties: a review. Environ Chem Lett [Internet]. 19(4):3009–3022. https://doi.org/10.1007/s10311-021-01215-w

Haleem A, Chen S-Q, Ullah M, Siddiq M, He W-D. 2021. Highly porous cryogels loaded with bimetallic nanoparticles as an efficient antimicrobial agent and catalyst for rapid reduction of water-soluble organic contaminants. J Environ Chem Eng [Internet]: 106510. https://doi.org/10.1016/j.jece.2021.106510

He T, Li B, Colombani T, Joshi-Navare K, Mehta S, Kisiday J, Bencherif SA, Bajpayee AG. 2021. Hyaluronic Acid-Based Shape-Memory Cryogel Scaffolds for Focal Cartilage Defect Repair. Tissue Eng - Part A. 27(11–12):748–760. https://doi.org/10.1089/ten.tea.2020.0264

He Y, Wang Chunhua, Wang Chenzhi, Xiao Y, Lin W. 2021. An overview on collagen and gelatin-based cryogels: Fabrication, classification, properties and biomedical applications. Polymers (Basel). 13(14). https://doi.org/10.3390/polym13142299

Hemeg HA. 2017. Nanomaterials for alternative antibacterial therapy. Int J Nanomedicine. 12:8211–8225. https://doi.org/10.2147/IJN.S132163

Hezarkhani M, Ustürk S, Özbilenler C, Yilmaz E. 2021. Pullulan/poly(N-vinylimidazole) cryogel: An efficient adsorbent for methyl orange. J Appl Polym Sci. 138(37). https://doi.org/10.1002/app.50958

Hixon KR, Lu T, Sell SA. 2017. A comprehensive review of cryogels and their roles in tissue engineering applications. Acta Biomater [Internet]. 62(August):29–41. https://doi.org/10.1016/j.actbio.2017.08.033

Hou S, Liu Y, Feng F, Zhou J, Feng X, Fan Y. 2020. Polysaccharide-Peptide Cryogels for Multidrug-Resistant-Bacteria Infected Wound Healing and Hemostasis. Adv Healthc Mater. 9(3):1–7. https://doi.org/10.1002/adhm.201901041

Huang Y, Zhao X, Zhang Z, Liang Y, Yin Z, Chen B, Bai L, Han Y, Guo B. 2020. Degradable gelatin-based ipn cryogel hemostat for rapidly stopping deep noncompressible hemorrhage and simultaneously improving wound healing. Chem Mater. 32(15):6595–6610. https://doi.org/10.1021/acs.chemmater.0c02030

Jamal M, Bukhari SMAUS, Andleeb S, Ali M, Raza S, Nawaz MA, Hussain T, Rahman S u., Shah SSA. 2019. Bacteriophages: An overview of the control strategies against multiple bacterial infections in different fields. J Basic Microbiol. 59(2):123–133. https://doi.org/10.1002/jobm.201800412

Joshi Navare K, Colombani T, Rezaeeyazdi M, Bassous N, Rana D, Webster T, Memic A, Bencherif SA. 2020. Needle-injectable microcomposite cryogel scaffolds with antimicrobial properties. Sci Rep [Internet]. 10(1):1–16. https://doi.org/10.1038/s41598-020-75196-1

Kang HK, Kim C, Seo CH, Park Y. 2017. The therapeutic applications of antimicrobial peptides (AMPs): a patent review. J Microbiol. 55(1):1–12. https://doi.org/10.1007/s12275-017-6452-1

Kangkamano T, Numnuam A, Limbut W, Kanatharana P, Thavarungkul P. 2017. Chitosan cryogel with embedded gold nanoparticles decorated multiwalled carbon nanotubes modified electrode for highly sensitive flow based non-enzymatic glucose sensor. Sensors Actuators, B Chem [Internet]. 246:854–863. https://doi.org/10.1016/j.snb.2017.02.105

Kumari J, Karande AA, Kumar A. 2016. Combined effect of cryogel matrix and temperature-reversible soluble-insoluble polymer for the development of *in vitro* human liver tissue. ACS Appl Mater Interfaces. 8(1):264–277. https://doi.org/10.1021/acsami.5b08607

Lin Z, Wu T, Wang W, Li B, Wang M, Chen L, Xia H, Zhang T. 2019. Biofunctions of antimicrobial peptide-conjugated alginate/hyaluronic acid/collagen wound dressings promote wound healing of a mixed-bacteria-infected wound. Int J Biol Macromol [Internet]. 140:330–342. https://doi.org/10.1016/j.ijbiomac.2019.08.087

Lozinsky VI, Shchekoltsova AO, Sinitskaya ES, Vernaya OI, Nuzhdina A V., Bakeeva I V., Ezernitskaya MG, Semenov AM, Shabatina TI, Melnikov MY. 2020. Influence of succinylation of a wide-pore albumin cryogels on their properties, structure, biodegradability, and release dynamics of dioxidine loaded in such spongy carriers. Int J Biol Macromol [Internet]. 160:583–592. https://doi.org/10.1016/j.ijbiomac.2020.05.251

Madduma-Bandarage USK, Madihally S V. 2021. Synthetic hydrogels: Synthesis, novel trends, and applications. J Appl Polym Sci. 138(19):1–23. https://doi.org/10.1002/app.50376

Mazer-Amirshahi M, Pourmand A, May L. 2017. Newly approved antibiotics and antibiotics reserved for resistant infections: Implications for emergency medicine. Am J Emerg Med [Internet]. 35(1):154–158. https://doi.org/10.1016/j.ajem.2016.10.034

Milakin KA, Acharya U, Hodan J, Trchová M, Stejskal J, Bober P. 2018. Effect of 1,3-phenylenediamine concentration on the properties of poly(aniline-co-1,3-phenylenediamine) cryogels. Mater Lett [Internet]. 229:68–70. https://doi.org/10.1016/j.matlet.2018.06.080

Milakin KA, Capáková Z, Acharya U, Vajďák J, Morávková Z, Hodan J, Humpolíček P, Bober P. 2020. Biocompatible and antibacterial gelatin-based polypyrrole cryogels. Polymer (Guildf). 197(February):2–8. https://doi.org/10.1016/j.polymer.2020.122491

Odonkor ST, Addo KK. 2018. Prevalence of multidrug-resistant escherichia coli isolated from drinking water sources. Int J Microbiol. 2018. https://doi.org/10.1155/2018/7204013

Pacelli S, Di Muzio L, Paolicelli P, Fortunati V, Petralito S, Trilli J, Casadei MA. 2021. Dextran-polyethylene glycol cryogels as spongy scaffolds for drug delivery. Int J Biol Macromol [Internet]. 166:1292–1300. https://doi.org/10.1016/j.ijbiomac.2020.10.273

Prasad R, Siddhardha B, Dyavaiah M (2020) Nanostructures for Antimicrobial and Antibiofilm Applications. Springer International Publishing (ISBN 978-3-030-40336-2) https://www.springer.com/gp/book/9783030403362

Priya SG, Gupta A, Jain E, Sarkar J, Damania A, Jagdale PR, Chaudhari BP, Gupta KC, Kumar A. 2016. Bilayer cryogel wound dressing and skin regeneration grafts for the treatment of acute skin wounds. ACS Appl Mater Interfaces. 8(24):15145–15159. https://doi.org/10.1021/acsami.6b04711

Ramesh S, Govender T, Kruger HG, de la Torre BG, Albericio F. 2016. Short AntiMicrobial Peptides (SAMPs) as a class of extraordinary promising therapeutic agents. J Pept Sci. (February):438–451. https://doi.org/10.1002/psc.2894

Razavi M, Qiao Y, Thakor AS. 2019. Three-dimensional cryogels for biomedical applications. J Biomed Mater Res - Part A. 107(12):2736–2755. https://doi.org/10.1002/jbm.a.36777

Rezaeeyazdi M, Colombani T, Memic A, Bencherif SA. 2018. Injectable hyaluronic acid-co-gelatin cryogels for tissue-engineering applications. Materials (Basel). https://doi.org/10.3390/ma11081374

Rodríguez-Dorado R, López-Iglesias C, García-González CA, Auriemma G, Aquino RP, Del Gaudio P. 2019. Design of aerogels, cryogels and xerogels of alginate: Effect of molecular weight, gelation conditions and drying method on particles' micromeritics. Molecules. 24(6):4–6. https://doi.org/10.3390/molecules24061049

Rosli NA, Teow YH, Mahmoudi E. 2021. Current approaches for the exploration of antimicrobial activities of nanoparticles. Sci Technol Adv Mater [Internet]. 22(1):885–907. https://doi.org/10.1080/14686996.2021.1978801

Rosselle L, Cantelmo AR, Barras A, Skandrani N, Pastore M, Aydin D, Chambre L, Sanyal R, Sanyal A, Boukherroub R, Szunerits S. 2020. An "on-demand" photothermal antibiotic release cryogel patch: Evaluation of efficacy on an: Ex vivo model for skin wound infection. Biomater Sci. 8(21): 5911–5919. https://doi.org/10.1039/d0bm01535k

Sahoo DR, Biswal T. 2021. Alginate and its application to tissue engineering. SN Appl Sci [Internet]. 3(1): 1–19. https://doi.org/10.1007/s42452-020-04096-w

Sanchez Ramirez DO, Varesano A, Carletto RA, Vineis C, Perelshtein I, Natan M, Perkas N, Banin E, Gedanken A. 2019. Antibacterial properties of polypyrrole-treated fabrics by ultrasound deposition. Mater Sci Eng C [Internet]. 102(October 2018):164–170. https://doi.org/10.1016/j.msec.2019.04.016

Şarkaya K, Bakhshpour M, Denizli A. 2019. Ag+ ions imprinted cryogels for selective removal of silver ions from aqueous solutions. Sep Sci Technol [Internet]. 54(18):2993–3004. https://doi.org/10.1080/01496395.2018.1556300

Şarkaya K, Akıncıoğlu G, Akıncıoğlu S. 2022. Investigation of tribological properties of HEMA-based cryogels as potential articular cartilage biomaterials. Polym Technol Mater [Internet]. 00(00):1–17. https://doi.org/10.1080/25740881.2022.2039190

Savina IN, Zoughaib M, Yergeshov AA. 2021. Design and assessment of biodegradable macroporous cryogels as advanced tissue engineering and drug carrying materials. Gels. 7(3). https://doi.org/10.3390/gels7030079

Scorciapino MA, Serra I, Manzo G, Rinaldi AC. 2017. Antimicrobial dendrimeric peptides: Structure, activity and new therapeutic applications. Int J Mol Sci. 18(3). https://doi.org/10.3390/ijms18030542

Shakya AK. 2022. Functionalized cryogel monoliths for fast and selective separation of nucleic acids directly from crude lysate. Biomed Chromatogr. 36(4). https://doi.org/10.1002/bmc.5333

Shirbin SJ, Karimi F, Chan NJA, Heath DE, Qiao GG. 2016. Macroporous hydrogels composed entirely of synthetic polypeptides: Biocompatible and enzyme biodegradable 3D cellular scaffolds. Biomacromolecules. 17(9):2981–2991. https://doi.org/10.1021/acs.biomac.6b00817

Sonawane JM, Rai AK, Sharma M, Tripathi M, Prasad R (2022) Microbial biofilms: Recent advances and progress in environmental bioremediation. Science of the Total Environment. 824(153843). https://doi.org/10.1016/j.scitotenv.2022.153843

Suner SS, Sahiner N. 2018. Humic acid particle embedded super porous gum Arabic cryogel network for versatile use. Polym Adv Technol. 29(1):151–159. https://doi.org/10.1002/pat.4097

Thakar H, Sebastian SM, Mandal S, Pople A, Agarwal G, Srivastava A. 2019. Biomolecule-conjugated macroporous hydrogels for biomedical applications. ACS Biomater Sci Eng. 5(12):6320–6341. https://doi.org/10.1021/acsbiomaterials.9b00778

Türkmen D, Bakhshpour M, Akgönüllü S, Aşır S, Denizli A. 2022. Heavy metal ions removal from wastewater using cryogels: A review. Front Sustain. 3(March):1–17. https://doi.org/10.3389/frsus.2022.765592

Tyshkunova I V., Poshina DN, Skorik YA. 2022. Cellulose cryogels as promising materials for biomedical applications. Int J Mol Sci. 23(4). https://doi.org/10.3390/ijms23042037

Wu T, Huang J, Jiang Y, Hu Y, Ye X, Liu D, Chen J. 2018. Formation of hydrogels based on chitosan/alginate for the delivery of lysozyme and their antibacterial activity. Food Chem [Internet]. 240:361–369. https://doi.org/10.1016/j.foodchem.2017.07.052

Xiao M. 2021. Advances and rational design of chitosan-based autonomic self-healing hydrogels for biomedical applications. J Polym Res [Internet]. 28(8). https://doi.org/10.1007/s10965-021-02688-7

Yang K, Han Q, Chen B, Zheng Y, Zhang K, Li Q, Wang J. 2018. Antimicrobial hydrogels: Promising materials for medical application. Int J Nanomedicine. 13:2217–2263. https://doi.org/10.2147/IJN.S154748

Yu Y, Li P, Zhu C, Ning N, Zhang S, Vancso GJ. 2019. Multifunctional and recyclable photothermally responsive cryogels as efficient platforms for wound healing. Adv Funct Mater. 29(35):1–11. https://doi.org/10.1002/adfm.201904402

Zare EN, Makvandi P, Ashtari B, Rossi F, Motahari A, Perale G. 2020. Progress in Conductive Polyaniline-Based Nanocomposites for Biomedical Applications: A Review. J Med Chem. 63(1):1–22. https://doi.org/10.1021/acs.jmedchem.9b00803

Zhang X, Hang Y, Ding Z, Xiao L, Cheng W, Lu Q. 2022. Macroporous Silk Nanofiber Cryogels with Tunable Properties. Biomacromolecules. https://doi.org/10.1021/acs.biomac.2c00222

Zhao X, Guo B, Wu H, Liang Y, Ma PX. 2018. Injectable antibacterial conductive nanocomposite cryogels with rapid shape recovery for noncompressible hemorrhage and wound healing. Nat Commun. https://doi.org/10.1038/s41467-018-04998-9

Zhao S, Zou Y, Liu X, Zhang H. 2019. Ecofriendly construction of enzyme reactor based on three-dimensional porous cryogel composites. Chem Eng J. https://doi.org/10.1016/j.cej.2018.12.101

Zhong Y, Xiao H, Seidi F, Jin Y. 2020. Natural Polymer-Based Antimicrobial Hydrogels without Synthetic Antibiotics as Wound Dressings. Biomacromolecules. 21(8):2983–3006. https://doi.org/10.1021/acs.biomac.0c00760

Zou X, Deng P, Zhou C, Hou Y, Chen R, Liang F, Liao L. 2017. Preparation of a novel antibacterial chitosan-poly(ethylene glycol) cryogel/silver nanoparticles composites. J Biomater Sci Polym Ed [Internet]. 28(13):1324–1337. https://doi.org/10.1080/09205063.2017.1321346

2 Natural Antimicrobial Therapeutic Peptides

Milk Lactoferricin and Spirulina platensis Peptides

*Nawal Abd El-Baky[1] and *Amro Abd Al Fattah Amara[2]

[1*]Protein Research Department, Genetic Engineering and Biotechnology Research Institute (GEBRI), City of Scientific Research and Technological Applications (SRTA-City), New Borg El-Arab City, Alexandria, Egypt
nelbaky@srtacity.sci.eg.

[2*]Protein Research Department, Genetic Engineering and Biotechnology Research Institute (GEBRI), City of Scientific Research and Technological Applications (SRTA-City), New Borg El-Arab City, Alexandria, Egypt
aamara@srtacity.sci.eg.

CONTENTS

DOI: 10.1201/9781003268932-2

TABLE OF ABBREVIATIONS

ACE	angiotensin I converting enzyme
FDA	Food and Drug Administration
GRAS	generally recognized as safe
GST	glutathione S tranferase
HS	heparan sulfate
MICs	minimum inhibitory concentrations
MRSA	methicillin-resistant *Staphylococcus aureus*
NAMP	natural antimicrobial peptides
NMR	nuclear magnetic resonance
RTE	ready-to-eat

2.1 INTRODUCTION

Natural antimicrobial peptides (NAMPs) are one of the successful mechanisms by which eukaryotic cells can perform innate chemical defenses against invading microbial pathogens (Hancock and Diamond 2000, Zasloff 2002). They have been isolated from natural sources and could exhibit broad-spectrum antimicrobial effects. Moreover, numerous bioactive synthetic variants of these peptides were synthesized and characterized. NAMPs have significant bactericidal and antiviral activities, and represent hope for novel antibiotic lead agents (Uddin et al. 2021). Most of NAMPs are cationic and amphipathic allowing their binding to and consequently disruption of lipid membranes of microbial cells.

Milk is rich in bioactive elements such as antibodies, lactoperoxidase, complex carbohydrates, and lactoferrin, which cause protecting activities of milk (Phelan and Kerins 2011, Lonnerdal 2014, Hill and Newburg 2015, Hsieh et al. 2015, Pammi and Abrams 2015, Demmelmair et al. 2017, Marcone, Belton, and Fitzgerald 2017, Sultan et al. 2018). Research has shown that milk lactoferrin and its related cationic peptides have significant and valuable antimicrobial activity (Redwan and Tabll 2007, El-Fakharany et al. 2008, Conesa, Calvo, and Sanchez 2010, Liao et al. 2012, Redwan et al. 2014, Theolier et al. 2014, Redwan et al. 2016, El-Baky, El-Fakharany, and Redwana 2017, Almehdar et al. 2019, 2020, El-Baky et al. 2021). Human milk contains the highest concentration of Lf (approximately 1.7 g/L), followed by camel milk (55–888 mg/L) and bovine milk (300–500 mg/L) (El-Agamy 2006, Farah et al. 2007, Konuspayeva et al. 2007). The Lf antimicrobial activity is attributed to two mechanisms, an iron sequestering-based bacteriostatic mechanism and an iron-independent bactericidal one (Arnold, Cole, and McGhee 1977, Arnold, Brewer, and Gauthier 1980, Ellison III et al. 1990, Ellison III and Giehl 1991, Nibbering et al. 2001, Legrand et al. 2005, Morgenthau et al. 2014, Jahani, Shakiba, and Jahani 2015, Redwan et al. 2016, Samaniego-Barron et al. 2016, Drago-Serrano et al. 2017, Almehdar et al. 2019, 2020, Kowalczyk et al. 2022). The iron-independent Lf bactericidal mechanism includes direct bacterial lipopolysaccharides (LPS)-binding (Ellison III and Giehl 1991), and damage of the bacterial outer membrane via binding to numerous bacterial membrane proteins (Ellison III et al. 1990, Morgenthau et al. 2014, Samaniego-Barron et al. 2016). The cationic N-terminal peptide of lactoferrin (LFcin) is responsible for these bactericidal effects of Lf parent protein through peptide binding to bacterial LPS and membrane proteins (Nibbering et al. 2001, Morgenthau et al. 2014, Theolier et al. 2014).

In our body, LFcin is released from N-terminal region of Lf parent protein by gastric pepsin digestion. This highly positively charged peptide was considered as the bactericidal domain of lactoferrin. The peptide shares many of the bioactivities performed by the intact protein and occasionally can exceed the potency of its parent protein (Bellamy, Takase, Yamauchi, et al. 1992, Shin et al. 1998, Rybarczyk et al. 2017). LFcin demonstrated antimicrobial, antiparasitic, antiviral, antitumor, and immunomodulatory effects (Gifford, Hunter, and Vogel 2005, Eliassen et al. 2006, Arias et al. 2014).

Spirulina platensis has been consumed as food and feed for centuries. It is marketed as either an active ingredient in functional foods or a food supplement. The high nutritional value of this filamentous blue-green microalga originates from its high protein content (60–70%), and its constituents of vitamins, essential amino acids and fatty acids, minerals, and carotenoids (Habib et al. 2008, Capelli and Cysewski 2010, Vo, Ngo, and Kim 2015). *Spirulina* also has pharmaceutical properties including antiviral, antibacterial, antioxidant, anti-allergic, antidiabetic, and anticancer activities (Mao, van de Water, and Gershwin 2005, Sharaf et al. 2010, Sharaf et al. 2013, Vaz et al. 2016, Wu et al. 2016, Amara 2017, Moreira et al. 2019, Shao et al. 2019).

Bioactive peptides were extracted from the hydrolysis of proteins of *Spirulina platensis*. These peptides have distinctive bioactivities such as antibacterial, antihypertensive, antidiabetic, antioxidant, antithrombotic, anticoagulant, immunomodulatory, and anticancer. These activities depend on the number, type, sequence, and characteristics of the amino acids in peptide (Aneiros and Garateix 2004, Lisboa, Pereira, and Costa 2016, Gogineni and Hamann 2018, Sadeghi et al. 2018, Srinivasan et al. 2021).

Until recently, the pharmaceutical industry rarely considered peptide therapeutics, mainly because of their costly production and their poor profile for absorption, distribution, metabolism, and excretion. Besides the alarming liability of peptides to enzymatic degradation in biological fluids, their problematic sequestration, and challenging transport across the host membranes, in addition to the difficulty to reach an effective concentration at deep organs or tissues. However, these limitations were recently addressed and appropriately solved. The solutions for these limitations included improved strategies for production of recombinant peptides to decrease their cost (Gaglione et al. 2019, Wibowo and Zhao 2019, Kaur et al. 2020, Sampaio de Oliveira et al. 2020, El-Baky et al. 2021), and engineering strategies to prevent enzymatic degradation and improve bioavailability of peptides via incorporating unnatural amino acids in peptide primary sequence (Yao et al. 2018), peptidomimetics (Kuppusamy et al. 2019), enantiomeric peptides (Miao et al. 2020), or by gaining a more stable conformation such as cyclation and stapling (Reguera and Rivera 2019). Furthermore, the application of nano-sized drug delivery systems to improve peptide bioavailability, targeting the right site, avoiding their degradation, as well as sustaining a gradual peptide delivery at the right site (Radaic, de Jesus, and Kapila 2020). Thus, FDA approved several peptide therapeutics (Fosgerau and Hoffmann 2015, Jad et al. 2019, Al Musaimi, de la Torre, and Albericio 2020, De la Torre and Albericio 2020).

Both milk lactoferricin and *Spirulina platensis* antibacterial peptides among other NAMPs are potential novel antibiotic lead agents that can be used as peptide drugs or in combination therapies to overcome the crisis of antimicrobial resistance and disturbing decline of antibiotic effectiveness as a result of the growing multidrug-resistance developed by microbial pathogens.

2.2 LACTOFERRICIN AND ITS PARENT PROTEIN

In 1991, it was reported that the antimicrobial activity of enzymatic hydrolysate of bovine lactoferrin (bLf), which was digested by porcine pepsin, was more potent than the intact protein effect against *Escherichia coli* 0111 isolate (Tomita et al. 1991). This active peptide fragment obtained by pepsin digestion of the parent protein was purified by reverse-phase high-performance liquid chromatography (Bellamy, Takase, Wakabayashi, et al. 1992, Bellamy, Takase, Yamauchi, et al. 1992). Sequence analysis of this peptide fragment revealed that it has a length of 25 amino acids (FKCRRWQWRMKKLGAPSITCVRRAF), which corresponds to amino acid segment 17–41 of N-terminal of parent Lf protein sequence (Goodman and Schanbacher 1991).

In 1993, Yamauchi and coworkers compared the antibacterial effect of both parent protein and the pepsin-derived peptide fragment, LFcin, and reported that LFcin peptide has an equivalent effect on the outer membrane of microbes to that of the parent molecule, which suggests that this N-terminal domain in lactoferrin is the exclusive site for disrupting microbial membrane (Yamauchi et al. 1993).

Lactoferricin displays a broad bactericidal spectrum against various Gram-positive and Gram-negative bacteria, and also exhibited virucidal, fungicidal, antiparasitic, and antitumor activities (Bellamy et al. 1992, Yamauchi et al. 1993, Bellamy et al. 1994, Turchany, Aley, and Gillin 1995, Yoo et al. 1997, Shin et al. 1998, Andersen et al. 2001, Omata et al. 2001, Eliassen et al. 2002, McCann et al. 2003, Gruden and Poklar-Ulrih 2021). So far, bovine lactoferricin (bLFcin) is the most studied peptide among other LFcin peptides from different animal species (Goodman and Schanbacher 1991, Tomita et al. 1991, Bellamy et al. 1992, Bellamy et al. 1992, Yamauchi et al. 1993, Bellamy et al. 1994, Turchany, Aley, and Gillin 1995, Yoo et al. 1997, Shin et al. 1998, Andersen et al. 2001, Omata et al. 2001, Eliassen et al. 2002, McCann et al. 2003, Gifford, Hunter, and Vogel 2005, Eliassen et al. 2006, Arias et al. 2014). LFcin shares common features with another antimicrobial peptide, magainin (Haukland et al. 2001, Huertas et al. 2017). Both peptides could cross the bacterial cytoplasmic membrane. Additionally, LFcin shares the amphipathic alpha-helical structure with magainins, dermaseptin, and other amphipathic tryptophan- and arginine-rich AMPs (Chan, Prenner, and Vogel 2006).

Moreover, LFcin peptide was confirmed to be the sole active domain obtained from pepsin digestion of bLf and human lactoferrin (hLf) (Tomita et al. 1991). The LFcin peptide domain is located in a glycosylation-free region of Lf parent protein and its sequence is lacking tyrosine and histidine residues. This peptide is unique among other Lf-derived peptides by the fact that it comprises a disulfide bond between amino acid residues Cys 19 and Cys 36 in bLf and Cys 20 and Cys 37 in hLf. Basic amino acid residues (e.g. lysine and arginine) are abundant in sequence of LFcin peptide, about eight basic residues of total 25 in bLFcin and seven basic residues in human lactoferricin (hLFcin) (Tomita et al. 1991). Hydrophobic amino acid residues such as tryptophan and phenylalanine are also abundant in sequence of LFcin.

2.3 STRUCTURAL CHARACTERISTICS OF LACTOFERRICIN PEPTIDE

Lactoferricin is a highly basic peptide, comprising several positively charged arginine and lysine residues. Even though both bLFcin and hLFcin are highly positively charged, there is a remarkable variance in their amino acid sequences and length. This variance was revealed by the relatively low sequence similarity of 69% between the aligned N-terminal regions of bLf and hLf (Sinha et al. 2013).

The well-established primary structure of bLFcin is a 25-residue peptide (amino acid segment 17–41 of N-terminal of parent bLf sequence). It forms into a looped structure via an intramolecular disulfide bond (Bellamy, Takase, Yamauchi, et al. 1992). The first amino acid sequence determination of hLFcin showed that it comprises two peptide chains linked by disulfide bond and includes the N-terminal 47 residues of hLf (Bellamy, Takase, Yamauchi, et al. 1992). Nevertheless, the more recent primary structure determined by mass spectrometry evidenced that hLFcin contains N-terminal 49 residues of hLf in a single continuous chain (Hunter et al. 2005). In 2016, Chahardooli et al. (Chahardooli et al. 2016) characterized camel lactoferricin (cLFcin) for the first time, as the amino acid region 17–41 of N-terminal of camel lactoferrin (cLf) that corresponds to previously characterized bLFcin and hLFcin.

Helical Wheel tool (https://heliquest.ipmc.cnrs.fr/) analysis for the primary structure (amino acid region 17—46 from N-terminal of intact Lf proteins) of bLFcin, cLFcin, and hLFcin, regarding hydrophilic, hydrophobic, and charge properties of peptides, demonstrates that cLFcin has the highest cationicity among the three peptides, while hLFcin shows the lowest cationicity (Figure 2.1). The analyzed sequences for bLFcin, cLFcin, and hLFcin were *FKCRRWQWRMKKLGAPSITCVR*

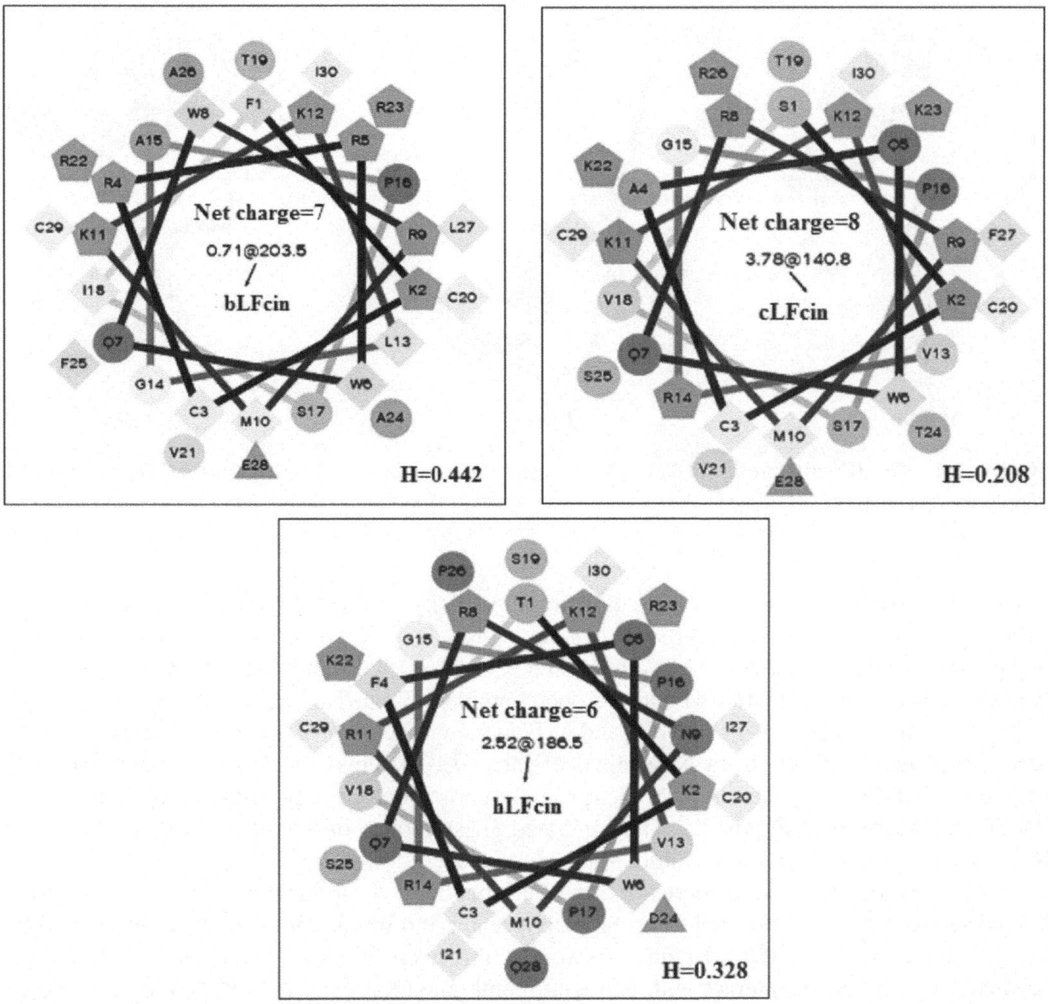

FIGURE 2.1 Helical Wheel diagrammatic illustration of primary structure of bLFcin, cLFcin, and hLFcin peptides. Hydrophilic residues are presented as circles, diamonds are hydrophobic residues, pentagons are potentially positively charged, while triangles are potentially negatively charged. Hydrophobicity is represented by diamond shapes, while hydrophilic residues are represented by dark grey circles. The potentially charged residues are represented by pentagons. H is the abbreviation for hydrophobicity (the image is original for this chapter).

RAFALECI, SKCAQWQRRMKKVRGPSVTCVKKTSRFECI, and *TKCFQWQRNMRKVRGPPVSCI KRDSPIQCI*, respectively.

Nuclear magnetic resonance (NMR) spectroscopy studies of the three-dimensional (3D) structure of bLFcin demonstrated that free bLFcin conformation in aqueous solution varies from that found in the intact protein (Hwang et al. 1998). In low salt solutions, bLFcin forms a twisted β-sheet instead of α-helix observed in intact bLf. This makes the released peptide turn into amphipathic form because almost all the hydrophobic residues lie on one face while the cationic residues lie on the opposite. On the other hand, hLFcin preserves its α-helixconformation observed in intact hLf (Hunter et al. 2005). The extra length and hydrogen bonds of hLFcin compared to bLFcin may be attributed to supporting the helix stabilization.

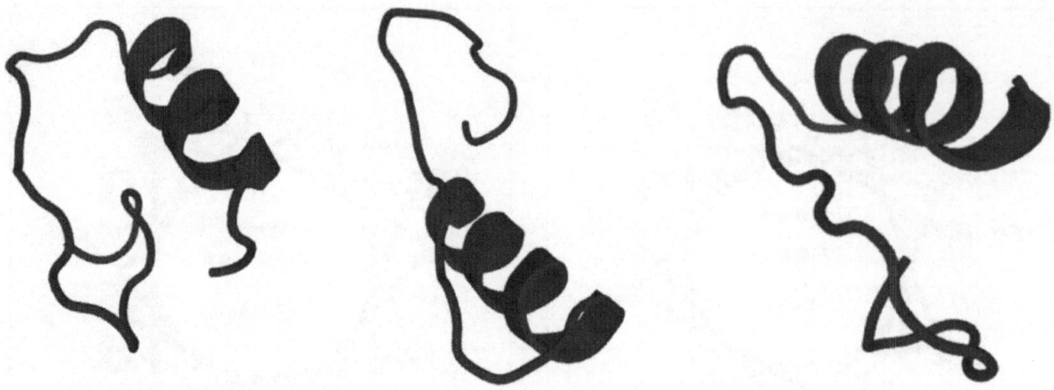

FIGURE 2.2 The 3D structures of bLFcin, cLFcin, and hLFcin peptides made by PEP-FOLD3 (the image is original for this chapter).

The released hLFcin resembles bLFcin in amphipathicity. Remarkably, the hLFcin hydrophobic surface is considerably larger than that in bLFcin, and in aqueous solution, this could be explained by the dimer structure behavior of hLFcin (Hunter et al. 2005). The amphipathic structure of bLFcin and hLFcin is a common feature of antimicrobial peptides (Zhou, Tieleman, and Vogel 2004).

The 3D structures of bLFcin, cLFcin, and hLFcin are predicted from their amino acid sequences (amino acid region 17–46 from N-terminal of intact Lf proteins) by *de novo* peptide structure predictor PEP-FOLD3 (https://mobyle.rpbs.univ-paris-diderot.fr/cgi-bin/portal.py#forms::PEP-FOLD3) as presented in Figure 2.2. The generated 3D structures of the three LFcin peptides show that they have a conserved amphipathic α-helical structure.

Analysis of the amino acid sequences of LFcin peptides via truncation and design of shorter derived sequences was performed to reveal active center and important residues of these peptides. As an example, the 11-residue bLFcin derivative (amino acids 20–30 of N-terminal of bLf) retains its bioactivity and could interact with bacterial membrane (Kang et al. 1996). Moreover, the 15-residue bLFcin derivative (amino acids 17–31 of N-terminal of bLf) with lower net charge compared to original 25-residue bLFcin (17–41) demonstrated enhanced antibacterial activity (Rekdal et al. 1999). In contrast, shorter bLFcin derivatives (18–31, 19–31, and 20–31) showed lower antibacterial effect on *Staphylococcus aureus* and *Escherichia coli* than the original 25-residue bLFcin (Tomita et al. 1994, Rekdal et al. 1999, Liu et al. 2011). Finally, the 6-residue bLFcin derivative sequence RRWQWR-NH2 (amino acids 20–25 of N-terminal of bLf) was identified to be the antimicrobial center of bLFcin (Tomita et al. 1994, Schibli, Hwang, and Vogel 1999).

In the case of hLFcin, the 16-residue hLFcin derivative (amino acids 21–36 of N-terminal of hLf, with a sequence of FQWQRNMRKVRGPPVS) and the 11-residue hLFcin derivative (amino acids 21–31) displayed significant antibacterial effects on *E. coli* and methicillin-resistant *S. aureus* (MRSA) (Odell et al. 1996). The hLFcin derivative fragment 19–31 was confirmed to be the optimal sequence and length for peptide antimicrobial activity. This fragment could bind to bacterial lipid A/LPS (Haversen et al. 2010).

Successive experiments were carried out to scan each individual amino acid in bLFcin derivative fragment 17–31 to detect the specific essential amino acids for antibacterial activity of the peptide. Obtained data revealed that the two tryptophan residues at positions 6 and 8 in bLFcin derivative fragment 17–31 could not be replaced by alanine, which resulted in slight loss of antibacterial activity (Strøm, Rekdal, and Svendsen 2000). This tryptophan effect was further confirmed by replacement of one, two, or three residues in bLFcin 17–31 by tryptophan, which improved its antibacterial activity (Strøm et al. 2002). The same result was observed with tryptophan-rich porcine lactoferricin derivative that has higher antibacterial activity than that of native sequence (Han et al. 2013). Another critical amino acid is arginine, when all arginine residues in bLFcin 17–30 and hLFcin 1–11 were replaced by lysine, the activity against *Mycobacterium avium* was significantly lower than the original peptide (Silva et al. 2014). Therefore, arginine- and tryptophan-rich peptides are often used as a good original sequence for designing novel antimicrobial peptides.

2.4 EXPRESSION SYSTEMS FOR RECOMBINANT LACTOFERRICIN PEPTIDES

Lactoferricin peptides resemble other NAMPs in their presence at very low concentrations in host organisms. For this reason, isolation of these peptides directly from natural sources is a time-consuming and high cost approach. Additionally, large-scale chemical synthesis of both natural and synthetic AMPs is costly, especially peptides with disulfide bonds (Li et al. 2010), and those require post-translational glycosylation or bromination (Otero-gonzalez et al. 2010).

Recombinant DNA technology has been considered the most effective method for NAMPs production regarding time and production costs (Xu et al. 2007, Amara 2013). The main hosts used for recombinant NAMPs production are *E. coli* (prokaryotic expression systems) and yeasts such as *Saccharomyces cerevisiae* and *Pichia pastoris* (eukaryotic expression systems) (Li and Chen 2008). Because of cytotoxicity of cationic NAMPs on *E. coli*, various fusion partners were used to neutralize these NAMPs to be successfully expressed in *E. coli* cells. In 2006, bLFcin was fused with glutathione S tranferase (GST) to be expressed in *E. coli*. The expression construct has a thrombin cleavage site in the vector located between GST and bLFcin to release recombinant peptide from the GST-bLFcin fused product by enzymatic digestion of thrombin (Feng et al. 2006). In 2007, Tian et al. (Tian et al. 2007) expressed tetramer of bLFcin derivative (LFcinB15-W4, 10) linked with thioredoxin fusion peptides (fusion tag in pET32a vector) in *E. coli*. About 10 mg/L of tetramer with 99% purity was obtained and exhibited comparable antimicrobial activity to that of synthetic LFcinB15-W4, 10 monomer.

Bovine lactoferricin peptide was produced in the *Pichia methanolica* expression system linked to the alpha factor signal peptide of *S. cerevisiae*. The signal peptide directed the secretion of bLFcin from the recombinant yeast cell into the culture medium (Wang, Zhao, and Lu 2007). Recombinant cLFcin was successfully expressed in *Pichia pastoris* and confirmed to have proper antimicrobial activity (Chahardooli et al. 2016).

Even transgenic plants were used for production of recombinant lactoferricin. As an example, recombinant chimera of bovine lactoferricin and lactoferrampin was expressed in *Nicotiana tabacum* hairy roots. The chimera accumulated at around 4.8 μg/g fresh weight of hairy root biomass (Chahardooli, Fazeli, and Ghabooli 2018). Recently, a cell-free or *in vitro* protein synthesis approach was used to produce four recombinant lactoferricin peptides, three natural (bLFcin, cLFcin, and hLFcin) and a synthetic consensus one (El-Baky et al. 2021). The four peptides were expressed in crude cell extract of *E. coli* S30 rather than living host cells to avoid any peptide-cell interaction.

2.5 *IN VITRO* BIOACTIVITIES ASSOCIATED WITH LACTOFERRICIN PEPTIDE

2.5.1 Antibacterial Activity

Lactoferricin has superior antibacterial effect to its parent protein (Arias et al. 2014, Tomita et al. 1991, Yamauchi et al. 1993). Both bLFcin and hLFcin peptides exhibited antibacterial effects on several Gram-negative and Gram-positive bacteria (Han et al. 2013, Hao et al. 2017, Chaparro et al. 2018) but bLFcin bactericidal activity exceeds that of hLFcin (Shin et al. 1998, Gifford, Hunter, and Vogel 2005, Silva et al. 2014). Camel LFcin was reported to achieve twice the antimicrobial activity of bLFcin against *E. coli*, *Salmonella typhi*, *Pseudomonas aeruginosa*, *S. aureus*, and MRSA and even four times higher than that of hLFcin against *S. typhi* (El-Baky et al. 2021). Moreover, it was found that the superior biological activities of cLf over other lactoferrins were partially correlated with difference in the sequence and levels of intrinsic disorder of cLFcin (Redwan et al. 2016).

The LPS in the outer membrane of Gram-negative bacteria and the layer of teichoic acid surrounding the cytoplasmic membrane of Gram-positive bacteria are negatively charged surfaces; accordingly, an electrostatic attraction occurs between the cationic LFcin peptides and these negatively charged surfaces. Thus, the predicted antibacterial mechanism of this peptide is that LFcin first binds to the outer bacterial cell surfaces and then crosses this barrier to interact with the bacterial cytoplasmic membrane (Chen, Shyu, and Mao 2003, Riahifard et al. 2018). Research reported that bLFcin and hLFcin could disturb membrane permeability, permitting the passage of small ions, and causing the loss of the electrochemical and pH gradients of bacterial and artificial membranes (Aguilera et al. 1999). Nevertheless, the antibacterial effect decreases if the positive charge of peptide is excessively high (Kohn et al. 2018). This is because when the number of positive charges exceeds a definite critical value, it will lead to electrostatic repulsion between the NAMPs, which is stronger than their electrostatic attraction to the bacterial membrane. This prevents the NAMPs molecules accumulation and the subsequent formation of membrane pores, decreasing membrane lysis and antimicrobial action (Sun et al. 2018).

2.5.2 Antifungal Activity

Bovine lactoferricin could efficiently inhibit the growth of numerous yeasts (e.g. *Candida albicans*), molds, and filamentous fungi (e.g. dermatophytes such as *Trichophyton mentagrophytes*) (Bellamy et al. 1994). Lactoferricin has two suggested antifungal mechanisms, either via a direct fungicidal activity (Bellamy et al. 1993) or by upregulating host defense. Both bLFcin and hLFcin interact with the plasma membrane of *Candida albicans* triggering the dispersion of the proton gradient across the cell membrane (Lupetti et al. 2000, Gifford, Hunter, and Vogel 2005, Basso 2020). Also, both LFcin peptides appear to affect the cytoplasm of fungal cells. Aggregation of cytoplasmic material was observed with bLFcin-exposed fungal cells and treatment of the blastoconidia life stage of *C. albicans* with hLFcin-derived peptide results in synthesis and secretion of ATP from the mitochondria. Interestingly, this ATP is released extracellularly, where it can interact with extracellular ATP binding sites on the plasma membrane, giving rise to pore formation and cell death (Bellamy et al. 1994, Ueta, Tanida, and Osaki 2001, Fernandes and Carter 2017). Bovine lactoferricin was showed to upregulate host defense through the stimulation of reactive oxygen species that induce the *Candida* cell-killing activity of polymorphonuclear leukocytes (Fernandes and Carter 2017).

2.5.3 Antiparasitic Activity

LFcin peptides are parasiticidal against a range of protozoans, however how LFcin exerts this effect is not known (Turchany, Aley, and Gillin 1995, Omata et al. 2001, Aguilar-Diaz et al. 2017, Zarzosa-Moreno et al. 2020). Probably, the cell surface of tachyzoites of *Toxoplasma gondii* is negatively charged, and LFcin peptide was proved to attract to this surface, resulting in the disruption of the membrane. It is also possible that this interaction releases parasitic structural components that subsequently activate the host defense system. Bovine lactoferricin could effectively reduce the sporozoites' infectivity of *Toxoplasma gondii* and *Eimeria stiedai* in mice and rabbits, respectively (Omata et al. 2001). Furthermore, LFcin peptides had *in vitro* giardicidal activity via interaction with *Giardia intestinalis* and causing an increase in the granularity and vacuolization of its cytoplasm. In addition, pore formation, membrane disruption, and programmed cell death were observed in trophozoites of *G. intestinalis* treated with LFcin peptides (Aguilar-Diaz et al. 2017).

2.5.4 Antiviral Activity

In contrast to other bioactivities, the antiviral activity of LFcin is less than that of the parent Lf protein. Though LFcin moderately inhibits *in vitro* multiplication of many viruses, the activity of intact

Lf against these same viruses is as much as seven-fold higher, suggesting that either the size of the molecule is important or that other regions of Lf contribute to the antiviral activity (Andersen et al. 2001, Berkhout et al. 2002, Di-Biase et al. 2003, McCann et al. 2003, Hu et al. 2021, Zhu 2022). LFcin peptides are highly positively charged peptides, which have a critical affinity to heparan sulfate (HS) or glycosaminoglycans (highly negatively charged carbohydrates that are the typical viral receptors on the host cell surface) which blocks the initial viral entry, thus inhibiting cell-to-cell spreading of an ongoing infection (Andersen et al. 2001, Jenssen et al. 2008, Hu et al. 2021). Although parent protein also prevents viral entry into the cell by binding to these carbohydrate molecules, there is evidence that Lf and LFcin act via different antiviral mechanisms (Jenssen et al. 2004). The structural characteristics of peptide such as hydrophobicity, molecular size, and the secondary structure are also involved in its antiviral activity (Jenssen et al. 2004). Both bLFcin and parent protein significantly reduced herpes simplex virus-1 (HSV-1) cellular uptake. Even internalized virus particles had delayed intracellular trafficking (Marr et al. 2009).

2.5.5 Anticancer Activity

LFcin resembles its parent protein in exhibiting cytotoxic effect against many cancer cell lines including leukemic, melanoma, fibrosarcoma, colon carcinoma, and neuroblastoma cells at a concentration that does not affect normal fibroblasts, lymphocytes, epithelial cells, endothelial cells, or erythrocytes (Yoo et al. 1997, Eliassen et al. 2002). It causes loss of tumor cell membrane integrity via membrane binding and formation of membrane pores, then enters the cytoplasm of the cancer cell and targets negatively charged mitochondria (Eliassen et al. 2006, Mader et al. 2007, Guerra JR et al. 2019). Another LFcin antitumor pathway against human leukemia and breast carcinoma is via killing these cells by the sequential generation of reactive oxygen species, impairing mitochondrial transmembrane potential, and inducing apoptosis (Mader et al. 2005). For high LFcin antitumor activity, both a net positive charge of +7 and amphipathic structure are required (Yang et al. 2004). Mader et al. (Mader et al. 2005) suggested that the cyclization of bLFcin by the disulfide bond between residues Cys 3 and Cys 20 is not required for its antitumor effect. However, another study reported an opposite concept, that *in vitro* anticancer activity of bLFcin against fibrosarcoma cells is critically dependent on its cyclic structure (Eliassen et al. 2002). Therefore, the role of the disulfide bridge in mediating the anticancer activity of bLFcin remains unclear.

2.6 SYNERGY OF LACTOFERRICIN WITH ANTIMICROBIAL DRUGS

The ability of LFcin to impair the integrity of bacterial inner membrane possibly will stimulate other antimicrobial agents' (e.g. other antibacterial peptides or antibiotics) uptake, which results in synergistic effects of LFcin with conventional antibiotics. An example of this antibacterial synergism is synergistic activity of bLFcin with ceftazidime or ciprofloxacin against *S. aureus* and *P. aeruginosa* (Oo et al. 2010, Vargas-Casanova et al. 2019). Another study evaluated the impact of bLFcin on the activity of fluoroquinolones against uropathogenic *E. coli* strains (Longhi et al. 2009). Obtained results demonstrated different effects including synergism, partial synergism, or indifference, which depend on the examined *E. coli* strains. Overall, combination of fluoroquinolones with bLFcin may reduce inhibitory concentrations of fluoroquinolones against some uropathogenic *E. coli* strains. Different combinations of bLFcin, bLf, and alphaS2-casein with nisin (a bacteriocin produced by *Lactocococcus lactis* spp. *Lactis* and used for food preservation) were analyzed against *E. coli*, *Salmonella enterica* Serotype *Choleraesuis*, *Staphylococcus epidermidis*, and *Listeria monocytogenes*. The combination of bLFcin and its parent protein showed a synergistic effect on *S. epidermidis* and *E. coli*, while a combination of bLFcin and nisin displayed an antagonistic effect against *E. coli* (López-Expósito et al. 2008).

In the same context, LFcin affects the physiology of fungal cell cytoplasm making some strains of azole-resistant *C. albicans* more susceptible to itraconazole or fluconazole in its presence even at comparatively low concentration. This synergistic effect was absent with antifungal agents like amphotericin B, which is a non-azole type agent (Wakabayashi et al. 1996, Wakabayashi et al. 1998).

2.7 LACTOFERRICIN: FROM MILK TO HUMAN AND VETERINARY MEDICINE

2.7.1 Preclinical Trials for Human Diseases Treatment

To our knowledge, until now, the *in vivo* trials of lactoferricin peptide were in mice and rabbits not humans. Those trials proved that lactoferricin represents a potential treatment as a bactericidal, fungicidal, virucidal, parasiticidal, anti-inflammatory, and antitumor agent.

Teraguchi et al. (1994) reported that bLf and its porcine pepsin hydrolysate (bLFcin) have bacteriostatic effects on intestinal *Enterobacteriaceae* in mice that were fed with bovine milk. This was the first report for *in vivo* bacteriostatic activities of both intact protein and its peptides. The same research team revealed the effectiveness of bLf and its porcine pepsin hydrolysate (bLFcin) to control the bacterial growth of diverse strains of *Clostridium* spp. in a mouse model (Teraguchi et al. 1995).

The first *in vivo* trial of bLFcin peptide as antitumor agent was in 1997 using established tumor models in mice (Yoo et al. 1997). The subcutaneous injection of 0.5 mg peptide/mouse after one day from tumor inoculation significantly inhibited liver, spleen, and lung metastasis. Bovine lactoferricin suppressed tumor growth on day 8 after tumor inoculation. Nevertheless, long-term analysis of tumor growth for up to 21 days after tumor inoculation revealed that single administration of bovine lactoferricin displayed inhibitory activity on tumor growth only during the early period (8 days) (Yoo et al. 1997).

Regarding parasiticidal activities of bLFcin, the peptide could effectively reduce the sporozoites infectivity of *Toxoplasma gondii* and *Eimeria stiedai* in mice and rabbits models, respectively (Omata et al. 2001).

In 2010, it was described that bLFcin acts synergistically with ciprofloxacin or ceftazidime against eight strains of *P. aeruginosa* isolated from ocular infection *in vitro*, counting multidrug-resistant ones. On the other hand, this synergism was absent between peptide and gentamicin. Furthermore, bLFcin acts synergistically with only ciprofloxacin against two out of ten test strains of *S. aureus* isolated from ocular infection *in vitro*. Combined treatment of peptide and ciprofloxacin to infected corneas of mice considerably enhanced the clinical results, as peptide acts as an adjuvant enhancing ciprofloxacin activity and consequently reducing bacterial growth and myeloperoxidase (MPO) levels. Interestingly, the peptide alone could decrease MPO levels in infected corneas resulting in reduction of *in vivo* inflammatory response (Oo et al. 2010).

During treatment of keratitis, biofilm eradication is crucial to solve the occurrence of drug resistance. In 2012, an *ex vivo* research was conducted to evaluate the combined efficacy of bLFcin with other antifungals in inhibiting formation of fungal biofilm on contact lens by three fungal pathogens that cause keratitis, *C. albicans*, *Fusarium solani*, and *Aspergillus fumigatus*. These fungal pathogens were isolated from keratitis patients (Sengupta et al. 2012). The combination of bLFcin with conventional antifungal agents such as amphotericin B, voriconazole, or fluconazole improved the sensitivity of fungal biofilms and planktonic cells to antifungals, and reduced minimum inhibitory concentrations (MICs) of antifungal agents more than eightfold. Moreover, the addition of peptide to solution of lens care has efficiently eradicated the mature fungal biofilm formed on contact lenses. Thus, bLFcin can be considered a promising therapeutic candidate that increases susceptibility of fungal biofilm to conventional antifungals. In addition, if a peptide is used in solution of lens care, it can act as an antibiofilm/antifungal additive (Sengupta et al. 2012).

Lactoferricin was reported to have therapeutic potential against genital herpes simplex virus 2 (HSV-2) infection in mice. It demonstrated a strong antiviral activity in female mice inoculated intravaginally with HSV-2 alone or with a mixture of HSV-2 and lactoferricin. Mice treated with lactoferricin did not develop any sign of the disease nor did they lose weight. The peptide delayed both disease onset and disease severity when given 24 h post-viral inoculation. Lactoferricin was also tested for its ability to stimulate the production of chemokines. It did not induce the production of CCL3, CCL5, CXCL1 or CXCL2 by mouse splenocytes *in vitro* but could induce the production of CCL5 and CXCL2 chemokines *in vivo* at 8 h post-administration when deposited on the genital mucosa (Shestakov et al. 2012).

Bovine LFcin has anti-inflammatory and anti-catabolic activities, which were studied in *ex vivo* tests on human articular cartilage and synovium (Yan et al. 2013). Results confirmed that bLFcin exhibited chondroprotective properties, hindering synovial fibroblasts inflammatory response, and could thus be described as a promising preventive agent and/or treatment of diseases of degenerative joint. This antimicrobial peptide may also have therapeutic effects on septic arthritis.

2.7.2 Lactoferricin for Veterinary Applications

When Lf alone was used as intramammary treatment for mastitis in dairy cattle caused by *Streptococcus dysgalactiae*, *Streptococcus uberis*, and *S. aureus*, the outcome was disappointing especially for control of infection by beta-lactam resistant *S. aureus*. Yet, co-administeration of Lf at low concentration with penicillin G efficiently reduced the activity of beta-lactamase of resistant *S. aureus* and elevated penicillin inhibitory activity up to four-fold, improving the rate of treatment from about 12.5% (penicillin alone) to 33.3% (combination of Lf and penicillin) (Lacasse et al. 2008, Sharun et al. 2021). Interestingly, bLFcin exhibited powerful activity against mastitis pathogens, which has attracted attention to its possible application in udder infections control. Kawai et al. (2003) carried out an *in vivo* trial in which lactoferrin hydrolysate (bLFcin) was infused in cows infected by subclinical mastitis. Obtained results revealed a significant reduction in numbers of *E. coli* and Staphylococci in the milk from infected cows on day one after infusion, and disappearance of the bacterial pathogens by the 14th day after lactoferrin hydrolysate administration. In a more recent study, the same research team reported that bovine lactoferrin hydrolysate (bLFcin) has an *in vitro* antimicrobial effect on *S. aureus*, *E. coli*, Enterococci, coagulase-negative Staphylococci, *Klebsiella pneumoniae*, *Prototheca zopfii*, and yeast-like fungi, all of which were isolated from bovine mastitis clinical cases (Kawai et al. 2007). Furthermore, bLFcin was proved to possess the ability to activate the production of bovine neutrophil superoxide. Nevertheless, these encouraging *in vitro* outcomes have to be further confirmed *in vivo* (Bruni et al. 2016).

In a step toward a new strategy to prevent bovine mastitis, bLFcin was expressed into the mammary gland of goat by an expression plasmid vector specific for mammary gland tissue and carrying the genes of bactericidal peptides (Zhang et al. 2007). This plasmid-mediated gene transfer approach enabled synthesis and secretion of bLFcin and bovine tracheal bactericidal peptides by mammary cells. Obtained results revealed that the expression of these antibacterial peptides in goat milk lasted for three to six days. Furthermore, milk collected from the injected mammary glands with a vector carrying the genes of bactericidal peptides displayed *in vitro* bacteriostatic effect on *E. coli* D12K31 and *S. aureus* ATCC 25923.

Tang et al. (2011) studied the effect of using bLFcin at a pharmacological level as a component of dietary supplementation in 21-day-weaned piglets that were challenged with enterotoxigenic *E. coli* and peptide role to replace colistin sulphate (an antibiotic that suppresses colonic flora and treats intestinal diseases). Dietary supplementation with cipB-LFcin, a fusion protein that releases the peptide in the piglets' stomach, improved gut microflora via decreasing the concentration of *E. coli* and increasing those of *Bifidobacterium* and *Lactobacillus* in the duodenum, jejunum, and ileum. The peptide also promoted piglet intestinal mucosal morphology via stimulating development of the jejunum and ileum crypt/villus architecture. Moreover, dietary supplementation with the peptide caused a decline in serum levels of cytokines and improved growth hormone. Overall, dietary supplementation with bLFcin could replace colistin sulphate in all of its effects on gut microflora, intestinal mucosal morphology, and circulating cytokines.

Vercelli et al. (2015) published an *in vivo* pilot study in dogs and recommended daily use for 14 days of an emulsion comprising bLFcin, verbascoside, and glycerophosphoinositole lysine for efficient treatment of recurrent otitis externa with clinical and bacterial and/or yeasts overgrowth. They also reported synergistic bactericidal and fungicidal activities of this therapeutic combination.

2.8 LACTOFERRICIN AS FOOD AND BEVERAGE PRESERVATIVE

In 2012, Quintieri et al. (2012) analyzed the antimicrobial effectiveness of bLf and its pepsin-digested hydrolysate (bLFcin) to control spoilage mesophilic bacteria, which contaminate traditional Mozzarella cheese during cold storage. They evidenced for the first time that bLFcin can inhibit the growth of cheese spoilage bacteria including coliforms and pseudomonads. In a more recent study, the same research group reported for the first time that bLf pepsin-digested hydrolysate (bLFcin) could prevent Mozzarella cheese blue discoloration via delaying the growth of psychrotrophic cheese pigmenting bacterial strains of *Pseudomonas fluorescens* (Caputo et al. 2015). Thus, bLFcin has the ability to counteract the chromatic spoilage of cheese during storage. Although bLFcin did not considerably affect the viable cell count of food beneficial microbes which is an advantage for its use as food preservative, its application has the drawback of peptide rapid degradation in food matrices, particularly foods comprising lactic acid bacteria with robust proteolytic activity (Quintieri et al. 2012, Caputo et al. 2015). Paul and Somkuti (2010) studied bLFcin hydrolytic breakdown by lactic acid bacteria including *Lactobacillus delbrueckii* sp. *bulgaricus* and *Streptococcus thermophilus* (strains of yogurt starter). They reported that the peptide was degraded by peptidases of lactic acid bacteria at neutral pH but compared to other bioactive peptides derived from milk, bLFcin is more resistant, with about 50% of the peptide lasting after 4 h of incubation. In addition, bLFcin was resistant to hydrolysis at acidic pH (pH 4.5).

Two synthetic derivatives from bLFcin (LFcinB (17–31) and LFcinB (20–25)) were found to possess antifungal activity against phytopathogenic filamentous fungus *Penicillium digitatum* that causes spoilage in mandarins, with a unique mode of action (Muñoz and Marcos 2006, Martínez-Culebras et al. 2021). LFcinB (17–31) showed superior inhibitory activity against *Penicillium digitatum* growth than LFcinB (20–25), nevertheless, LFcinB (17–31) was not fungicidal to quiescent fungal conidia, whereas LFcinB (20–25) was. Baruzzi et al. (2015) provided for the first time direct evidence for bLFcin ability to control microbial spoilage caused by various strains of *Pseudomonas* spp. in ready-to-eat (RTE) leafy vegetables (escarole chicory and Iceberg and Trocadero lettuce) throughout cold storage. These reports present an alternative green strategy to fungicides in the production of vegetables and fruits.

The *in vitro* antibacterial activity of bLf, pepsin-digested bLf (bLFcin), and amidated bLf was evaluated against *Listeria monocytogenes*, *S. aureus*, and *Enterococus faecalis*. The latter showed the highest resistance to these bactericidal agents, whereas *Listeria monocytogenes* was the most sensitive, and *S. aureus* displayed a moderate resistance. These bactericidal agents reduced *L. monocytogenes* to less than 0.5 log CFU/g in inoculated chicken breast fillets. When antimicrobials were combined with high-pressure treatment for 10 min at 400 MPa, their bactericidal effect slightly improved to less than 1 logCFU/g. Treatment of inoculated chicken breast fillets with antimicrobials either 18 h before or 1 h after high-pressure handling mostly gave the best results (reduced *L. monocytogenes* to less than 1.5 log CFU/g). Thus, both parent bLf protein and its peptide derivatives have a restricted potential for controlling meat pathogens (Del Olmo, Calzada, and Nunez 2012).

Enrique et al. (2009) examined the bactericidal effects of bLf pepsin hydrolysate (bLFcin) and a synthetic derivative from bLFcin (LFcinB (17–31)) against wine-related spoilage lactic acid bacteria responsible for wine spoilage during processes of winemaking. Both peptide and its synthetic derivative could inhibit growth of wine-related spoilage lactic acid bacteria *in vitro*. The bactericidal effect of LFcinB (17–31) against *Oenococcus oeni*, *Pediococcus damnosus*, and *Lactobacillus brevis* was demonstrated in most fermentation experiments in which these three LAB co-existed with the winemaking *Saccharomyces cerevisiae* T73 in the presence of the peptide. Last, LFcinB (17–31) at inhibitory concentrations did not modify enological parameters of wine must or fermentation kinetics.

Table 2.1 summarizes antimicrobial and antiviral activities of milk lactoferricin peptides and their different potential applications.

TABLE 2.1
Antimicrobial and Antiviral Activities of Milk Lactoferricin Peptides

Peptide	Bioactivity and applications	Target pathogens	Reference
Porcine pepsin hydrolysate of bLf	Antibacterial	*Escherichia coli* 0111	(Tomita et al. 1991)
Lactoferricin B (bLFcin)	Antibacterial	*Escherichia coli* *Salmonella enteritidis Klebsiella pneumoniae* *Proteus vulgaris* *Yersinia enterocolitica* *Pseudomonas aeruginosa* *Campylobacter jejuni* *Staphylococcus aureus* *Streptococcus mutans* *Corynebacterium diphtheriae* *Listeria monocytogenes* *Clostridium perfringens*	(Bellamy, Takase, Wakabayashi, et al. 1992)
Antimicrobial domain near the N-terminus of bLf and hLf in a region separate from sites of iron-binding	Antibacterial	*Escherichia coli* IID-861 *Klebsiella pneumoniae* JCM1662T *Pseudomonas aeruginosa IFO-3446* *Staphylococcus aureus* JCM-2151 *Listeria monocytogenes* IDF-1b	(Bellamy, Takase, Yamauchi, et al. 1992)
Lactoferricin B (bLFcin)	Antifungal	*Candida albicans*	(Bellamy et al. 1993)
Pepsin-derived bLf peptide (lactoferricin)	Antibacterial	*Escherichia coli* CL99 1-2 *Salmonella typhimurium* SL696 *Salmonella montevideo* SL5222	(Yamauchi et al. 1993)
Lactoferricin B (bLFcin)	Antifungal	*Candida albicans Trichophyton mentagrophytes*	(Bellamy et al. 1994)
Porcine pepsin hydrolysate (bLFcin)	Orally administered bacteriostatic agent	Intestinal Enterobacteriaceae in SPF mice	(Teraguchi et al. 1994)
Porcine pepsin hydrolysate (bLFcin)	Orally administered bacteriostatic agent	*Clostridium* species in the gut of mice	(Teraguchi et al. 1995)
N-terminal peptides derived from hLf and bLf	Antiparasitic	*Giardia lamblia*	(Turchany, Aley, and Gillin 1995)
LFcin combined with itraconazole or fluconazole	Antifungal	Some strains of azole-resistant *Candida albicans*	(Wakabayashi et al. 1996, Wakabayashi et al. 1998)

(continued)

TABLE 2.1 (Continued)
Antimicrobial and Antiviral Activities of Milk Lactoferricin Peptides

Peptide	Bioactivity and applications	Target pathogens	Reference
Three bLFcin peptides; the sequence corresponds to residues 17 to 42, two sequences residues 1 to 16 and 43 to 48 linked by a single disulfide bond, and disulfide-linked heterodimer corresponds to residues 1 to 48 cleaved between residues 42 and 43	Antibacterial	A number of pathogenic and food spoilage microorganisms, *Listeria monocytogenes*	(Dionysius and Milne 1997)
Pepsin hydrolysate of bLf (bLFH) and bLFcin	Antibacterial	*Escherichia coli* O157:H7	(Shin et al. 1998)
Synthetic peptides derived from N-terminus of hLf; hLF (1-11) and hLF (21-31)	Antifungal	*Candida albicans*	(Lupetti et al. 2000)
Cyclic lactoferricin	Inhibiting viral entry into human fibroblasts	Human cytomegalovirus	(Andersen et al. 2001)
bLFcin	Reduction of the infectivity of sporozoites in mice and rabbit models	*Toxoplasma gondii* *Eimeria stiedai*	(Omata et al. 2001)
A novel bLf peptide, FKCRRWQWRM	Antifungal	Two strains of *Candida albicans*	(Ueta, Tanida, and Osaki 2001)
Lactoferrin- and other milk proteins-derived peptides	Antiviral	HIV	(Berkhout et al. 2002)
N-terminal peptide of bLf (bLFcin)	Antiviral	Adenovirus	(Di-Biase et al. 2003)
bLf hydrolysate (bLFcin)	Infused antibacterial agent into mammary glands of cows infected with subclinical mastitis	*Escherichia coli* and Staphylococci	(Kawai et al. 2003)
Lactoferricin B (bLFcin)	Inhibiting the viral ability to infect cell cultures	Feline calici virus Poliovirus	(McCann et al. 2003)
Cyclic and linear human and bovine LFcin derivatives	Antiviral	HSV-1 and HSV-2	(Jenssen et al. 2004)
Two synthetic derivatives from bLFcin (LFcinB (17-31) and LFcinB (20-25))	Control spoilage in mandarins	*Penicillium digitatum*	(Muñoz and Marcos 2006)
bLf hydrolysate (bLFcin)	*In vitro* antibacterial effect on mastitis pathogens	*Staphylococcus aureus* *Escherichia coli* Enterococci Coagulase-negative Staphylococci *Klebsiella pneumoniae*	(Kawai et al. 2007)
hLFcin and bLFcin	Antiviral	Herpes simplex virus 1 and 2	(Jenssen et al. 2008)
Combination of bLFcin and bLf	Synergistic antibacterial effect	*Staphylococcus epidermidis* *Escherichia coli*	(López-Expósito et al. 2008)
bLf pepsin hydrolysate (bLFcin) and a synthetic derivative from bLFcin (LFcinB (17-31))	Antibacterial action against wine lactic acid bacteria	*Oenococcus oeni* *Pediococcus damnosus* *Lactobacillus brevis*	(Enrique et al. 2009)
bLFcin combined with fluoroquinolones	Synergistic antibacterial effect	Uropathogenic *Escherichia coli* strains	(Longhi et al. 2009)

TABLE 2.1 (Continued)
Antimicrobial and Antiviral Activities of Milk Lactoferricin Peptides

Peptide	Bioactivity and applications	Target pathogens	Reference
bLf and bLFcin	Interfering with intracellular viral trafficking	Herpes simplex virus-1	(Marr et al. 2009)
bLFcin combined with ceftazidime or ciprofloxacin	Synergistic antibacterial effect	*Staphylococcus aureus* *Pseudomonas aeruginosa*	(Oo et al. 2010)
bLFcin	Antibacterial component of dietary supplementation in 21-day-weaned piglets	Enterotoxigenic *Escherichia coli*	(Tang et al. 2011)
bLFcin	*In vitro* antibacterial effect combined with high pressure on spoilage of chicken breast fillets	*Listeria monocytogenes* *Staphylococcus aureus*	(Del Olmo, Calzada, and Nunez 2012)
bLf and its pepsin-digested hydrolysate (bLFcin)	Control spoilage mesophilic bacteria that contaminate traditional Mozzarella cheese during cold storage	Coliforms and pseudomonads	(Quintieri et al. 2012)
bLFcin combined with amphotericin B, voriconazole, or fluconazole	Inhibiting formation of fungal biofilm on contact lens by keratitis-causing fungi	*Candida albicans* *Fusarium solani* *Aspergillus fumigatus*	(Sengupta et al. 2012)
Lactoferricin	Inhibiting viral infection in mice	Herpes simplex virus type 2	(Shestakov et al. 2012)
bLFcin	Control microbial spoilage in escarole chicory and Iceberg and Trocadero lettuce throughout cold storage	Various strains of *Pseudomonas* spp.	(Baruzzi et al. 2015)
bLFcin	Prevent Mozzarella cheese blue discoloration	Psychrotrophic cheese pigmenting bacterial strains of *Pseudomonas fluorescens*	(Caputo et al. 2015)
bLFcin combined with verbascoside and glycerophosphoinositol	Bactericidal and fungicidal emulsion in treatment of otitis externa in dogs	Some bacterial strains and/or *Malassezia* spp. Yeasts	(Vercelli et al. 2015)
Bovine lactoferrin peptides	Antiparasitic	*Giardia intestinalis*	(Aguilar-Diaz et al. 2017)
cLFcin, bLFcin, and hLFcin	Antibacterial	*Escherichia coli* *Salmonella typhi* *Pseudomonas aeruginosa* *Staphylococcus aureus* MRSA	(El-Baky et al. 2021)

2.9 RECENT RESEARCH ON LACTOFERRICIN

Recent research studies focused on modifying the natural lactoferricin peptide to become more potent. Björn et al. (2016) developed HLR1r peptide from hLFcin by substitution of amino acids based on the hydrophobicity and the addition of an arginine-rich motif to the hLFcin peptide. This peptide was designed to facilitate the peptide-bacterial membrane interactions. Another approach was taken involving synthetic generation offusion or chimeric peptides from the combination of multiple bioactive peptides such as LFchimera, a fusion of bLFcin and lactoferrampin (Silva et al. 2013, Chahardooli, Fazeli, and Ghabooli 2018, Sarhadi, Jahandar, and Tanhaeian 2020), and LFT33, a fusion of bLFcin and the insect-derived antimicrobial peptide thanatin (Feng et al. 2012). Moreover, new functions and capabilities of LFcin may be uncovered in future studies, leading to additional therapeutic applications of LFcin.

2.10 *SPIRULINA PLATENSIS* AS SOURCE OF NUTRACEUTICAL AND PHARMACEUTICAL INGREDIENTS

The marine ecosystem is characterized by unique and diverse environmental conditions including different levels of temperature, illumination, pressure, and salinity, which increase the availability and chemical diversity of nutraceutical and pharmaceutical ingredients derived from marine organisms (Freitas et al. 2012). *Spirulina platensis* is a prokaryotic multicellular cyanobacterium microalga with photosynthesis capability that was isolated from soda lakes, seawater, as well as fresh water (Agustini et al. 2015). This microalga has been commercially produced for more than three decades. *Spirulina* is mostly consumed in vitamin supplements, aquaculture, fish food, food dyes, nutraceuticals, and pharmaceuticals (Belay et al. 1993).

Spirulina platensis is an excellent source of protein (percent composition of 60–70 by wet weight and includes eight essential amino acids and ten non-essential amino acids), carbohydrates (percent composition of 13.5–15 by wet weight and comprise galactose, glucose, rhamnose, mannose, xylose in addition to two unusual sugars (3-O-methyl-L-rhamnose and 2-O-methyl-L-rhamnose)), lipids (percent composition of 5–6 by wet weight and include polyunsaturated fatty acids), vitamins (percent composition of below 1 by wet weight), essential minerals (percent composition of about 7 by wet weight), and pigments (percent composition of below 1 by wet weight) (Becker et al. 1986, Yücetepe and Özçelik 2016).

Many scientific reports listed *Spirulina* health benefits such as its use in treatment of obesity and diabetes, reduction of high blood pressure, and its anticancer as well as antiviral effects (Hosoyamada, Takai, and Kato 1991, Aneiros and Garateix 2004, Mao, van de Water, and Gershwin 2005, Capelli and Cysewski 2010, Sharaf et al. 2010, Tang and Suter 2011, Sharaf et al. 2013, Vo, Ngo, and Kim 2015, Lisboa, Pereira, and Costa 2016, Vaz et al. 2016, Wu et al. 2016, Amara 2017, Gogineni and Hamann 2018, Sadeghi et al. 2018, Moreira et al. 2019, Shao et al. 2019, Bilal and Altiner 2020). In 2010, Sharaf et al. (2010) confirmed that phosphate buffer and hot water extracts of an American strain of *Spirulina platensis* could inhibit multiplication of herpes simplex virus prior to and after virus infection of host cells. On the other hand, they reported that the same extracts from an Egyptian *Spirulina platensis* strain directly inactivated virus particles before infection of host cells. They suggested application of *Spirulina* crude extracts as treatment for recurrent herpetic infection. In a more recent study, the same researchers demonstrated that crude extracts of *Spirulina platensis* strain from Chad lake also have antiviral effects on herpes virus before and during virus infection of host cells (Sharaf et al. 2013). Moura et al. (2011) reported that *Spirulina platensis* could effectively reduce total circulating levels of hepatic lipids and LDL-cholesterol, either alone or combined with physical exercise in diabetic Wistar rats. Recently, *Spirulina* was proved to be an efficient inhibitor to oxidative stress and hyperglycemia induced by diabetes and its administration was suggested to prevent diabetic complications (Gargouri et al. 2018). The therapeutic effect on diabetic complications was even more notable with *Spirulina* than that caused by insulin injection.

The protein content of *Spirulina platensis* is considered as a complete protein with high nutritional quality, comprising numerous essential amino acids, such as isoleucine, leucine, and valine, but with lower amounts of cysteine, methionine, and lysine compared to standard proteins (e.g. eggs, soybean, etc.) (Benelhadj et al. 2016). *Spirulina platensis* contains predominantly two phycobiliproteins (water soluble and highly fluorescent pigment-protein complexes responsible for capturing of light in cyanobacterial photosynthetic apparatus), namely C-phycocyanin and allophycocyanin (Patil et al. 2008). Both phycobiliproteins were used as natural dyes for cosmetics and food (Yoshida, Takagaki, and Nishimune 1996), fluorescent markers with various applications in biomedical research (Glazer 1994), in addition to potential pharmaceuticals in oxidative stress-induced diseases (Bhat and Madyastha 2001, Dagnino-Leone et al. 2022). Recently, Braune et al. (2021) reviewed current understanding of molecular mechanisms of potential phycocyanin-based anticancer treatments and approaches to improve their efficacy.

2.11 ANTIBACTERIAL PEPTIDES FROM *SPIRULINA PLATENSIS*

Bioactive peptides derived from natural sources are mostly found in enzymatic hydrolysate of protein. Peptides are released from proteins in food during digestion by the gut hydrolytic enzymes then affect health and cellular functions. Sun et al. (2016) isolated an antibacterial peptide of 18 amino acid residues, namely SP-1 from the papain and alkaline protease hydrolysate of *Spirulina platensis* by a series of chromatographic procedures. SP-1 was the first bactericidal peptide derived from protein hydrolysate of *S. platensis*. Its amino acid sequence was KLVDASHRLATGDVAVRA and has a molecular weight of 1878.97 Da. The peptide showed antibacterial activity against *E. coli* and *S. aureus* at MICs of 8 mg/ml and 16 mg/ml, respectively.

They reported that the antibacterial activity of SP-1 peptide was dependent on its α-helical structure, hydrophobicity that allows partitioning of the peptide into lipid bilayers of the bacterial membrane, and cationicity that allows peptide binding to negatively charged bacterial LPS, leading to perturbation of the integrity of bacterial cytoplasmic membrane or translocation of peptide into the bacterial cytoplasm (Sun et al. 2016).

Helical Wheel tool analysis for primary structure of SP-1 is presented in Figure 2.3. This analysis demonstrates that the peptide has a net charge of 1, and hydrophobicity of 0.228, comprising polar residues + glycine with number/percent of 9/50, uncharged residues + glycine including histidine

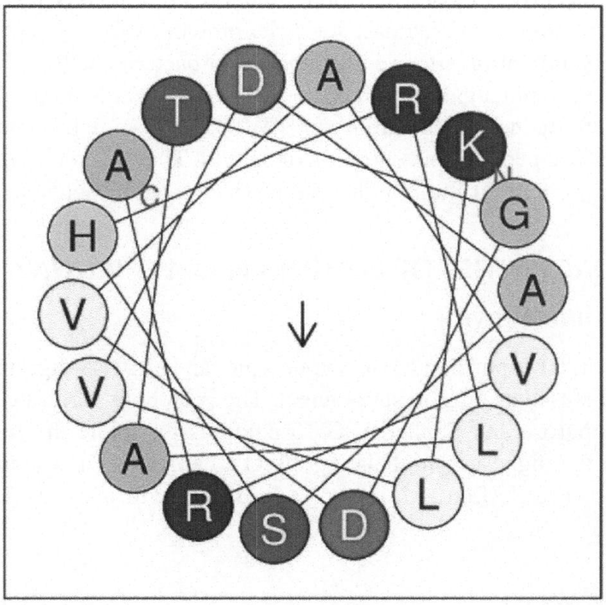

FIGURE 2.3 Helical Wheel diagrammatic illustration of primary structure of SP-1 peptide from protein hydrolysate of *Spirulina platensis*. D residue is acidic hydrophilic, V and L are nonpolar hydrophobic residues, T and S are polar and uncharged hydrophilic resides, R and K are basic hydrophilic residues.

FIGURE 2.4 The 3D structure of SP-1 peptide made by PEP-FOLD3 (the image is original for this chapter).

1, serine 1, threonine 1, and glycine 1, charged residues including lysine 1, arginine 2, and aspartic acid 2, nonpolar residues with number/percent of 9/50, and no special residues (cysteine 0, proline 0) or aromatic residues.

The 3D structure of SP-1 peptide is predicted from its amino acid sequence by PEP-FOLD3 as presented in Figure 2.4. The generated 3D structure of the peptide shows that it has a conserved α-helical structure.

Sadeghi et al. (2018) evaluated antibacterial and anticancer activities of chymotrypsin- and trypsin-digested protein hydrolysate of *S. platensis* against *E. coli*, *S. aureus*, and human colon adenocarcinoma cell line SW480. Hydrolyzed protein of *S. platensis* has approximately 22 kDa molecular weight and its derived bioactive peptide fraction has a molecular weight of below 3 kDa. This <3 kDa peptide fraction could decrease bacterial growth of *E. coli* and *S. aureus* with 15.2% and 19.6%, respectively, inhibition after 16 h incubation of bacterial cells with peptide fraction. The MIC values for peptide fraction against both bacterial species were 625 μg/ml. The <3 kDa peptide fraction also significantly reduced cell viability of SW480 in a dose-dependent manner with percent of cell viability of 44.9% at peptide concentration of 8 μg/μl. They suggested that *Spirulina plantesis* is a potential source of protein for future industrial production of bioactive peptides.

2.12 OTHER BIOACTIVITIES OF PEPTIDES FROM *SPIRULINA PLATENSIS*

2.12.1 Iron-Chelating Activity

An iron-chelating peptide was purified from flavourzyme- and alcalase-digested protein hydrolysate of *Spirulina* that can be used as an iron supplement. The protein hydrolysate was ultra-filtered and fractionated on Q-Sepharose and Sephadex G-15 columns to isolate the fractions below 3 kDa, then the peptide with the highest iron-chelating effect was isolated and identified. Its amino acid sequence was Thr-Asp-Pro-Ile(Leu)-Ala-Ala-Cys-Ile(Leu), with a molecular weight of 802 Da (Kim et al. 2014).

2.12.2 Anticancer Activity

An anticancer peptide was identified from papain-digested protein hydrolysate of *Spirulina platensis* that can be considered as a promising pharmaceutical or food ingredient. The protein hydrolysate was fractionated by gel filtration chromatography, and a new peptide with inhibitory effects on five cancer cell lines (A549, HepG-2, HT-29, MCF-7, and SGC-7901) was isolated. The peptide sequence was YGFVMPRSGLWFR, and has values of half-maximal inhibitory concentrations (IC_{50}) that range between 104.05 and 336.57 µg/ml against test cancer cells (Wang and Zhang 2016).

2.12.3 Antioxidant Activity

Yu et al. (2016) isolated a novel antioxidant peptide from an enzymatic hydrolysate of *Spirulina platensis* that can be applied in food and pharmaceutical industries. Its sequence was Pro-Asn-Asn and has a molecular mass of 343.15 Da. This peptide exhibited free radicals scavenging activity of 81.44 ± 0.43% for the antioxidant test of DPPH at 100 µg/ml peptide concentration, which was comparable to that of glutathione (82.63 ± 0.56%), 54.01 ± 0.82% for hydroxyl radicals and 47.84 ± 0.49% for hydrogen peroxide at 10 mg/ml peptide concentration.

2.12.4 Antihypertensive Activity

Heo et al. (2017) purified a heptameric peptide from chymotrypsin-, pepsin-, and trypsin-digested protein hydrolysate of *Spirulina*. Its sequence was TMEPGKP and has molecular weight of 759 Da. It has a non-competitive inhibitory effect on the angiotensin I converting enzyme (ACE). It also blocked the phosphorylation of p38 mitogen-activated protein kinase, and inhibited angiotensin II-induced vascular dysfunction and reactive oxygen species in human endothelial cells. Altogether, this heptameric peptide is a promising candidate as antihypertensive agent.

2.13 FUTURE PROSPECTS

The alarming rise in the numbers of pathogens that developed multidrug-resistance has turned the attention of scientific and health communities toward using alternative therapeutics as a substitute to conventional antibiotics and exploration of natural resources for novel, unconventional or complementary antimicrobial therapeutics for the future. During their exploration, scientists luckily found the treasure multifunctional protein, milk Lf, which plays a key role in various physiological processes. Aside from the significant and crucial bioactivities performed by intact Lf protein, this protein is an excellent source for derivatives or fragments with potential pharmaceutical applications. These derivatives or fragments are still being prepared by different enzymatic and chromatography processes, and explored for bioactivities (Rastogi et al. 2014, Bruni et al. 2016).

Until now, the highest bactericidal, fungicidal, and parasiticidal bioactivities against human and animal pathogens were achieved by intact protein and LFcin, nevertheless, extensive research remains to be carried out to explore their antimicrobial mechanisms. In fact, antimicrobial mechanisms of LFcin are still not fully discovered but promising and interesting outcomes begin to be obtained.

Antibacterial and antifungal effects of bLf and bLFcin, combined with their safety profile, have made them potential food preservatives that preserve food from pathogens and spoilage. This recent approach may decrease both food loss caused by spoilage and the use of chemical preservatives. It can also lead to developing new food products in the future for industry and markets.

Furthermore, the established protective roles of bLFcin against pathogens, cancer, and inflammation, its safety, as well as its natural origin make the peptide a promising candidate for therapeutics and functional foods.

Extensive research should be done in the future concerning synergy between LFcin and conventional antibiotics, via testing various combinations of LFcin-antibiotic and different pathogens, besides systematic studies on *in vivo* models, and selecting different administration routes of peptide (Sanchez-Gomez et al. 2011). Another underexplored approach is applying LFcin ability to penetrate cells as vectors of drugs for intracellular targets (Durzynska et al. 2015).

The construction of chimera peptides from LFcin and other AMPs is a promising approach for obtaining improved activity. Yet, such chimera may suffer from some drawbacks like difficulty of their large-scale production, in addition to stability and toxicity issues (Flores-Villasenor et al. 2010, Bolscher et al. 2012).

Moreover, besides LFcin derived from lactoferrin, there is a need to pursue research on enzymatic hydrolysate of other whey proteins from different species of mammals, which may result in constant discovery of novel peptides with antimicrobial effects. Therefore, evaluating both the *in vitro* and *in vivo* synergy profile of these peptides with conventional antimicrobials and their antimicrobial spectrum are the key targets for the future development of novel antimicrobial peptides.

Spirulina platensis is a famous marketed food supplement with so many health benefits. Various reports studied these benefits and the nutraceutical and pharmaceutical ingredients. Although, this microalga is well studied, only two antibacterial peptides were isolated and identified from its protein hydrolysate. Thus, more effort should be done in the future to isolate more antibacterial peptides from *Spirulina platensis* and explore their antimicrobial mechanisms.

2.14 CONCLUSION

Recently, peptides as a novel therapeutic approach have gained increasing attention. Peptides can play a crucial role in several biochemical processes because of their broad-spectrum activity and wide availability. The number of therapeutic peptides available in the market already exceeds 60, while many hundreds of novel peptide pharmaceuticals are currently under development in preclinical and clinical trials. The *in vitro* and *in vivo* reports detailed in this chapter regarding bioactivities of milk lactoferricin and *Spirulina platensis* antibacterial peptides present these NAMPs as promising candidates for novel antibiotic lead agents, antiviral and anticancer therapeutics, veterinary medicine applications, or even their utilization as functional food ingredients, food preservatives, and so many other applications. However, improving their cellular permeability and stability via appropriate modifications on the physical and chemical properties of peptides remains a challenge to certify success of drugs derived from these peptides. As soon as this improvement will be achieved in future, the new peptides with high cell permeability and stability, combined with their specificity, high efficacy, and safety as natural molecules, will be transformed into effective peptide therapeutics.

2.15 ACKNOWLEDGMENTS

We are grateful to Hoda Reda Kholef, MSc. in Microbiology, Protein Research Department, Genetic Engineering and Biotechnology Research Institute, City of Scientific Research and Technological Applications, for her support and help in collecting data of milk lactoferricin.

REFERENCES

Aguilar-Diaz, H., A. Canizalez-Roman, T. Nepomuceno-Mejia, F. Gallardo-Vera, Y. Hornelas-Orozco, K. Nazmi, J. G. M. Bolscher, J. C. Carrero, C. Leon-Sicairos, and N. Leon-Sicairos. 2017. Parasiticidal effect of synthetic bovine lactoferrin peptides on the enteric parasite *Giardia intestinalis*. *Biochem. Cell Biol.* 95 (1):82–90.

Aguilera, O., H. Ostolaza, L. M. Quiros, and J. F. Fierro. 1999. Permeabilizing action of an antimicrobial lactoferricin-derived peptide on bacterial and artificial membranes. *FEBS Lett.* 462 (3):273–277.

Agustini, T. W., M. Suzery, D. Sutrisnanto, W. F. Ma'rufa, and Hadiyanto. 2015. Comparative studyof bioactive substances extracted from fresh and dried *Spirulina sp. Procedia Environmental Sciences* 23:282–289.

Al Musaimi, O., B. G.. de la Torre, and F. Albericio. 2020. Greening Fmoc/t Bu solid-phase peptide synthesis. *Green Chem* 22 (4):996–1018.

Almehdar, H. A., N. A. El-Baky, A. A. Alhaider, S. A. Almuhaideb, A. A. Alhaider, R. S. Albiheyri, V. N. Uversky, and E. M. Redwan. 2019. Synergistic killing of pathogenic *Escherichia coli* using camel lactoferrin from different Saudi camel clans and various antibiotics. *Protein J.* 38 (4):479–496.

Almehdar, H. A., N. A. El-Baky, A. A. Alhaider, S. A. Almuhaideb, A. A. Alhaider, R. S. Albiheyri, V. N. Uversky, and E. M. Redwan. 2020. Bacteriostatic and bactericidal activities of camel lactoferrins against *Salmonella enterica Serovar Typhi. Probiotics Antimicrob. Proteins* 12 (1):18–31.

Amara, A. A. 2013. Pharmaceutical and Industrial protein engineering: Where we are? *Pak. J. Pharm. Sci.* 26 (1):217–232.

Amara, A. A. 2017. The Antisickling Effect of the *Arthrospira platensis* bilins for liver protection: A modeling, hypothesis, and food for thought. *SOJ Biochem* 3 (1):1–12.

Andersen, J. H., S. A. Osbakk, L. H. Vorland, T. Traavik, and T. J. Gutteberg. 2001. Lactoferrin and cyclic lactoferricin inhibit the entry of human cytomegalovirus into human fibroblasts. *Antiviral Res.* 51 (2):141–149.

Aneiros, A., and A. Garateix. 2004. Bioactive peptides from marine sources: Pharmacological properties and isolation procedures. *J. Chromatogr. B Analyt. Technol. Biomed. Life Sci.* 803 (1):41–53.

Arias, M., L. J. Mcdonald, E. F. Haney, K. Nazmi, J. G. Bolscher, and H. J. Vogel. 2014. Bovine and human lactoferricin peptides: Chimeras and new cyclic analogs. *Biometals* 27 (5):935–948.

Arnold, R. R., M. F. Cole, and J. R. McGhee. 1977. A bactericidal effect for human lactoferrin. *Science* 197 (4300):263–265.

Arnold, R. R., M. Brewer, and J. J. Gauthier. 1980. Bactericidal activity of human lactoferrin: Sensitivity of a variety of microorganisms. *Infect. Immun.* 28 (3):893–898.

Baruzzi, F., L. Pinto, L. Quintieri, A. Carito, N. Calabrese, and L. Caputo. 2015. Efficacy of lactoferricin B in controlling ready-to-eat vegetable spoilage caused by *Pseudomonas spp. Int. J. Food Microbiol.* 215:179–186.

Basso, V., Tran, D. Q., Ouellette, A. J., and Selsted, M. E. 2020. Host defense peptides as templates for antifungal drug development. *J. Fungi (Basel).* 6 (4):241.

Becker, E., B. Jakober, D. Luft, and R. M. Schmulling. 1986. Clinical and biochemical evaluations of the alga *spirulina* with regard to its application in the treatment of obesity. A double-blind cross-over study. *Nutr. Rep. Int.* 33 (4):565–574.

Belay, A., Y. Ota, K. Miyakawa, and H. Shimamatsu. 1993. Current knowledge on potential health benefits of *spirulina. J.Appl.Phycol* 5 (2):235–241.

Bellamy, W., M. Takase, H. Wakabayashi, K. Kawase, and M. Tomita. 1992. Antibacterial spectrum of lactoferricin B, a potent bactericidal peptide derived from the N-terminal region of bovine lactoferrin. *J. Appl. Bacteriol* 73 (6):472–479.

Bellamy, W., M. Takase, K. Yamauchi, H. Wakabayashi, K. Kawase, and M. Tomita. 1992. Identification of the bactericidal domain of lactoferrin. *Biochim. Biophys. Acta* 1121 (1–2):130–136.

Bellamy, W., H. Wakabayashi, M. Takase, K. Kawase, S. Shimamura, and M. Tomita. 1993. Killing of *Candida albicans* by lactoferricin B, a potent antimicrobial peptide derived from the N-terminal region of bovine lactoferrin. *Med. Microbiol. Immunol.* 182 (2):97–105.

Bellamy, W., K. Yamauchi, H. Wakabayashi, M. Takase, N. Takakura, S. Shimamura, and M. Tomita. 1994. Antifungal properties of lactoferricin B, a peptide derived from the N-terminal region of bovine lactoferrin. *Lett. Appl. Microbiol.* 18 (4):230–233.

Benelhadj, S., A. Gharsallaoui, P. Degraeve, H. Attia, and D. Ghorbel. 2016. Effect of pH on the functional properties of *Arthrospira* (*Spirulina*) *platensis* protein isolate. *Food Chem.* 194:1056–1063.

Berkhout, B., J. L. B. van Wamel, L. Beljaars, D. K. F. Meijer, S. Visser, and R. Floris. 2002. Characterization of the anti-HIV effects of native lactoferrin and other milk proteins and protein-derived peptides. *Antiviral Res.* 55 (2):341–355.

Bhat, V. B., and K. M. Madyastha. 2001. Scavenging of peroxynitrite by phycocyanin and phycocyanobilin from *Spirulina platensis*: Protection against oxidative damage to DNA. *Biochem. Biophys. Res. Commun.* 285 (2):262–266.

Bilal, T., and A. Altiner. 2020. Effect of *Spirulina Platensis* on some serum markers, performance metrics and organ weights, in rats fed with hydrogenated vegetable oil and/or cholesterol. *VMPH* 1 (2):60–69.

Björn, C., M. Mahlapuu, I. Mattsby-Baltzer, and J. Håkansson. 2016. Anti-infective efficacy of the lactoferrin-derived antimicrobial peptide HLR1r. *Peptides* 81:21–28.

Bolscher, J., K. Nazmi, J. van Marle, W. van Hof, and E. Veerman. 2012. Chimerization of lactoferricin and lactoferrampin peptides strongly potentiates the killing activity against *Candida albicans*. *Biochem. Cell Biol.* 90 (3):378–388.

Braune, S., A. Krüger-Genge, S. Kammerer, F. Jung, and J.-H. Küpper. 2021. Phycocyanin from *Arthrospira platensis* as potential anti-cancer drug: Review of *in vitro* and in vivo studies. *Life (Basel)* 11 (2):91.

Bruni, N., M. T. Capucchio, E. Biasibetti, E. Pessione, S. Cirrincione, L. Giraudo, A. Corona, and F. Dosio. 2016. Antimicrobial activity of lactoferrin-related peptides and applications in human and veterinary medicine. *Molecules* 21 (6):752.

Capelli, B., and G. R. Cysewski. 2010. Potential health benefits of *Spirulina* microalgae: A review of the existing literature. *Nutra foods* 9 (2):19–26.

Caputo, L., L. Quintieri, D. M. Bianchi, L. Decastelli, L. Monaci, A. Visconti, and F. Baruzzi. 2015. Pepsin-digested bovine lactoferrin prevents mozzarella cheese blue discoloration caused by *Pseudomonas fluorescens*. *Food Microbiol.* 46:15–24.

Chahardooli, M., A. Niazi, F. Aram, and S. M. Sohrabi. 2016. Expression of recombinant Arabian camel lactoferricin related peptide in *Pichia pastoris* and its antimicrobial identification. *J. Sci. Food Agric.* 96 (2):569–575.

Chahardooli, M., A. Fazeli, and M. Ghabooli. 2018. Recombinant production of bovine lactoferrin-derived antimicrobial peptide in tobacco hairy roots expression system. *Plant physiology et biochemistry* 123:414–421.

Chan, D. I., E. J. Prenner, and H. J. Vogel. 2006. Tryptophan- and arginine-rich antimicrobial peptides: Structures and mechanisms of action. *Biochim. Biophys. Acta* 1758 (9):1184–1202.

Chaparro, S. C. V., J. T. V. Salguero, D. A. M. Baquero, and J. E. R. Pérez. 2018. Effect of polyvalence on the antibacterial activity of a synthetic peptide derived from bovine lactoferricin against healthcare-associated infectious pathogens. *Biomed Res. Int.* 2018:5252891.

Chen, P.-W., C.-L. Shyu, and F. C. Mao. 2003. Antibacterial activity of short hydrophobic and basic-rich peptides. *Am. J. Vet. Res.* 64 (9):1088–1092.

Conesa, C., M. Calvo, and L. Sanchez. 2010. Recombinant human lactoferrin: A valuable protein for pharmaceutical products and functional foods. *Biotechnol. Adv.* 28 (6):831–838.

Dagnino-Leone, J., C. P. Figueroa, M. L. Castañeda, A. D. Youlton, A. Vallejos-Almirall, A. Agurto-Muñoz, J. Pavón Pérez, and C. Agurto-Muñoz. 2022. Phycobiliproteins: Structural aspects, functional characteristics, and biotechnological perspectives. *Comput. Struct. Biotechnol. J.* 20:1506–1527.

De la Torre, B. G., and F. Albericio. 2020. Peptide Therapeutics 2.0. *Molecules* 25 (10):2293.

Del Olmo, A., J. Calzada, and M. Nunez. 2012. Effect of lactoferrin and its derivatives against gram-positive bacteria *in vitro* and, combined with high pressure, in chicken breast fillets. *Meat Sci.* 90 (1):71–76.

Demmelmair, H., C. Prell, N. Timby, and B. Lonnerdal. 2017. Benefits of lactoferrin, osteopontin and milk fat globule membranes for infants. *Nutrients* 9 (8):817.

Di-Biase, A. M., A. Pietrantoni, A. Tinari, R. Siciliano, P. Valenti, G. Antonini, L. Seganti, and F. Superti. 2003. Heparin-interacting sites of bovine lactoferrin are involved in anti-adenovirus activity. *J. Med. Virol.* 69 (4):495–502.

Dionysius, D. A., and J. M. Milne. 1997. Antibacterial peptides of bovine lactoferrin: Purification and characterization. *J. Dairy Sci.* 80 (4):667–674.

Drago-Serrano, M. E., R. Campos-Rodríguez, J. C. Carrero, and M. de la Garza. 2017. Lactoferrin: Balancing ups and downs of inflammation due to microbial infections. *Int. J. Mol. Sci.* 18 (3):501.

Durzynska, J., L. Przysiecka, R. Nawrot, J. Barylski, G. Nowicki, A. Warowicka, O. Musidlak, and A. Gozdzicka-Jozefiak. 2015. Viral and other cell-penetrating peptides as vectors of therapeutic agents in medicine. *J. Pharmacol. Exp. Ther.* 354 (1):32–42.

El-Agamy, E.-S. I. 2006. Camel milk (Chapter 6). In: Y. W. Park and G. F. W. Haenlein (eds.) Handbook of Milk of Non-Bovine Mammals. Blackwell Publishers, Iowa, USA:297–344.

El-Baky, N. A., E. M. El-Fakharany, and E. M. Redwana. 2017. Lactoferrin and hepatitis C virus: Flexible fitting and invasive mechanism. In: T. B. Ng, J. H. Wong, R. Tse, T. G. Tse, and H. Chan (eds.) Lactoferrin: Structure, biological functions, health benefits and clinical applications. New York: Nova Science Publishers, Inc.:49–71.

El-Baky, N. A., M. A. Elkhawaga, E. S. Abdelkhalek, M. M. Sharaf, E. M. Redwan, and H. R. Kholef. 2021. *De novo* expression and antibacterial potential of four lactoferricin peptides in cell-free protein synthesis system. *Biotechnol. Rep. (Amsterdam, Netherlands)* 29:e00583.

El-Fakharany, E. M., A. Tabll, A. A. El-Wahab, B. M. Haroun, and E. M. Redwan. 2008. Potential activity of camel milk amylase and lactoferrin against hepatitis C virus infectivity in HepG2 and lymphocytes. *Hepat Mon* 8 (2):101–109.

Eliassen, L. T., G. Berge, B. Sveinbjornsson, J. S. Svendsen, L. H. Vorland, and O. Rekdal. 2002. Evidence for a direct antitumor mechanism of action of bovine lactoferricin. *Anticancer Res.* 22 (5):2703–2710.

Eliassen, L. T., G. Berge, A. Leknessund, M. Wikman, I. Lindin, C. Lokke, F. Ponthan, J. I. Johnsen, B. Sveinbjornsson, P. Kogner, T. Flaegstad, and O. Rekdal. 2006. The antimicrobial peptide, lactoferricin B, is cytotoxic to neuroblastoma cells *in vitro* and inhibits xenograft growth *in vivo*. *Int. J. Cancer* 119 (3):493–500.

Ellison III, R. T., F. M. LaForce, T. J. Giehl, D. S. Boose, and B. E. Dunn. 1990. Lactoferrin and transferrin damage of the Gram-negative outer membrane is modulated by Ca^{2+} and Mg^{2+}. *J Gen Microbiol* 136 (7):1437–1446.

Ellison III, R. T., and T. J. Giehl. 1991. Killing of gram-negative bacteria by lactoferrin and lysozyme. *J. Clin. Invest.* 88 (4):1080–1091.

Enrique, M., P. Manzanares, M. Yuste, M. Martínez, S. Vallés, and J. F. Marcos. 2009. Selectivity and antimicrobial action of bovine lactoferrin derived peptides against wine lactic acid bacteria. *Food Microbiol.* 26 (3):340–346.

Farah, Z., M. Mollet, M. Younan, and R. Dahir. 2007. Camel dairy in Somalia: Limiting factors and development potential. *Livestock Sci* 110 (1–2):187–191.

Feng, X.-J., J.-H. Wang, A.-S. Shan, D. Teng, Y.-L. Yang, Y. Yao, G.-P. Yang, Y.-C. Shao, S. Liu, and F. Zhang. 2006. Fusion expression of bovine lactoferricin in *Escherichia coli*. *Protein Expr. Purif.* 47 (1):110–117.

Feng, X., C. Liu, J. Guo, X. Song, J. Li, W. Xu, and Z. Li. 2012. Recombinant expression, purification, and antimicrobial activity of a novel hybrid antimicrobial peptide LFT33. *Appl. Microbiol. Biotechnol.* 95 (5):1191–1198.

Fernandes, K. E., and D. A. Carter. 2017. The antifungal activity of lactoferrin and its derived peptides: Mechanisms of action and synergy with drugs against fungal pathogens. *Front. Microbiol.* 8:2.

Flores-Villasenor, H., A. Canizalez-Roman, M. Reyes-Lopez, K. Nazmi, M. de la Garza, J. Zazueta-Beltran, N. Leon-Sicairos, and J. G. Bolscher. 2010. Bactericidal effect of bovine lactoferrin, LFcin, LFampin and LFchimera on antibiotic-resistant *Staphylococcus aureus* and *Escherichia coli*. *Biometals* 23 (3):569–578.

Fosgerau, K., and T. Hoffmann. 2015. Peptide therapeutics: Current status and future directions. *Drug Discov. Today* 20 (1):122–128.

Freitas, A. C., D. Rodrigues, T. A. P. Rocha-Santos, A. M. P. Gomes, and A. C. Duarte. 2012. Marine biotechnology advances towards applications in new functional foods. *Biotechnol. Adv.* 30 (6):1506–1515.

Gaglione, R., K. Pane, E. Dell'Olmo, V. Cafaro, E. Pizzo, G. Olivieri, E. Notomista, and A. Arciello. 2019. Cost-effective production of recombinant peptides in *Escherichia coli*. *New Biotechnol.* 51:39–48.

Gargouri, M., H. Hamed, A. Akrouti, X. Dauvergne, C. Magne, and A. El Feki. 2018. Effects of Spirulina platensis on lipid peroxidation, antioxidant defenses, and tissue damage in kidney of alloxan-induced diabetic rats. *Appl. Physiol. Nutr. Metab.* 43 (4):345–354. doi: doi: 10.1139/apnm-2017-0461.

Gifford, J. L., H. N. Hunter, and H. J. Vogel. 2005. Lactoferricin: A lactoferrin-derived peptide with antimicrobial, antiviral, antitumor and immunological properties. *Cell. Mol. Life Sci.* 62 (22):2588–2598.

Glazer, A. N. 1994. Phycobiliproteins—A family of valuable, widely used fluorophores. *J. Appl. Phycol.* 6 (2):105–112.

Gogineni, V., and M. T. Hamann. 2018. Marine natural product peptides with therapeutic potential: Chemistry, biosynthesis, and pharmacology. *Biochim. Biophys. Acta Gen. Subj.* 1862 (1):81–196.

Goodman, R. E., and F. L. Schanbacher. 1991. Bovine lactoferrin mRNA: Sequence, analysis, and expression in the mammary gland. *Biochem. Biophys. Res. Commun.* 180 (1):75–84.

Gruden, Š., and N. Poklar-Ulrih. 2021. Diverse mechanisms of antimicrobial activities of lactoferrins, lactoferricins, and other lactoferrin-derived peptides. *Int. J. Mol. Sci.* 22 (20):11264.

Guerra, J. R., A. B. Cárdenas, A. Ochoa-Zarzosa, J. L. Meza, A. Umaña Pérez, R. Fierro-Medina, Z. J. Rivera Monroy, and J. E. García Castañeda. 2019. The tetrameric peptide LfcinB (20–25)4 derived from bovine lactoferricin induces apoptosis in the MCF-7 breast cancer cell line. *RSC Adv.* 9 (36):20497–20504.

Habib, M. A. B., M. Parvin, T. C. Huntington, and M. R. A. Hasan. 2008. A review on culture, production and use of spirulina as food for humans and feeds for domestic animals and fish. In Food and Agriculture Organization (FAO) of the United Nations Rome. *FAO Fisheries and Aquaculture Circular* No. 1034 FIMA/C1034 (En). 1–33.

Han, F. F., Y. H. Gao, C. Luan, Y. G. Xie, Y. F. Liu, and Y. Z. Wang. 2013. Comparing bacterial membrane interactions and antimicrobial activity of porcine lactoferricin derived peptides. *J. Dairy Sci.* 96 (6):3471–3487.

Hancock, R. E., and G. Diamond. 2000. The role of cationic antimicrobial peptides in innate host defences. *Trends Microbiol.* 8 (9):402–410.

Hao, Y., N. Yang, X. Wang, D. Teng, R. Mao, X. Wang, Z. Li, and J. Wang. 2017. Killing of *Staphylococcus aureus* and *Salmonella enteritidis* and neutralization of lipopolysaccharide by 17-residue bovine lactoferricins: Improved activity of Trp/Ala-containing molecules. *Sci. Rep.* 7:44278–44278.

Haukland, H. H., H. Ulvatne, K. Sandvik, and L. H. Vorland. 2001. The antimicrobial peptides lactoferricin B and magainin 2 cross over the bacterial cytoplasmic membrane and reside in the cytoplasm. *FEBS Lett.* 508 (3):389–393.

Haversen, L., N. Kondori, L. Baltzer, L. A. Hanson, G. T. Dolphin, K. Duner, and I. Mattsby-Baltzer. 2010. Structure microbicidal activity relationship of synthetic fragments derived from the antibacterial α-helix of human lactoferrin. *Antimicrob. Agents Chemother.* 54 (1):418–425.

Heo, S. Y., S. C. Ko, C. S. Kim, G. W. Oh, B. Ryu, Z. J. Qian, G. Kim, W. S. Park, I. W. Choi, T. T. Phan, S. J. Heo, D. H. Kang, M. Yi, and W. K. Jung. 2017. A heptameric peptide purified from *Spirulina sp.* Gastrointestinal hydrolysate inhibits angiotensin I-converting enzyme- and angiotensin II-induced vascular dysfunction in human endothelial cells. *Int. J. Mol. Med.* 39 (5):1072–1082.

Hill, D. R., and D. S. Newburg. 2015. Clinical applications of bioactive milk components. *Nutr. Rev.* 73 (7):463–476.

Hosoyamada, Y., Y. Takai, and T. Kato. 1991. Effect of water soluble and water insoluble fractions of *Spirulina* over serum lipids and glucose resistance of rats. *J. Jpn. Soc. Nutr. Food Sci.* 44 (4):273–277.

Hsieh, C. C., B. Hernandez-Ledesma, S. Fernandez-Tome, V. Weinborn, D. Barile, and J. M. de Moura Bell. 2015. Milk proteins, peptides, and oligosaccharides: Effects against the 21st century disorders. *Biomed Res. Int.* 2015:146840.

Hu, Y., X. Meng, F. Zhang, Y. Xiang, and J. Wang. 2021. The *in vitro* antiviral activity of lactoferrin against common human coronaviruses and SARS-CoV-2 is mediated by targeting the heparan sulfate co-receptor. *Emerg. Microbes Infect.* 10 (1):317–330.

Huertas, N. D. J., Z. J. R. Monroy, R. F. Medina, and J.E.G. Castañeda. 2017. Antimicrobial activity of truncated and polyvalent peptides derived from the FKCRRQWQWRMKKGLA sequence against *Escherichia coli* ATCC 25922 and *Staphylococcus aureus* ATCC 25923. *Molecules* 22:987.

Hunter, H. N., A. R. Demcoe, H. Jenssen, T. J. Gutteberg, and H. J. Vogel. 2005. Human lactoferricin is partially folded in aqueous solution and is better stabilized in membrane mimetic solvent. *Antimicrob. Agents Chemother.* 49 (8):3387–3395.

Hwang, P. M., N. Zhou, X. Shan, C. H. Arrowsmith, and H. J. Vogel. 1998. Three-dimensional solution structure of lactoferricin b, an antimicrobial peptide derived from bovine lactoferrin. *Biochemistry* 37 (12):4288–4298.

Jad, Y. E., A. Kumar, A. El-Faham, B. G. de la Torre, and F. Albericio. 2019. Green transformation of solid-phase peptide synthesis. *ACS Sustain. Chem. Eng.* 7 (4):3671–3683.

Jahani, S., A. Shakiba, and L. Jahani. 2015. The antimicrobial effect of lactoferrin on Gram-negative and Gram-positive Bacteria. *Int J Infect* 2 (3):e27954.

Jenssen, H., J. H. Andersen, L. Uhlin-Hansen, T. J. Gutteberg, and O. Rekdal. 2004. Anti-HSV activity of lactoferricin analogues is only partly related to their affinity for heparan sulfate. *Antiviral Res.* 61 (2):101–109.

Jenssen, H., K. Sandvik, J. H. Andersen, R. E. W. Hancock, and T. J. Gutteberg. 2008. Inhibition of HSV cell-to-cell spread by lactoferrin and lactoferricin. *Antiviral Res.* 79 (3):192–198.

Kang, J. H., M. K. Lee, K. L. Kim, and K. S. Hahm. 1996. Structure-biological activity relationships of 11-residue highly basic peptide segment of bovine lactoferrin. *Int. J. Pept. Protein Res.* 48 (4):357–363.

Kaur, N., R. Dilawari, A. Kaur, G. Sahni, and P. Rishi. 2020. Recombinant expression, purification and PEGylation of Paneth cell peptide (cryptdin-2) with value added attributes against *Staphylococcus aureus*. *Sci. Rep.* 10 (1):12164.

Kawai, K., H. Nagahata, N. Y. Lee, A. Anri, and K. Shimazaki. 2003. Effect of infusing lactoferrin hydrolysate into bovine mammary glands with subclinical mastitis. *Vet. Res. Commun.* 27 (7):539–548.

Kawai, K., K. Shimazaki, H. Higuchi, and H. Nagahata. 2007. Antibacterial activity of bovine lactoferrin hydrolysate against mastitis pathogens and its effect on superoxide production of bovine neutrophils. *Zoonoses Public Health* 54 (3–4):160–164.

Kim, N. H., S. H. Jung, J. Kim, S.-H. Kim, H.-J. Ahn, and K. B. Song. 2014. Purification of an iron-chelating peptide from *Spirulina* protein hydrolysates. *J. Korean Soc. Appl. Biol. Chem.* 57 (1):91–95.

Kohn, E. M., D. J. Shirley, L. Arotsky, A. M. Picciano, Z. Ridgway, M. W. Urban, B. R. Carone, and G. A. Caputo. 2018. Role of cationic side chains in the antimicrobial activity of C18G. *Molecules* 23 (2):329.

Konuspayeva, G., B. Faye, G. Loiseau, and D. Levieux. 2007. Lactoferrin and immunoglobulin contents in camel's milk (*Camelus bactrianus*, *Camelus dromedarius*, and Hybrids) from Kazakhstan. *J. Dairy Sci.* 90 (1):38–46.

Kowalczyk, P., K. Kaczyńska, P. Kleczkowska, I. Bukowska-Ośko, K. Kramkowski, and D. Sulejczak. 2022. The lactoferrin phenomenon—A Miracle molecule. *Molecules* 27 (9):2941.

Kuppusamy, R., M. Willcox, D. S. Black, and N. Kumar. 2019. Short cationic peptidomimetic antimicrobials. *Antibiotics* 8 (2):44.

Lacasse, P., K. Lauzon, M. S. Diarra, and D. Petitclerc. 2008. Utilization of lactoferrin to fight antibiotic-resistant mammary gland pathogens. *J. Anim. Sci.* 86 (Suppl–13):66–71.

Legrand, D., E. Elass, M. Carpentier, and J. Mazurier. 2005. Lactoferrin: A modulator of immune and inflammatory responses. *Cell. Mol. Life Sci.* 62 (22):2549–2559.

Li, C., H. M. Blencke, V. Paulsen, T. Haug, and K. Stensvag. 2010. Powerful workhorses for antimicrobial peptide expression and characterization. *Bioeng. Bugs* 1 (3):217–220.

Li, Y. F., and Z. X. Chen. 2008. RAPD: A database of recombinantly-produced antimicrobial peptides. *FEMS Microbiol. Lett.* 289 (2):126–129.

Liao, Y., E. El Fakharany, B. Lönnerdal, and E. M. Redwan. 2012. Inhibitory effects of native and recombinant full length camel lactoferrin and its N and C lobes on hepatitis C virus infection of Huh7.5 Cells. *J. Med. Microbiol.* 61 (Pt–3):375–383.

Lisboa, C. R., A. M. Pereira, and J. A. V. Costa. 2016. Biopeptides with antioxidant activity extracted from the biomass of *Spirulina sp.* LEB 18. *Afr. J. Microbiol. Res.* 10 (3):79–86.

Liu, Y., F. Han, Y. Xie, and Y. Wang. 2011. Comparative antimicrobial activity and mechanism of action of bovine lactoferricin-derived synthetic peptides. *Biometals* 24 (6):1069–1078.

Longhi, C., M. Marazzato, M. P. Conte, V. Iebba, S. Schippa, L. Seganti, and A. Comanducci. 2009. Effect of lactoferricin on fluoroquinolone susceptibility of uropathogenic *Escherichia coli*. *J. Antibiot. (Tokyo)* 62 (2):109–111.

Lonnerdal, B. 2014. Infant formula and infant nutrition: Bioactive proteins of human milk and implications for composition of infant formulas. *Am. J. Clin. Nutr.* 99 (3):712S–717S.

López-Expósito, I., A. Pellegrini, L. Amigo, and I. Recio. 2008. Synergistic effect between different milk-derived peptides and proteins. *J. Dairy Sci.* 91 (6):2184–2189.

Lupetti, A., A. Paulusma-Annema, M. M. Welling, S. Senesi, J. T. Van Dissel, and P. H. Nibbering. 2000. Candidacidal activities of human lactoferrin peptides derived from the N-terminus. *Antimicrob. Agents Chemother.* 44 (12):3257–3263.

Mader, J. S., J. Salsman, D. M. Conrad, and D. W. Hoskin. 2005. Bovine lactoferricin selectively induces apoptosis in human leukemia and carcinoma cell lines. *Mol. Cancer Ther.* 4 (4):612–624.

Mader, J. S., A. Richardson, J. Salsman, D. Top, R. de Antueno, R. Duncan, and D. W. Hoskin. 2007. Bovine lactoferricin causes apoptosis in Jurkat T-leukemia cells by sequential permeabilization of the cell membrane and targeting of mitochondria. *Exp. Cell Res.* 313 (12):2634–2650.

Mao, T. K., J. van de Water, and M. E. Gershwin. 2005. Effects of a *Spirulina*-based dietary supplement on cytokine production from allergic rhinitis patients. *J. Med. Food* 8 (1):27–30.

Marcone, S., O. Belton, and D. J. Fitzgerald. 2017. Milk-derived bioactive peptides and their health promoting effects: A potential role in atherosclerosis. *Br. J. Clin. Pharmacol.* 83 (1):152–162.

Marr, A. K., H. Jenssen, M. R. Moniri, R. E. Hancock, and N. Pante. 2009. Bovine lactoferrin and lactoferricin interfere with intracellular trafficking of herpes simplex virus-1. *Biochimie* 91 (1):160–164.

Martínez-Culebras, P. V., M. Gandía, S. Garrigues, J. F. Marcos, and P. Manzanares. 2021. Antifungal peptides and proteins to control toxigenic fungi and mycotoxin biosynthesis. *Int. J. Mol. Sci.* 22 (24):13261.

McCann, K. B., A. Lee, J. Wan, H. Roginski, and M. J. Coventry. 2003. The effect of bovine lactoferrin and lactoferricin B on the ability of feline calici virus (a norovirus surrogate) and poliovirus to infect cell cultures. *J. Appl. Microbiol.* 95 (5):1026–1033.

Miao, X., T. Zhou, J. Zhang, J. Xu, X. Guo, H. Hu, X. Zhang, M. Hu, J. Li, W. Yang, J. Xie, Z. Xu, and L. Mou. 2020. Enhanced cell selectivity of hybrid peptides with potential antimicrobial activity and immunomodulatory effect. *Biochim. Biophys. Acta Gen. Subj.* 1864 (4):129532.

Moreira, J. B., L.-T. Lim, E. D. R. Zavareze, A. R. G. Dias, J. A. V. Costa, and M. G. de Morais. 2019. Antioxidant ultrafine fibers developed with microalga compounds using a free surface electrospinning. *Food Hydrocolloids* 93:131–136.

Morgenthau, A., S. K. Partha, P. Adamiak, and A. B. Schryvers. 2014. The specificity of protection against cationic antimicrobial peptides by lactoferrin binding protein B. *Biometals* 27 (5):923–933.

Moura, L. P., G. M. Puga, W. R. Beck, I. P. Teixeira, A. C. Ghezzi, G. A. Silva, and M. A. R. Mello. 2011. Exercise and *Spirulina* control non-alcoholic hepatic steatosis and lipid profile in diabetic Wistar rats. *Lipids Health Dis.* 10:77.

Muñoz, A., and J. F. Marcos. 2006. Activity and mode of action against fungal phytopathogens of bovine lactoferricin-derived peptides. *J. Appl. Microbiol.* 101 (6):1199–1207.

Nibbering, P. H., E. Ravensbergen, M. M. Welling, L. A. van Berkel, P. H. van Berkel, E. K. Pauwels, and J. H. Nuijens. 2001. Human lactoferrin and peptides derived from its N terminus are highly effective against infections with antibiotic-resistant bacteria. *Infect. Immun.* 69 (3):1469–1476.

Odell, E. W., R. Sarra, M. Foxworthy, D. S. Chapple, and R. W. Evans. 1996. Antibacterial activity of peptides homologous to a loop region in human lactoferrin. *FEBS Lett.* 382 (1–2):175–178.

Omata, Y., M. Satake, R. Maeda, A. Saito, K. Shimazaki, K. Yamauchi, Y. Uzuka, S. Tanabe, T. Sarashina, and T. Mikami. 2001. Reduction of the infectivity of *Toxoplasma gondii* and *Eimeria stiedai* sporozoites by treatment with bovine lactoferricin. *J. Vet. Med. Sci.* 63 (2):187–190.

Oo, T. Z., N. Cole, L. Garthwaite, M. D. P. Willcox, and H. Zhu. 2010. Evaluation of synergistic activity of bovine lactoferricin with antibiotics in corneal infection. *J. Antimicrob. Chemother.* 65 (6):1243–1251.

Otero-gonzalez, A. J., B. S. Magalhaes, M. Garcia-villarino, C. Lopez-Abarrategui, D. A. Sousa, S. C. Dias, and O. L. Franco. 2010. Antimicrobial peptides from marine invertebrates as a new frontier for microbial infection control. *FASEB J.* 24 (5):1320–1334.

Pammi, M., and S. A. Abrams. 2015. Oral lactoferrin for the prevention of sepsis and necrotizing enterocolitis in preterm infants. *Cochrane Database Syst. Rev.* (2):CD007137.

Patil, G., S. Chethana, M. C. Madhusudhan, and K. S. M. S. Raghavarao. 2008. Fractionation and purification of the phycobiliproteins from *Spirulina platensis*. *Bioresour. Technol.* 99 (15):7393–7396.

Paul, M., and G. A. Somkuti. 2010. Hydrolytic breakdown of lactoferricin by lactic acid bacteria. *J. Ind. Microbiol. Biotechnol.* 37 (2):173–178.

Phelan, M., and D. Kerins. 2011. The potential role of milk-derivedpeptides in cardiovascular disease. *Food Funct.* 2 (3–4):153–167.

Quintieri, L., L. Caputo, L. Monaci, D. Deserio, M. Morea, and F. Baruzzi. 2012. Antimicrobial efficacy ofpepsin-digested bovine lactoferrin on spoilage bacteria contaminating traditional Mozzarella cheese. *Food Microbiol.* 31 (1):64–71.

Radaic, A., M. B. de Jesus, and Y. L. Kapila. 2020. Bacterial anti-microbial peptides and nano-sized drug delivery systems: The state of the art toward improved bacteriocins. *J. Control. Release* 321:100–118.

Rastogi, N., N. Nagpal, H. Alam, S. Pandey, L. Gautam, M. Sinha, K. Shin, N. Manzoor, J. S. Virdi, P. Kaur, S. Sharma, and T. P. Singh. 2014. Preparation and antimicrobial action of three tryptic digested functional molecules of bovine lactoferrin. *PLoS ONE* 9 (3):e90011.

Redwan, E. M., and A. Tabll. 2007. Camel lactoferrin markedly inhibits hepatitis C virus genotype 4 infection of human peripheral blood leukocytes. *J. Immunoassay Immunochem.* 28 (3):267–277.

Redwan, E. M., E. M. El-Fakharany, V. N. Uversky, and M. H. Linjawi. 2014. Screening the anti-infectivity potentials of native N- and C-lobes derived from the camel lactoferrin against hepatitis C virus. *BMC Complement. Altern. Med.* 14:219.

Redwan, E. M., N. A. El-Baky, A. M. Al-Hejin, M. N. Baeshen, H. A. Almehdar, A. Elsaway, A. M. Gomaa, S. B. Al-Masaudi, F. A. Al-Fassi, I. E. AbuZeid, and V. N. Uversky. 2016. Significant antibacterial activity and synergistic effects of camel lactoferrin with antibiotics against methicillin-resistant *Staphylococcus aureus* (MRSA). *Res. Microbiol.* 167 (6):480–491.

Reguera, L., and D. G. Rivera. 2019. Multicomponent reaction toolbox for peptide macrocyclization and stapling. *Chem. Rev.* 119 (17):9836–9860.

Rekdal, O., J. Andersen, L. H. vorland, and J. S. Svendsen. 1999. Construction and synthesis of lactoferricin derivatives with enhanced antibacterial activity. *J. Pept. Sci.* 5 (1):32–45.

Riahifard, N., Mozaffari, S., Aldakhil, T., Nunez, F., Alshammari, Q., Alshammari, S., Yamaki, J., Parang, K., and Tiwari, R. K. 2018. Design, synthesis, and evaluation of amphiphilic cyclic and linear peptides composed of hydrophobic and positively-charged amino acids as antibacterial agents. *Molecules* 23 (10):2722.

Rybarczyk, J., E. Kieckens, D. Vanrompay, and E. Cox. 2017. *In vitro* and *in vivo* studies on the antimicrobial effect of lactoferrin against *Escherichia coli* O157:H7. *Vet. Microbiol.* 202:23–28.

Sadeghi, S., H. Jalili, S. O. Ranaei Siadat, and M. Sedighi. 2018. Anticancer and antibacterial properties in peptide fractions from hydrolyzed *Spirulina protein. J. Agric. Sci. Technol* 20 (4):673–683.

Samaniego-Barron, L., S. Luna-Castro, C. Pina-Vazquez, F. Suarez-Guemes, and M. de la Garza. 2016. Two outer membrane proteins are bovine lactoferrin-binding proteins in *Mannheimia haemolytica,* A1. *Vet. Res.* 47 (1):93.

Sampaio de Oliveira, K. B., M. L. Leite, G. R. Rodrigues, H. M. Duque, R. A. da Costa, V. A. Cunha, L. S. de Loiola Costa, N. B. da Cunha, O. L. Franco, and S. C. Dias. 2020. Strategies for recombinant production of antimicrobial peptides with pharmacological potential. *Expert Rev. Clin. Pharmacol.* 13 (4):367–390.

Sanchez-Gomez, S., B. Japelj, R. Jerala, I. Moriyon, M. Fernandez Alonso, J. Leiva, S. E. Blondelle, J. Andra, K. Brandenburg, K. Lohner, and G. Martínez de Tejada. 2011. Structural features governing the activity of lactoferricin-derived peptides that act in synergy with antibiotics against *Pseudomonas aeruginosa in vitro* and i*n vivo*. *Antimicrob. Agents Chemother.* 55 (1):218–228.

Sarhadi, H., M. H. Jahandar, and A. Tanhaeian. 2020. Evaluation of antibacterial properties of chimeric bovine lactoferrin peptide for inhibition of food and plant pathogens. *Int. J. Infect.* 7 (2):e104594.

Schibli, D. J., P. M. Hwang, and H. J. Vogel. 1999. The structure of the antimicrobial active center of lactoferricin b bound to Sodium dodecyl sulfate micelles. *FEBS Lett.* 446 (2–3):213–217.

Sengupta, J., S. Saha, A. Khetan, S. K. Sarkar, and S. M. Mandal. 2012. Effects of lactoferricin B against keratitis-associated fungal biofilms. *J. Infect. Chemother.* 18 (5):698–703.

Shao, W., R. Ebaid, M. El-Sheekh, A. Abomohra, and H. Eladel. 2019. Pharmaceutical applications and consequent environmental impacts of *Spirulina* (*Arthrospira*): An overview. *Grasas y Aceites* 70 (1):e292.

Sharaf, M., A. Amara, A. Aboul-Enein, S. Helmi, A. Ballot, A. Astani, and P. Schnitzler. 2010. Molecular authentication and characterization of the antiherpetic activity of the cyanobacterium *Arthrospira fusiformis*. *Pharmazie* 65 (2):132–136.

Sharaf, M., A. Amara, A. Aboul-Enein, S. Helmi, A. Ballot, and P. Schnitzler. 2013. Antiherpetic efficacy of aqueous extract of the cyanobacterium *Arthrospira fusiformis* from Chad. *Pharmazie* 68 (5):376–380.

Sharun, K., K. Dhama, R. Tiwari, M. B. Gugjoo, Y. M. Iqbal, S. K. Patel, M. Pathak, K. Karthik, S. K. Khurana, R. Singh, B. Puvvala, Amarpal, R. Singh, K. P. Singh, and W. Chaicumpa. 2021. Advances in therapeutic and managemental approaches of bovine mastitis: A comprehensive review. *Vet. Q.* 41 (1):107–136.

Shestakov, A., H. Jenssen, I. Nordström, and K. Eriksson. 2012. Lactoferricin but not lactoferrin inhibit herpes simplex virus type 2 infection in mice. *Antiviral Res.* 93 (3):340–345.

Shin, K., K. Yamauchi, S. Teraguchi, H. Hayasawa, M. Tomita, and Y. Otsuka. 1998. Antibacterial activity of bovine lactoferrin and its peptides against enterohaemorrhagic *Escherichia col*i O157:H7. *Lett. Appl. Microbiol.* 26 (6):407–411.

Silva, T., R. Adao, K. Nazmi, J. G. Bolscher, S. S. Funari, D. Uhrikova, and M. Bastos. 2013. Structural diversity and mode of action on lipid membranes of three lactoferrin candidacidal peptides. *Biochim. Biophys. Acta* 1828 (5):1329–1339.

Silva, T., B. Magalhaes, S. Maia, P. Gomes, K. Nazmi, J. G. M. Bolscher, P. N. Rodrigues, M. Bastos, and M. S. Gomes. 2014. Killing of *Mycobacterium avium* by lactoferricin peptides: Improved activity of arginine- and D-amino-acid-containing molecules. *Antimicrob. Agents Chemother.* 58 (6):3461–3467.

Sinha, M., S. Kaushik, P. Kaur, S. Sharma, and T. P. Singh. 2013. Antimicrobial lactoferrin peptides: The hidden players in the protective function of a multifunctional protein. *Int. J. Pept.* 2013:390230.

Srinivasan, R., A. Kannappan, C. Shi, and X. Lin. 2021. Marine bacterial secondary metabolites: A treasure house for structurally unique and effective antimicrobial compounds. *Mar. Drugs* 19 (10):530.

Strøm, M. B., O. Rekdal, and J. S. Svendsen. 2000. Antibacterial activity of 15-residue lactoferricin derivatives. *J. Pept. Res.* 56 (5):265–274.

Strøm, M. B., B. E. Haug, O. Rekdal, M. L. Skar, W. Stensen, and J. S. Svendsen. 2002. Important structural features of 15-residue lactoferricin derivatives and methods for improvement of antimicrobial activity. *Biochem. Cell Biol.* 80 (1):65–74.

Sultan, S., N. Huma, M. S. Butt, M. Aleem, and M. Abbas. 2018. Therapeutic potential of dairy bioactive peptides: A contemporary perspective. *Crit. Rev. Food Sci. Nutr.* 58 (1):105–115.

Sun, C., Y. Li, S. Cao, H. Wang, C. Jiang, S. Pang, M. A. Hussain, and J. Hou. 2018. Antibacterial activity and mechanism of action of bovine lactoferricin derivatives with symmetrical amino acid sequences. *Int. J. Mol. Sci.* 19 (10):2951.

Sun, Y., R. Chang, Q. Li, and B. Li. 2016. Isolation and characterization of an antibacterial peptide from protein hydrolysates of *Spirulina platensis*. *Eur. Food Res. Technol.* 242:685–692.

Tang, G., and P. M. Suter. 2011. Vitamin A, nutrition, and health values of algae: *Spirulina*, *Chlorella*, and *Dunaliella*. *J. Pharm. Nutr. Sci.* 1 (2):111–118.

Tang, Z. R., Z. H. Sun, X. S. Tang, Z. M. Feng, D. Zhou, D. F. Xiao, B. Zhang, and L. L. Li. 2011. Dietary bovine lactoferricin improved gut microflora, benefited intestinal mucosal morphology and inhibited circular cytokines delivery in 21-day weaned piglets. *J. Appl. Anim. Res.* 39 (2):153–157.

Teraguchi, S., K. Ozawa, S. Yasuda, K. Shin, Y. Fukuwatari, and S. Shimamura. 1994. The bacteriostatic effects of orally administered bovine lactoferrin on intestinal Enterobacteriaceae of SPF mice fed bovine milk. *Biosci. Biotechnol. Biochem.* 58 (3):482–487.

Teraguchi, S., K. Shin, K. Ozawa, S. Nakamura, Y. Fukuwatari, S. Tsuyuki, H. Namihira, and S. Shimamura. 1995. Bacteriostatic effect of orally administered bovine lactoferrin on proliferation of *Clostridium* species in the gut of mice fed bovine milk. *Appl. Environ. Microbiol.* 61 (2):501–506.

Theolier, J., I. Fliss, J. Jean, and R. Hammami. 2014. MilkAMP: A comprehensive database of antimicrobial peptides of dairy origin. *Dairy Sci. Technol.* 94 (2):181–193.

Tian, Z.-G., D. Teng, Y.-L. Yang, J. Luo, X. Feng, Y. Fan, F. Zhang, and J. Wang. 2007. Multimerization and fusion expression of bovine lactoferricin derivative LFcinB15-W4, 10 in *Escherichia coli*. *Appl. Microbiol. Biotechnol.* 75 (1):117–124.

Tomita, M., W. Bellamy, M. Takase, K. Yamauchi, H. Wakabayashi, and K. Kawase. 1991. Potent antibacterial peptides generated by pepsin digestion of bovine lactoferrin. *J. Dairy Sci.* 74 (12):4137–4142.

Tomita, M., M. Takase, W. Bellamy, and S. Shimamura. 1994. A review: The active peptide of lactoferrin. *Acta Paediatr. Jpn.* 36 (5):585–591.

Turchany, J. M., S. B. Aley, and F. D. Gillin. 1995. *Giardicidal* activity of lactoferrin and N-terminal peptides. *Infect. Immun.* 63 (11):4550–4552.

Uddin, S. J., J. A. Shilpi, L. Nahar, S. D. Sarker, and U. Göransson. 2021. Editorial: Natural antimicrobial peptides: Hope for new antibiotic lead molecules. *Front. Pharmacol.* 12:640938.

Ueta, E., T. Tanida, and T. Osaki. 2001. A novel bovine lactoferrin peptide, FKCRRWQWRM, suppresses *Candida* cell growth and activates neutrophils. *J. Pept. Res* 57 (3):240–249.

Vargas-Casanova, Y., A. V. Rodríguez-Mayor, K. J. Cardenas, A. L. Leal-Castro, L Muñoz-Molina, C., R. Fierro-Medina, Z. J. Rivera-Monroy, and J. E.. García-Castañeda. 2019. Synergistic bactericide and antibiotic effects of dimeric, tetrameric, or palindromic peptides containing the RWQWR motif against Gram-positive and Gram-negative strains. *RSC Adv.* 9 (13):7239–7245.

Vaz, B. D. S., J. B. Moreira, M. G. deMorais, and J. A. V. Costa. 2016. Microalgae as a new source of bioactive compounds in food supplements. *Curr. Opin. Food Sci.* 7:73–77.

Vercelli, A., N. Fanton, N. Bruni, D. Vergnano, M. Bigliati, and L. Cornegliani. 2015. Use of lactoferricin, verbascoside and glycerophosphoinositol emulsion in otitis externa treatment: A pilot study. *Veterinaria* 29 (3):49–57.

Vo, T.-S., D.-H. Ngo, and S.-K. Kim. 2015. Nutritional and pharmaceutical properties of microalgal *Spirulina* (Chapter 19), In: S.-K. Kim (ed.) Handbook of Marine Microalgae, Academic Press:299–308.

Wakabayashi, H., S. Abe, T. Okutomi, S. Tansho, K. Kawase, and H. Yamaguchi. 1996. Cooperative anti-Candida effects of lactoferrin or its peptides in combination with azole antifungal agents. *Microbiol. Immunol.* 40 (11):821–825.

Wakabayashi, H., S. Abe, S. Teraguchi, H. Hayasawa, and H. Yamaguchi. 1998. Inhibition of hyphal growth of azole-resistant strains of *Candida albicans* by triazole antifungal agents in the presence of lactoferrin-related compounds. *Antimicrob. Agents Chemother* 42 (7):1587–1591.

Wang, H., X. Zhao, and F. Lu. 2007. Heterologous expression of bovine lactoferricin in *Pichia methanolica*. *Biochemistry Mosc.* 72 (6):640–643.

Wang, Z., and X. Zhang. 2016. Inhibitory effects of small molecular peptides from *Spirulina* (*Arthrospira*) platensis on cancer cell growth. *Food Funct.* 7 (2):781–788.

Wibowo, D., and C.-X. Zhao. 2019. Recent achievements and perspectives for large-scale recombinant production of antimicrobial peptides. *Appl. Microbiol. Biotechnol.* 103 (2):659–671.

Wu, Q., L. Liu, A. Miron, B. Klimova, D. Wan, and K. Kuca. 2016. The antioxidant, immunomodulatory, and anti-inflammatory activities of *Spirulina*: An overview. *Arch. Toxicol.* 90 (8):1817–1840.

Xu, X., F. Jin, X. Yu, S. Ji, J. Wang, H. Cheng, C. Wang, and W. Zhang. 2007. Expression and purification of a recombinant antibacterial peptide, cecropin, from *Escherichia coli*. *Protein Expr. Purif.* 53 (2):293–301.

Yamauchi, K., M. Tomita, T. J. Giehl, and R. T. Ellison III. 1993. Antibacterial activity of lactoferrin and a pepsin-derived lactoferrin peptide fragment. *Infect. Immun.* 61 (2):719–728.

Yan, D., D. Chen, J. Shen, G. Xiao, A. J. van Wijnen, and H. J. Im. 2013. Bovine lactoferricin is anti-inflammatory and anti-catabolic in human articular cartilage and synovium. *J. Cell. Physiol.* 228 (2):447–456.

Yang, N., M. B. Strøm, S. M. Mekonnen, J. S. Svendsen, and Ø. Rekdal. 2004. The effects of shortening lactoferrin derived peptides against tumour cells, bacteria and normal human cells. *J. Pept. Sci.* 10 (1):37–46.

Yao, J.-F., H. Yang, Y.-Z. Zhao, and M. Xue. 2018. Metabolism of peptide drugs and strategies to improve their metabolic stability. *Curr. Drug Metab.* 19 (11):892–901.

Yoo, Y. C., S. Watanabe, R. Watanabe, K. Hata, K. Shimazaki, and I. Azuma. 1997. Bovine lactoferrin and lactoferricin, a peptide derived from bovine lactoferrin, inhibit tumor metastasis in mice. *Jpn. J. Cancer Res.* 88 (2):184–190.

Yoshida, A., Y. Takagaki, and T. Nishimune. 1996. Enzyme immunoassay for phycocyanin as the main component of *Spirulina* color in foods. *Biosci. Biotechnol. Biochem.* 60 (1):57–60.

Yu, J., Y. Hu, M. Xue, Y. Dun, S. Li, N. Peng, Y. Liang, and S. Zhao. 2016. Purification and identification of antioxidant peptides from enzymatic hydrolysate of *Spirulina platensis*. *J. Microbiol. Biotechnol.* 26 (7):1216–1223.

Yücetepe, A., and B. Özçelik. 2016. Bioactive peptides isolated from microalgae *Spirulina platensis* and their biofunctional activities. *Akademik Gıda* 14 (4):412–417.

Zarzosa-Moreno D., Avalos-Gómez C., Ramírez-Texcalco L. S., Torres-López E., Ramírez-Mondragón R., Hernández-Ramírez J. O., Serrano-Luna J., and de la Garza M. 2020. Lactoferrin and its derived peptides: An alternative for combating virulence mechanisms developed by pathogens. *Molecules* 25 (24):5763.

Zasloff, M. 2002. Antimicrobial peptides of multicellular organisms. *Nature* 415 (6870):389–395.

Zhang, J. X., S. F. Zhang, T. D. Wang, X. J. Guo, and R. L. Hu. 2007. Mammary gland expression of antibacterial peptide genes to inhibit bacterial pathogens causing mastitis. *J. Dairy Sci.* 90 (11):5218–5225.

Zhou, N., D. P. Tieleman, and H. J. Vogel. 2004. Molecular dynamics simulations of bovine lactoferricin: Turning a helix into A sheet. *Biometals* 17 (3):217–223.

Zhu, Y. 2022. Human Papillomavirus (HPV) Entry Inhibitors. *Adv. Exp. Med. Biol.* 1366:223–239.

3 Plant-Based Antimicrobials for Sepsis Management

What Progress Have We Made?

*Shikha Joon,[1] Enty Tjoa,[2] Rajeev K. Singla,[1] Rakesh Bhatnagar,[**3,4] and Bairong Shen[*1]*

[1]Institutes for Systems Genetics, Frontiers Science Center for Disease-related Molecular Network, West China Hospital, Sichuan University, Chengdu, Sichuan, China.

[2]Department of Microbiology, School of Medicine and Sciences, Atma Jaya Catholic University of Indonesia, Jakarta, Indonesia.

[3]Molecular Biology and Genetic Engineering Laboratory, School of Biotechnology, Jawaharlal Nehru University, New Delhi, India.

[4]Amity University of Rajasthan, Jaipur, Rajasthan, India.

[*]**Corresponding author**: Bairong Shen bairong.shen@scu.edu.cn

[**]**Co-corresponding author**: Rakesh Bhatnagar rakeshbhatnagar@jnu.ac.in

CONTENTS

DOI: 10.1201/9781003268932-3

3.1 INTRODUCTION

Sepsis is an ill-regulated host inflammatory response to microbial intrusion that leads to massive and fatal tissue damage [1]. Depending upon the magnitude of severity, sepsis is categorized into severe sepsis, septic shock, and multi-organ failure [1]. A life-threatening clinical syndrome affects the health status of the population from all age groups and is among the leading cause of mortality worldwide [2]. In particular, the elderly (aged more than 65 years), children (aged less than 4 years), immune-compromised individuals, and patients with chronic ailments are primarily victimized [2]. Notably, the clinical manifestations of sepsis comprise a pronounced fluctuation in the cardiac, blood pressure and respiratory rates, and body temperature (fever with chills) of the affected host [3–5]. At present, the therapeutic approaches for sepsis rely on the empirical administration of an antibiotic or a combination of broad-spectrum antimicrobials [6–14]. Quite dismally, the therapeutic efficacy of these chemically synthesized antimicrobials is limited by the highly adaptive resistance mechanisms for myriad drugs amongst these pathogens [6–14]. This, in turn, raises serious environmental concerns about the vicious cycle of reckless administration of antibiotics and the emergence of multidrug-resistant microbial strains [6–14]. Moreover, these are accompanied by side effects that include dizziness, diarrhea, nausea, skin rashes, severe allergic reactions, and might predispose the host to other microbial infections, such as those caused by Clostridium difficile [15]. This emphasizes the exploration of naturally occurring low-cost, plant-based herbal formulations and therapeutics for sepsis management. Undoubtedly, the phytoconstituents obtained from medicinal plants offer innumerable clinical and economic advantages as a treatment alternative for sepsis over commercial antibiotics, including negligible side effects (after accurate dose-determination) and low cost of production [16]. These phytocompounds are biologically active secondary metabolites that display diverse pharmacological properties such as antibacterial, antidiabetic, antifungal, anti-inflammatory, antioxidant, and antitumor activities [16].

In light of the above facts, this review aimed at highlighting the therapeutic potential of traditionally used medicinal plants for sepsis management and presenting important findings on the anti-septic properties of their crude extracts or bioactive phytoconstituents *in vitro* and/or *in vivo*.

3.2 TRADITIONAL MEDICINAL PLANTS IN SEPSIS TREATMENT

3.2.1 *Angelica sinensis*

Angelica sinensis, a Chinese herb, is commonly known as Dang gui or Dong quai or female ginseng and belongs to the family Apiaceae. The roots of *A. sinensis* harbor pharmacologically important phytoconstituents such as ferulic acid, Z-ligustilide, and polysaccharide sulphate [17, 18]. It was demonstrated that the aqueous root fraction (low molecular weight, >10 kDa) from *A. sinensis*, significantly affected the release of high mobility group box 1 (HMGB1) protein *in vitro*. Further, an intraperitoneal administration of the same aqueous root extract from *A. sinensis* was shown to significantly attenuate the systemic accumulation of HMGB1 protein in BALB/c septic mice models and impart protection against endotoxemia (highly lethal) in a dose-dependent manner [19]. HMGB1 is a ubiquitous late pro-inflammatory mediator cytokine, which is released by the activated immune cells (macrophages and monocytes) during endotoxemia and sepsis [20–25]. In this study, ferulic acid was not found to be inhibitory against HMGB1 *in vitro* (Figure 3.1). Finally, the authors suggested exploring the pharmacological significance of this herb in human inflammatory diseases, such as sepsis [19].

3.2.2 *Salvia miltiorrhiza*

Salvia miltiorrhiza is commonly known as Chinese sage or Red sage, Danshen or Tan Shen, and belongs to the family Lamiaceae. The roots of *S. miltiorrhiza* harbor pharmacologically important phytoconstituents such as salvianolic acid B (Sal B) [26, 27] and steroid-like pigments called tanshinones (cryptotanshinone, tanshinones I, II, and IV) [28]. Sal B is reported to impart protection against chronic renal ailments and treatment of cerebrovascular disorders via anti-oxidation and cell-protective attributes (Figure 3.1) [26, 27], while the tanshinones possess anti-inflammatory properties *in vitro* [29–31]. Tanshinone II-A sodium sulfonate (TSNIIA-SS) from *S. miltiorrhiza* was shown to confer protection to the murine model against highly lethal endotoxemia (lipopolysaccharide, LPS-induced) and sepsis (Figure 3.1). Notably, there was a significant reduction in the systemic HMGB1 levels in experimental murine septic models. These findings strengthen the involvement of HMGB1 protein in the sepsis lethality and suggest TSNIIA-SS as a potential therapeutic candidate for sepsis management [32].

3.2.3 *Camellia sinensis*

Camellia sinensis is commonly known as a tea plant or tea shrub or tea tree or green tea and belongs to the family Theaceae. *C. sinensis* contains catechins, which are a group of polyphenols. These include clinically important catechins such as epicatechin, epigallocatechin, and (-)-epigallocatechin-3-gallate (EGCG) (Figure 3.1) [33]. These are documented to display antimicrobial, anti-oxidative, and antitumor activities [34, 35]. Notably, EGCG is the key active catechin present in the natural *C. sinensis* extracts [33]. The administration of natural EGCG extract from *C. sinensis* post-inception of induced sepsis successfully rescued the mouse models. Noteworthy, a dose-dependent EGCG inhibition of LPS (macrophage or monocyte cultures) and CLP-induced sepsis (intraperitoneal administration in male, BALB/c mice) was observed. There was an increase of 29 percent in the survival rate of sepsis mouse models in comparison to the control group. This effect was seen with a delay of 24 hours in the treatment of the mice models post-inception of induced sepsis. At a concentration of 10 mM EGCG, there occurred a complete annulment of HMGB1 release. This could be ascribed to the anti-HMGB1 activity of EGCG, which attenuated the systemic accumulation of HMGB1. Eventually, this averted the clustering of HMGB1 on the macrophage surfaces. Finally, the inflammatory responses mediated by HMGB1 were also suppressed [20]. In another study, there occurred a complete HMGB1 inhibition *in vivo* at a dose of 10 μl/ml or 1ml/kg of leaf extracts of EGCG (*C. sinensis* folium). EGCG was not observed to be cytotoxic *in vitro* (macrophage cell cultures) [21]. Conclusively, the authors proposed that EGCG in the *C. sinensis* extract significantly reduces HMGB1 release in induced sepsis conditions (*in vitro* and *in vivo*) by novel mechanisms and imparts protection against sepsis lethality and endotoxemia [36, 37].

3.2.4 *Panax ginseng* C.A. Meyer

Panax ginseng is commonly known as Korean ginseng and belongs to the family Araliaceae [28]. The diverse biological activities of *P. ginseng* are attributed to its rich root reserves of ginsenosides [39–41]. Amongst these, ginsenoside-Rh1 is the key active component with reported anti-inflammatory [42], antioxidant [43], anti-stress [44], and immunomodulatory [42] activities.

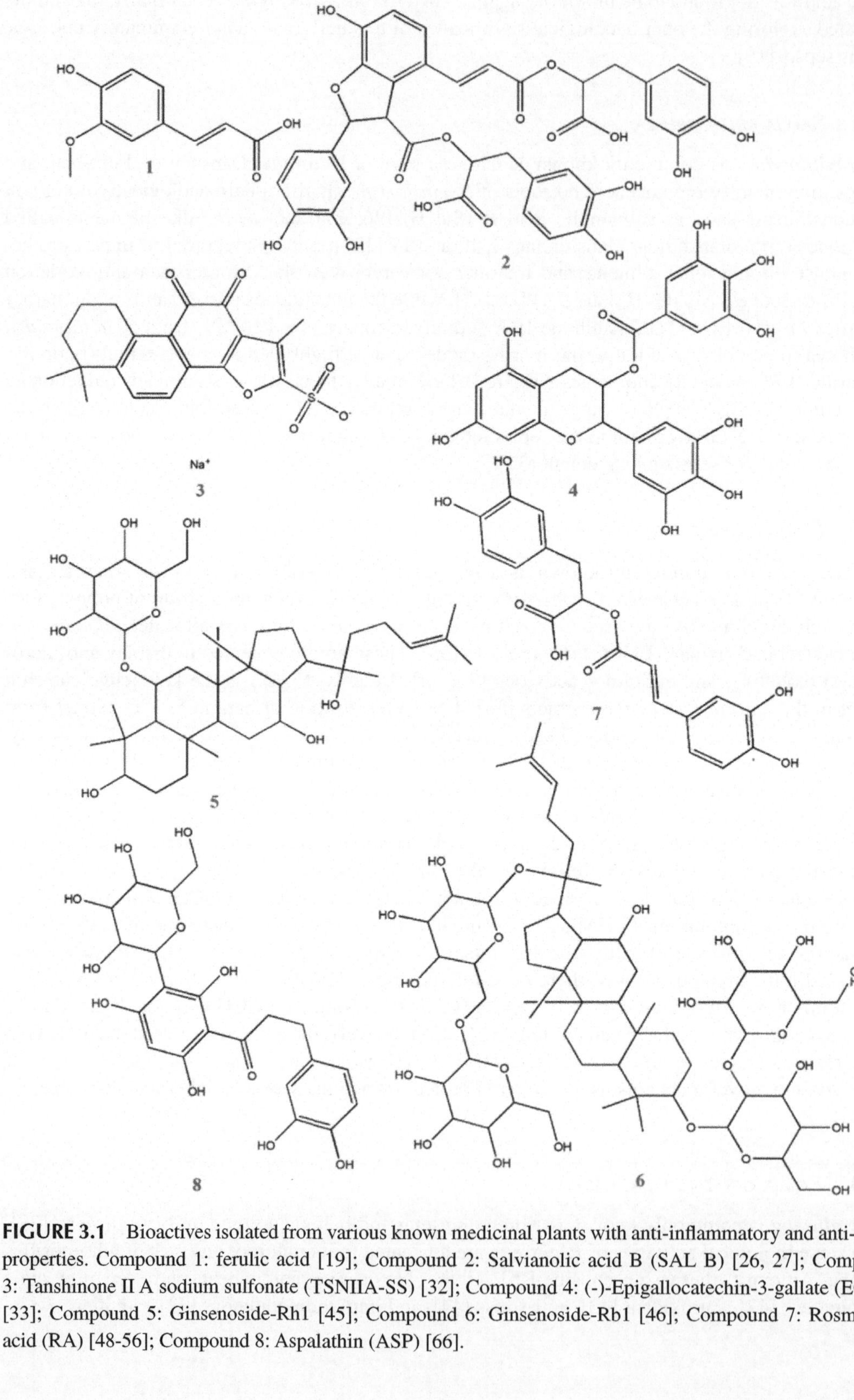

FIGURE 3.1 Bioactives isolated from various known medicinal plants with anti-inflammatory and anti-septic properties. Compound 1: ferulic acid [19]; Compound 2: Salvianolic acid B (SAL B) [26, 27]; Compound 3: Tanshinone II A sodium sulfonate (TSNIIA-SS) [32]; Compound 4: (-)-Epigallocatechin-3-gallate (EGCG) [33]; Compound 5: Ginsenoside-Rh1 [45]; Compound 6: Ginsenoside-Rb1 [46]; Compound 7: Rosmarinic acid (RA) [48-56]; Compound 8: Aspalathin (ASP) [66].

In an interesting study, ginsenoside-Rh1 exhibited *in vitro* (HUVECs, human umbilical vein endothelial cells) and *in vivo* (CLP-induced male C57BL/6 mouse model of sepsis) anti-septic effects against HMGB1. Precisely, there was a significant ginsenoside-Rh1 dose-dependent reduction in the release of HMGB1 (LPS-induced HUVECs), while the survival rate in the mice increased (Figure 3.1) [45]. Another ginsenoside, that is, ginsenoside-Rb1 from *P. ginseng*, led to a reduction in the *in vitro* (human uterine fibroid cells) levels of HMGB1. This occurred as ginsenoside Rb1 inhibited uterine fibroid cell proliferation and promoted their apoptosis as a part of its HMGB1 regulatory activity (Figure 3.1) [46]. These observations strengthen the candidature of ginsenosides Rb1 and Rh1 for the management of severe inflammatory diseases, including sepsis.

3.2.5 *Perilla frutescens* (L.) Britt. var. *acuta*

Perilla frutescens var. acuta is commonly known as 'beef steak plant' or 'Chinese basil' and belongs to the family Labiatae [47]. The leaves of *P. frutescens* contain Rosmarinic acid (RA) as a key active phytocompound with well-documented anti-inflammatory [48–50], antimicrobial [51], antioxidant [52], and neuroprotective [53, 54] properties (Figure 3.1). As a proof of concept, there was diminished mortality in the mice group (BALB/c mice) that were infected with Japanese encephalitis virus (JEV), but RA-treated. In this group, the levels of viral load ($P < 0.001$) and pro-inflammatory cytokines ($P < 0.001$) decreased significantly in comparison to the JEV-infected, RA-untreated mice. These effects were observed after 8 to 9 days of JEV infection in mice. Hence, RA could successfully inhibit the replication of JEV and the induced inflammation *in vivo* [55]. In another study, the leaf extract of RA from *P. frutescens* displayed anti-HMGB1 activity. Specifically, the HMGB1 release was inhibited together with the down-regulation of the inflammatory responses dependent on HMGB1 *in vitro* (in HUVECs). Further, RA also inhibited the hyper-permeability and leukocyte migration mediated by HMGB1 *in vivo* (in male C57BL/6 mice). In addition, there was a reduction in the release of HMGB1 upon induction by CLP as well as diminished mortality related to sepsis in the mouse models. The authors suggested RA as a potential HMGB1-inhibitor and therapeutic compound with anti-inflammatory attributes [56].

3.2.6 *Prunella vulgaris* var. *lilacina*

Prunella vulgaris var. lilacina is commonly known as 'self-heal' and belongs to the family Labiatae [57]. *P. vulgaris* contains pharmacologically important constituents, namely, flavonoids, phenols, and polysaccharides, and possess anti-allergic, anti-inflammatory, anti-oxidative, antiviral, and anxiolytic-like properties [58–63]. *P. vulgaris* ethanol extract-induced heme oxygenase-1 (HO-1) protein and inhibit the release of HMGB1 in LPS-activated macrophages *in vitro* (RAW 264.7 cells). Similarly, there was a reduction in the serum levels of HMGB1 and its expression in the lungs in the septic mice models (CLP-induced BALB/c mice). This strengthens the clinical relevance of *P. vulgaris* for sepsis treatment [64].

3.2.7 *Aspalathus linearis* (Rooibos)

Aspalathus linearis belongs to the family Fabaceae. It contains bioactive polyphenols, namely, dihydrochalcones, flavanones, flavones, and flavonols [65]. The key active dihydrochalcones, namely, aspalathin (Asp), 8 and nothofagin (Not), 9 from Rooibos (green tea) abrogated HMGB1 release in the LPS-induced HUVECs (Figures 2.1 and 2.2). Also, Asp and Not could effectively inhibit HMGB1 release along with concomitant suppression of leukocyte migration and adhesion, hyper-permeability, and cell adhesion molecule (CAM) expression in CLP-induced animal septic models (male C57BL/6 mice). All these sepsis-related responses are mediated by HMGB1. Thus, Asp and Not are recommended to be explored for their therapeutic potential in treating severe inflammatory diseases including sepsis [66].

3.2.8 *Cyclopia subternata*

Cyclopia subternata is commonly known as 'honey bush' and belongs to the family Fabaceae.

The key active flavonoids, namely, vicenin-2, 10 (VCN, Apigenin-6,8-di-C-glucoside) and scolymoside, 11 (SCL, Luteolin 7-O-rutinoside) [67–69] from *C. subternata* abrogated HMGB1 release in the LPS-induced HUVECs (Figure 3.2). Also, Asp and Not could effectively inhibit HMGB1 release along with concomitant suppression of leukocyte migration and adhesion, hyperpermeability, and CAM expression in CLP-induced animal septic models (male C57BL/6 mice). Additionally, Tumor Necrosis Factor Alpha (TNFα) and Interleukin 6 (IL-6) production as well as Nuclear Factor Kappa B (NF-κB) and extracellular signal-regulated protein kinase (ERK1/2) were suppressed. All these sepsis-related responses are mediated by HMGB1. Thus, VCN and SCL are recommended to be explored for their therapeutic potential in treating severe inflammatory diseases including sepsis [70].

3.2.9 *Rhodiola rosea* L.

Rhodiola rosea is commonly known as 'golden root' or 'rose root' and belongs to the family Crassulaceae. The treatment with an ethanolic root extract from *R. rosea* led to a significant attenuation of tumor necrosis factor-α-inducible protein 8-like 2 (TIPE2), Fas and Fas ligand (FasL) expression in the CLP-induced sepsis mouse models (male, BALB/c mice). In this treated mice group, there was an increase in the B-cell lymphoma 2 (Bcl-2) expression ($P<0.05$). In comparison to the CLP-induced treated mice group, a significant elevation in the count of T-lymphocytes was observed in the CLP-induced untreated mice group ($P<0.05$). On the contrary, the rate of apoptosis of T-lymphocytes was significantly lower in the CLP-induced treated mice group in comparison to the CLP-induced untreated mice group ($P<0.05$). Further, treatment with an ethanolic root extract from *R. rosea* caused a significant decline in the cluster of differentiation 3 (CD3+) and the cluster of differentiation 4 (CD4+) T-lymphocytes without affecting the elevated IL-4 and IL-10 cytokine levels in the CLP-induced septic mice. Of note, the otherwise enhanced IL-12, IL-2, and IFN-γ cytokine levels were further elevated significantly in the CLP-induced *R. rosea* extract-treated mice group ($P<0.05$). In comparison to the CLP-induced untreated mice group, there was a marked improvement in the thymus index and survival rate of the extract-treated group. These key observations are suggestive of the anti-septic properties of *R. rosea* attributed to its apoptotic, cytokine, and host immune-regulatory activities [71].

Histological examination revealed the administration of salidroside, 12 from *R. rosea* (SAL, p-hydroxyphenethyl-β-D-glucoside) in the CLP-induced septic rats (Wistar rats) alleviated pulmonary injury along with a decreased HMGB1 serum levels (Figure 3.2). Further, SAL pre-treatment caused a significant dose-dependent (50, 100, and 200 μg/ml) inhibition of HMGB1 release in LPS-induced RAW267.4 cells. Furthermore, *in vitro* LPS-induced HMGB1 translocation into the nucleocytoplasm was inhibited by SAL pre-treatment (200 μg/ml). Moreover, SAL pre-treatment (200 μg/ml) inhibited HMGB1 acetylation (activated by LPS-induction) *in vitro* through Adenosine Mono Phosphate-activated protein kinase (AMPK)-pathway-mediated up-regulation of sirtuin 1 (SirT1). As observed, the over-expression of SirT1 led to an attenuation of HMGB1 acetylation and translocation into nucleocytoplasm *in vitro* (in the LPS-induced RAW267.4 cells). SAL treatment reversed HMGB1 acetylation (due to LPS-activation) and its translocation into nucleocytoplasm in RAW267.4 cells transfected with SirT1 shRNA (small hair-pin ribonucleic

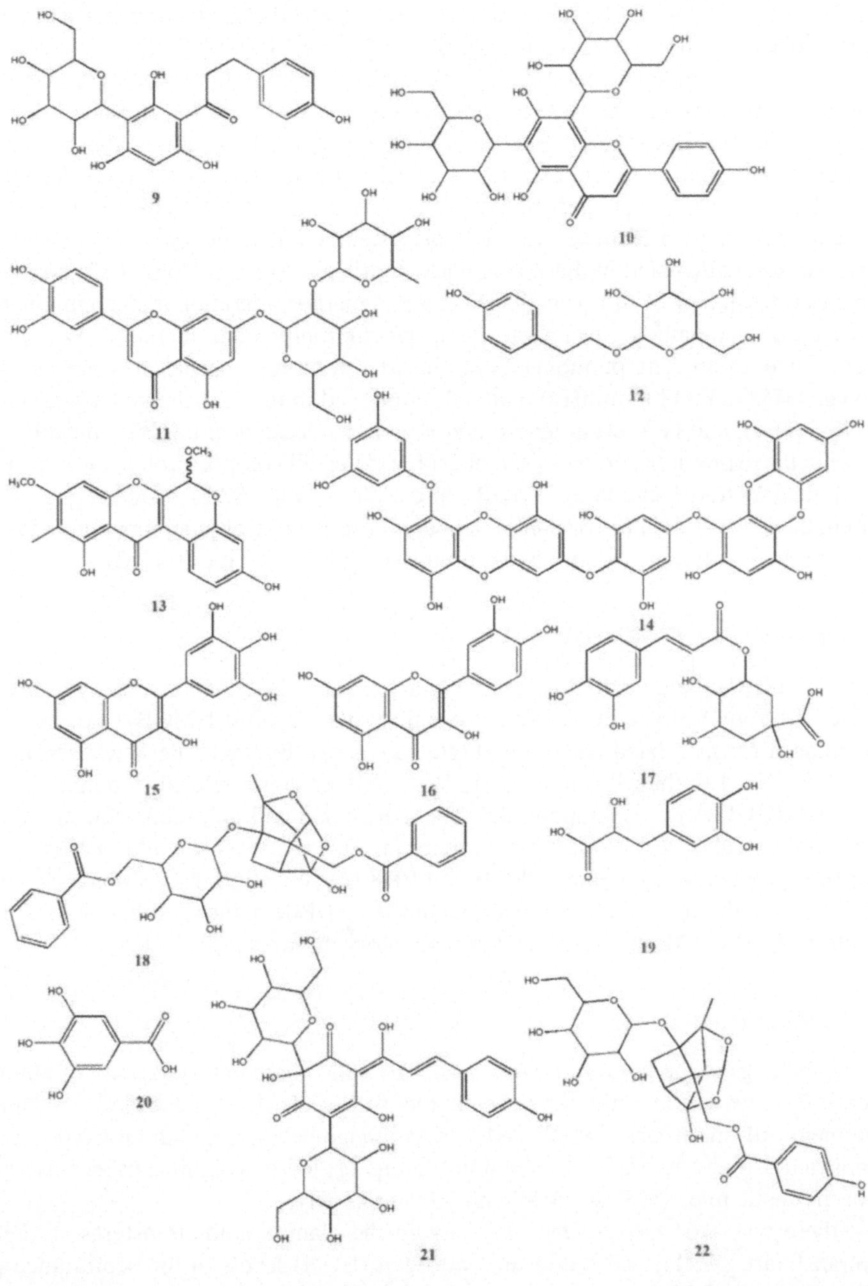

FIGURE 3.2 Bioactives isolated from various known medicinal plants with anti-inflammatory and anti-septic properties. Compound 9: Nothofagin (Not) [66]; Compound 10: Vicenin-2 (VCN, Apigenin-6,8-di-C-glucoside) [67–70]; Compound 11: Scolymoside (SCL, Luteolin 7-O-rutinoside) [67–70]; Compound 12: Salidroside (SAL, p-hydroxyphenethyl-β-D-glucoside) [72, 73]; Compound 13: Boeravinone X (C-methylrotenoid) [74]; Compound 14: Dieckol (phlorotannin) [75]; Compound 15: Myricetin [79]; Compound 16: Quercetin [79]; Compound 17: Chlorogenic acid (CGA) [87]; Compound 18: Benzoylpaeoniflorin [96–98]; Compound 19: Danshensu [96–98]; Compound 20: Gallic acid [96–98]; Compound 21: Hydroxysafflor yellow A [96–98]; Compound 22: Oxypaeoniflorin [96–98].

acid) plasmid. In conclusion, SAL causes attenuation of HMGB1 acetylation and its translocation into nucleocytoplasm possibly by up-regulating the expression of SirT1 [72].

In another study, a dose-dependent treatment with SAL caused a significant reversal in the otherwise enhanced expression of iNOS protein, phosphorylation of NF-κB p65, and declined expression of SIRT1 protein in the LPS-treated RAW264.7 macrophages. In addition, there was a significant inhibition of elevated cytosolic expression of HMGB1 protein with a dose-dependent SAL treatment of RAW264.7 macrophages. Similar effects of SAL were observed in the LPS-induced sepsis mouse models (male, ICR mice). In brief, the systemic inflammatory responses and release of HMGB1 protein were alleviated in the SAL-treated sepsis mice group. Here, SAL led to a reversal of diminished expression of SIRT1 protein. There was a marked alteration in the pulmonary edema, lipid peroxidation, histopathological parameters, and mortality rates in the CLP-induced SAL-treated septic mice group. The pulmonary ratio of arterial oxygen partial pressure and fractional inspired oxygen (PaO2/FiO2 in mmHg) markedly improved in the SAL-treated septic mice group. Further, TNF-α, IL-6, and NO serum levels, and HMGB1 release declined significantly. A significant decrease in the expressions of iNOS and phosphorylated NF-κB p65 proteins was also observed together with nuclear translocation of HMGB1 in the lungs in the SAL-treated septic mice. These findings strengthen the candidature of SAL as a potential anti-septic phytocompound with protective effects against severe pulmonary injury and mortality in mice septic models [73].

3.2.10 *Abronia nana S. Watson*

Abronia nana S. Watson belongs to the family Nyctaginaceae. Boeravinone X, 13, a rare and new C-methylrotenoid from *A. nana*, possesses anti-septic activity against HMGB-1 protein, which is a known mediator of sepsis-related responses (Figure 3.2) [74]. Boeravinone X was found to inhibit the release of HMGB-1 in the LPS-induced HUVECs. Other sepsis-related responses known to be mediated by HMGB-1 were also suppressed. These included cell adhesion molecule expression, hyper-permeability, leukocyte adhesion, and migration. Boeravinone X could efficiently suppress TNF-α and IL-6 production as well as NF-κB and ERK1/2 activation in CLP-induced septic mice models (C57BL/6 male mice). The authors concluded that Boeravinone X could be a promising treatment alternative for HMGB-1-mediated inflammatory diseases [74].

3.2.11 *Ecklonia cava*

Ecklonia cava belongs to the family Lessoniaceae. The anti-inflammatory activity of phlorotannin-rich extract of *E. cava* was investigated in septic mouse models (LPS-induced C57BL male mice) and macrophage cell line (murine RAW 264.7 macrophage cells). Oral administration of *E. cava* phlorotannin-rich extract (before LPS-induction (25 mg/kg) led to a significant increase in the survival rate of the septic mice (50% survival with 100 mg/kg/day).

Notably, there was a dose-dependent decrease in the alanine aminotransferase (ALT), aspartate aminotransferase (AST), and blood urea nitrogen (BUN) levels in the septic mice group as compared to the control LPS-induced group. Thus, the phlorotannin-rich extract demonstrated hepatic and nephroprotective activities as it could efficiently protect mice from lethal septic effects, namely, endotoxemia and septic shock. Additionally, there was an attenuation of nitric oxide (NO), Prostaglandin E2 (PGE2), and HMGB1 serum levels in the septic mice group [75].

Pre-treatment of primary murine peritoneal macrophages and RAW 264.7 cells with various concentrations of *E. cava* extract (25, 50, 100 μg/ml for 60 minutes) before induction with LPS (1 μg/ml for 24 hours) led to a marked inhibition of NO and PGE2 production. The results were compared with the positive controls, that is NS398 (5 μM) and L-NIL (10 μM). Further, *E. cava* phlorotannin-rich extract pre-treatment (for 1 hour) caused suppression of HMGB-1, IL-6, and TNF-α, and down-regulation of COX-2 and iNOS mRNA. The corresponding COX-2 and iNOS enzymes are responsible for NO and PGE2 synthesis. Thus, the anti-inflammatory activity of *E. cava* phlorotannin-rich extract is attributable to its transcriptional regulation of important pro-inflammatory cytokine sepsis mediators (HMGB-1, IL-6, and TNF-α), and NO and PGE2 synthesizing enzymes (COX-2 and iNOS). Furthermore, pre-treatment with *E. cava* phlorotannin-rich extract led to significant suppression of the NIK/TAK1/IKK/IκBα-mediated NFκB pathway *in vitro*. Here, LPS-induction of primary murine peritoneal macrophages and RAW 264.7 cells was done for 8 hours (NIK: Nuclear Factor-κB-inducing kinase, TAK1: Transforming growth factor beta-activated kinase 1, IKK: inhibitory κ-kinase, IκBα: Inhibitor of nuclear factor κB alpha). *E. cava* phlorotannin-rich extract led to a marked induction of HO-1 expression and stimulation of the nuclear factor erythroid-2-related factor 2 (Nrf2) translocation to the nucleus. In an HO-1 knockdown, specific small interfering ribonucleic acid (siRNA) partially attenuated *E. cava* extract-induced suppression of NO and PGE2. This ascertains the association between the anti-inflammatory activity of *E. cava* extract and Nrf2/HO-1 pathway activation, which, in turn, led to the production of NO and PGE2 in RAW 264.7 cells [75].

Dieckol, 14, a key phlorotannin in *E. cava* extract was found to inhibit the production of NO and PGE2 in macrophages (primary macrophages) (Figure 3.2). Administration of 100 mg/kg of dieckol conferred protection to the septic mice models against LPS-induced hyper-inflammatory responses leading to a 75% increase in their survival. Additionally, it was shown to exhibit hepatic and nephroprotective effects with a dose-dependent attenuation of ALT, AST, and BUN. A significant attenuation of NO and PGE2 serum levels, and production of HMGB1 was also observed upon dieckol administration to septic mice models. The authors concluded that *E. cava* negatively regulates pro-inflammatory factors and has the potential to suppress severe sepsis [75].

3.2.12 *Inula helenium* L.

Inula helenium L. is commonly known as 'horse-heal' or 'elf-dock' and belongs to the family Asteraceae. The ethanolic root extract from *I. helenium* was found to up-regulate p38 MAPK/Nrf2 -mediated HO-1 expression in a concentration- and time-dependent manner *in vitro* (LPS-induced RAW 264.6 macrophage cells). Notably, there occurred a down-regulation of COX-2 and iNOS expression in LPS-induced RAW 264.6 cells. Here, NO and PGE2 production was inhibited together with the HMGB1 release. Further, the expression of intercellular adhesion molecule 1 (ICAM-1) and vascular cell adhesion molecule 1 (VCAM-1) was also inhibited in HUVECs stimulated by TNF-α. Furthermore, there was a marked reduction in the levels of TNF-α and IL-1β in the LPS-induced macrophages. Of note, the ethanolic root extract led to an induction of HO-1 expression via translocation of Nrf2 and inhibition of NF-κB activation in a dose-dependent manner in the LPS-induced macrophages [76].

In vivo studies showed that the ethanolic root extract administration (5 and 10 mg/kg of intraperitoneal dose) caused a significant reduction in the ALT, AST, BUN, and creatinine plasma levels in septic mouse models (CLP-induced, male, BALB/c mice). This indicates attenuation of multiple organ injury in the septic mice by an ethanolic root extract from *I. helenium*. Similarly, there occurred a significant reduction in HMGB1 concentration in the blood of septic mice. Conclusively, ethanolic root extract from *I. helenium* possesses HO-1-pathway-dependent anti-inflammatory activity [76].

3.2.13 *Alpinia katsumadai* Hayata

Alpinia katsumadai Hayata is commonly known as 'shell ginger' and belongs to the family Zingiberaceae. The oral administration of ethanol seed extracts from *A. katsumadai* (EAKH) conferred protection to the CLP-induced septic mice models (male ICR mice) against polymicrobial sepsis. The oral dose of ethanol seed extract was administered both pre (1 hour) and post (2 hours)

CLP-induction in the mice. A significant improvement in the mouse survival rate at an EAKH dosage of 100 mg/kg and 200 mg/kg) was observed. In addition, the mean arterial pressure markedly improved with EAKH in a dose-dependent manner with a nearly complete restoration at 200 mg/kg. The septic mice were protected from tissue injury and organ dysfunction (liver and lungs) as confirmed by the evaluation of ALT/AST serum levels and histological assessment. Notably, the ALT and AST serum levels were significantly reduced by EAKH in a dose-dependent manner (50 mg/kg, 100 mg/kg, and 200 mg/kg). Even though EAKH did not exert bacteriostatic and bactericidal effects directly, bacterial clearance (from the peritoneal cavity) and enhanced migration of leukocyte (into the peritoneal cavity) was facilitated in the septic mice group. There was a marked decrease in the serum levels of IL-1β, NO, and TNF-α (pro-inflammatory cytokines) in the experimental septic mice models. Thus, EAKH exerts its septic preventive effects in CLP-induced mice through a heightened local defense mechanism [77].

3.2.14 *Syzygium jambolanum*

Syzygium jambolanum is commonly known as 'Malabar plum' or 'Java plum' or 'black plum' and belongs to the family Myrtaceae. The prophylactic administration (subcutaneous) of hydroalcoholic crude leaf extract (HCE) of *S. jambolanum* conferred protection to the CLP-induced mouse models (male C57Bl/6 mice) against polymicrobial sepsis. There was an enhanced survival of mice with increased migration of neutrophils to the site of infection (peritoneal cavity) and release of hydrogen peroxide. A decrease in the TNF and nitrite serum levels was observed. The bacterial colony forming units (CFU) did not decline even when there was an increase in the peritoneal cell migration and activation. The number of bone marrow cells was also significantly reduced by HCE treatment without causing any alteration in the spleen and lymph node cells. Thus, *S. jambolanum* exhibit anti-septic properties via activation and recruitment of immune cells (neutrophils) to the site of infection, and reduction in the systemic inflammatory responses. The authors recommend the characterization of phytocompounds responsible for this observed anti-septic activity of *S. jambolanum* [78].

3.2.15 *Eugenia uniflora*

Eugenia uniflora is commonly known as 'Pitanga cherry shrub' and belongs to the family Myrtaceae. A hydroalcoholic (butanol) crude leaf extract from *E. uniflora* (HE-Bu) caused a reduction in the sepsis lethality in CLP-induced septic mice models (male Swiss albino mice) when administered orally. There occurred a marked decline in the rate of mortality in the septic mice models by 30%. Further, the accumulation of neutrophils in the lungs was thwarted followed by a decrement in the serum levels of TNF-α and IL-1β. A significant down-regulation of cellular expression (ileum cells) of iNOS and COX-2 was also observed. Notably, the Hu-Bu fraction was found to contain myricetin, 15 and quercetin, 16 rhamnosides as the major flavonoid phytoconstituents (Figure 3.2). Thus, Hu-Bu from *E. uniflora* possesses anti-septic properties that could efficiently protect mice from the lethal sepsis effects. The authors recommended the exploration of phytocompounds from *E. uniflora* to exploit their anti-septic adjuvant properties [79].

3.2.16 *Carum carvi* L.

Carum carvi L. is commonly known as 'caraway' or 'meridian fennel' or 'Persian cumin'. It belongs to the family Apiaceae. The intraperitoneal administration of essential oils (post-sepsis induction) from gamma-irradiated caraway seeds (at 100 mg/kg dose) demonstrated hepato-protective effects against polymicrobial sepsis lethality on the CLP-induced septic rat models (male Wistar albino rats). The administration of oils from fresh and gamma-irradiated seeds led to a reversal in the CLP-induced increased myeloperoxidase (MPO) activity along with increased hepatic levels of thiobarbituric acid reactive substances (TBARS). The hepatic levels of Glutathione (GST) were also replenished. There was significant suppression of ALT and AST plasma concentrations in the septic rats treated with the caraway essential oils evident from the liver function test. Histopathological studies revealed that the ALP, albumin, total bilirubin, and total protein remained unaltered in the CLP-induced treated rats. This implies that the anti-oxidative effects of essential oils from caraway seeds remain unaffected with gamma-irradiation [80].

In another study, intraperitoneal treatment with the essential oils (50 and 100 mg/kg body weight) from caraway seeds altered the otherwise elevated lipid peroxidation (LP) (kidney) and urea/creatinine levels (plasma) in the CLP-induced sepsis rat models (male Wistar albino rats). This effect was not observed in the sepsis rats treated with hydroethanolic seed extract from caraway (50 and 100 mg/kg body weight). GST levels (in the kidney and heart) remained unaffected in the sepsis rats treated with either essential oils or hydroethanolic seed extract from caraway. These findings strengthen the protective efficacy of essential oils from caraway against sepsis-induced oxidative stress [81].

3.2.17 *Nigella sativa*

Nigella sativa is commonly known as 'black cumin' or 'kalanji' or 'kalonji' and belongs to the family Ranunculaceae. Treatment with *N. sativa* (NS) ethanolic seed extract caused a significant decline in the serum levels of pro-inflammatory cytokines in CLP-induced sepsis rats. Also, alterations in the hepatic histopathology, lipid peroxides (LPO) level, and MPO activity were observed. On the contrary, there was a significant elevation in the hepatic levels of GSH and superoxide dismutase (SOD) activity. In addition, there was a reduction in the mortality rate in the sepsis rats that received NS post-CLP-induction. This was probably due to the reduced oxidative stress in the hepatic tissues and decreased cytokine levels in the serum. This suggests that NS has the potential to mitigate the occurrence of multiple organ failure by altering the host systemic inflammatory responses against sepsis. The authors concluded that NS harbors anti-oxidative and anti-inflammatory properties that prevent CLP-induced sepsis formation in rats [82].

3.2.18 *Melilotus suaveolens Ledeb*

Melilotus suaveolens Ledeb belongs to the family Fabaceae. A prior oral administration of Melilotus extract (2 hours) from *M. suaveolens* conferred protection to the CLP-induced septic rat models (Sprague-Dawley rats, male) against acute lung injury (ALI) [83]. This protective effect was attributed to the up-regulation of cannabinoid (CB) 2 receptor expression on the peripheral monocytes by *Melilotus* extract. Further, there was an abrogation of NF-κβ65 activity with a significant reduction in the inflammatory (neutrophils and lymphocytes) and total cells in the *Melilotus* extract-treated mice over the untreated control mice group. Furthermore, *Melilotus* extract caused a significant decrease in the levels of TNF-α and IL-6. Notably, *Melilotus* extract ameliorated the pulmonary inflammation in the CLP-induced sepsis mice group as revealed by histological studies. In conclusion, CB2 was found to be negatively correlated to the mRNA and protein expression of NF-κβ [84].

The administration of *Melilotus* extract (25 mg/kg) in the CLP-induced sepsis mouse models had a pronounced effect on the pulmonary pathophysiology and the inflammatory response. Accordingly, the expression of TIPE2, HO-1, and inhibitor of κB (IκB) kinase was up-regulated. Further, the activities of NF-κB and toll-like receptor-4 (TLR-4) were inhibited. The Melilotus extract-treated septic mice models showed a decline in the MPO and malondialdehyde (MDA) levels. Contrastingly, SOD levels were elevated in this group of mice. The authors concluded that *Melilotus* extract possesses anti-inflammatory and anti-oxidative properties that can be exploited for treating pulmonary injuries evident from *in vivo* studies [84].

3.2.19 *Astragalus membranaceus*

Astragalus membranaceus is commonly known as the 'yellow leader' (Huang Qi) and belongs to the family Fabaceae [85]. The oral administration of saponin fraction from roots of *A. membranaceus* (SAM) imparted protection to CLP-induced septic mice models (male ICR mice) against polymicrobial sepsis. There was a significant increase in the survival rate (33%) of the SAM-treated mice group. SAM attributed this effect to the up-regulation of the protein C pathway and inhibition of inflammation. There was a reversal in the reduced expression of the mRNA for protein C receptors, namely thrombomodulin (TM) and endothelial PC receptor (EPCR) by SAM *in vitro* (LPS-induced HUVECs) [86].

3.2.20 *Lonicera japonica*

Lonicera japonica is commonly known as 'Japanese honeysuckle' or 'golden-and-silver honey' or 'suckle honeysuckle' and belongs to the family Caprifoliaceae. The intravenous administration of *Lonicera* flos extract (HS-23) conferred protection against polymicrobial sepsis in LPS-induced septic mouse models (C57BL/6J mice). There was a decline in the blood cytokine levels of TNF-α, IL-1β, and HMGB-1 in the septic mice. Further, it was found that Chlorogenic acid, 17 (CGA) and its isomers were the key phytocompounds responsible for imparting anti-endotoxemia protection to the mice intoxicated with LPS (Figure 3.2). Precisely, HS-23 or CGA or its isomers (crypto-CGA, neo-CG) abrogated the IL-1 receptor-associated kinase 4 (IRAK-4) auto-phosphorylation *ex vivo* (murine peritoneal macrophages). The phosphor-transfer (kinase) function of IRAK-4 was also inhibited in cell-free reactions. Consequently, the downstream signal-transduction pathways were also suppressed by HS-23. These pathways are indispensable for NF-κB and activating protein 1 (AP-1) activation *ex vivo* (post-LPS-induction). In addition, NO production induced by toll-like receptor agonists was inhibited by HS-23. Furthermore, the expression of inflammatory genes targets concerning NF-κB or AP-1 was also down-regulated *ex vivo*. The authors concluded that the anti-inflammatory effects of HS-23 or CGA (or its isomers) are attributed to its IRAK-4 inhibitory function as observed in the septic mouse models [87].

The intravenous administration of HS-23 (at 0 and 24 h post-CLP-induction) in sepsis mouse models (male, ICR mice) ameliorated survival in mice. Also, there was a reduced bacterial load and attenuation of sepsis-induced multiple organ failure in septic mice. Further, HS-23 down-regulated the hepatic and pulmonary expression of TLR-4 protein as well as mRNA. HS-23 attenuated the CLP-induced elevated hepatic and pulmonary expression of myeloid differentiation primary response protein 88 (MyD88). Likewise, the elevated hepatic and pulmonary expression of p38 and c-Jun N-terminal kinase (JNK1/2) was also suppressed by HS-23. The sepsis-induced elevation in the hepatic and pulmonary expression of TRIF and IFN-β protein, and IRF3 phosphorylation was attenuated by HS-23. Of note, this attenuation was observed only in the liver. Further, the CLP-induced hepatic and pulmonary increase in NF-κB nuclear localization and phospho-c-jun (p-c-jun) nuclear translocation were attenuated by HS-23. Moreover, the sepsis-induced increment in the bacterial load (CFU) was reduced significantly by HS-23 in the lung, peritoneal fluid, and spleen. Conclusively, the anti-septic activity of HS-23 is attributed to its TLR-pathway suppression activity, which makes it a potential therapeutic candidate for sepsis treatment [88].

In another study, the intravenous administration of HS-23 in CLP-induced sepsis mouse models (male ICR mice) conferred protection against sepsis. HS-23 attenuated the CLP-induced decline in splenocytes and lymphocyte (B cells and natural killer T cells) populations along with T cell (CD4(+) and CD8(+) cells) apoptosis. Moreover, the splenic apoptotic pathways (intrinsic and extrinsic pathways) were also inhibited by HS-23 together. Furthermore, there occurred an attenuation of the otherwise reduced production of IL-17, TNF-α and IL-2 (due to CLP-induction) by HS-23. HS-23 attenuated the CLP-induced decrease in interleukin (IL)-17 production. CLP significantly decreased splenic production of tumor necrosis factor-α and IL-2, and these effects were attenuated by HS-23.

Thus, HS-23 has immunosuppressive reversal activity observed in the later sepsis phase attributed to its lymphocyte apoptotic inhibition and elevated cytokine (Th1) production properties. This makes HS-23 a potential therapeutic candidate for sepsis treatment [89].

3.2.21 Aloe Vera

Aloe vera belongs to the family Asphodelaceae. Reportedly, the intravenous administration of PBS (Phosphate buffered saline) -dissolved Aloe vera gel (25 mg/kg and 50 mg/kg) conferred protection against the polymicrobial sepsis in the CLP-induced mouse models. It could therapeutically relieve sepsis lethality (multiple organ dysfunction syndromes, MODS) as there was a rapid bacterial clearance in the septic mice models [90]. There was a significant reduction in the serum levels of TNF-α and IL-1β with complete attenuation of IL-6 upon therapeutic administration of Aloe vera in septic mouse models. The pro-inflammatory cytokines, namely, TNF-α and IL-1β, are important mediators of sepsis-related inflammatory hepatic responses, while IL-6 arbitrates organ dysfunction and sepsis mortality [91]. The authors were convinced by their findings and recommended Aloe vera as an attractive candidate for sepsis treatment (See Figure 3.3).

3.2.22 *Chenopodium ambrosioides L.*

Chenopodium ambrosioides L. is commonly known as Jesuit's or Mexican-tea and belongs to the family Amaranthaceae. The subcutaneous administration of the hexane fraction (HEX, 5 mg/kg) of *C. ambrosioides* extracts in the septic mice models (CLP-induced) effectively retarded bacterial growth (reduction in the CFUs) possibly via phagocyte activation (macrophages and neutrophils) with a simultaneous check on the sepsis inflammatory responses in this group. This prophylactic treatment was given 6 hours before the CLP-induction *in vivo* and the assessment was made in comparison to the effects observed in the control mice group and sodium chloride (0.9% NaCl, control). Although no change was observed in the bacterial CFU with hydroalcoholic crude extract (HCE, 5

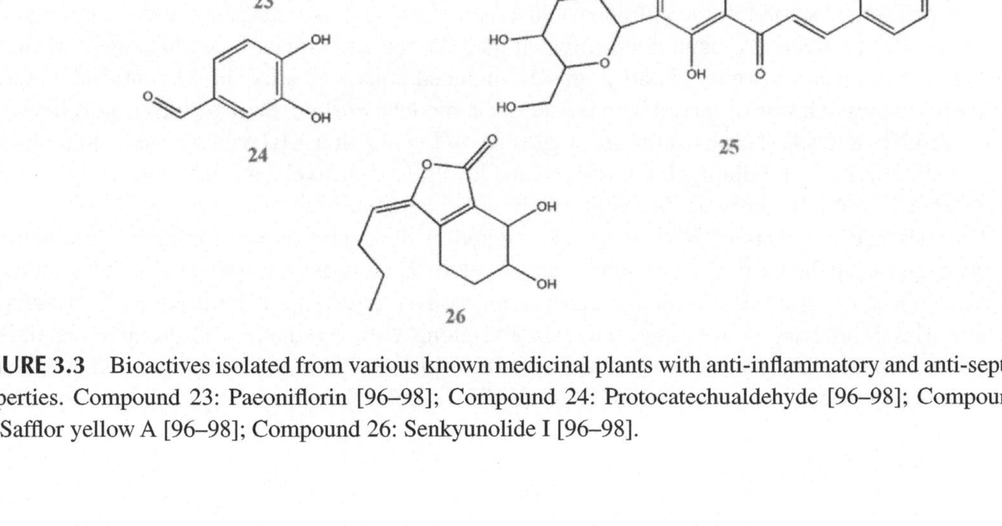

23 24 25 26

FIGURE 3.3 Bioactives isolated from various known medicinal plants with anti-inflammatory and anti-septic properties. Compound 23: Paeoniflorin [96–98]; Compound 24: Protocatechualdehyde [96–98]; Compound 25: Safflor yellow A [96–98]; Compound 26: Senkyunolide I [96–98].

mg/kg) prophylactically, there occurred an elevation in the phagocyte secretion of hydrogen peroxide and production of NO *ex vivo* with both HCE and HEX. This was evaluated 12 hours post-CLP-induction. In addition, there were diminished serum levels of pro-inflammatory cytokines. These observations support the anti-inflammatory activities of HCE and HEX from *C. ambrosioides* [92].

3.2.23 *Toona sinensis*

Toona sinensis Roem is commonly known as Chinese cedar or Chinese toona and belongs to the family Meliaceae. A pre-treatment with the aqueous leaf extracts of *T. sinensis* exhibited anti-septic activity in LPS-induced murine macrophage cells (RAW 264.7 cell line) and CLP-induced septic rat model (male adult Sprague-Dawley rats) [93]. Notably, there was an attenuation of sepsis-induced damage in the lungs as observed histologically. In addition, there was an improvement in the survival of septic animals upon oral administration of *T. sinensis* leaf extracts (pre-treatment) for 30 days. Moreover, there occurred an attenuation of LPS-induced iNOS expression, which, in turn, abrogated the production, as well as the release of NO *in vitro*. Other cytokines, that is, TNF-α and IL-1β, remained unaffected. However, there was an increase in the expression of HO-1 protein. Conclusively, the authors recommended further rigorous experimentation for these preliminary findings [93].

3.2.24 *Attalea speciosa* (synonym *Orbignya phalerata Mart. Babassu*)

Attalea speciosa belongs to the family Arecaceae. The babassu mesocarp extract (powdered or flour) exhibited *in vitro* antimicrobial activity against *Enterococcus faecalis*, *Staphylococcus aureus*, and methicillin-resistant *Staphylococcus aureus* (MRSA). Further, there was an enhanced survival in the CLP-induced septic mice models (Swiss mice) that received a subcutaneous dosage of the mesocarp extract (125 mg/kg and 250 mg/kg) 6 hours post-CLP-induction. The cytokine production, mainly TNF-α, and IL-6, was significantly inhibited in this mice group. Thus, babassu mesocarp extract harbors antimicrobial and immunomodulatory attributes, which makes it an attractive anti-septic agent for the management of sepsis and related inflammatory diseases [94].

3.2.25 *Xuebijing* (XBJ)

This traditional Chinese herbal formula comprises five herbs (Radix *Angelicae Sinensis*, Radix *Salviae Miltiorrhizae*, *Flos Carthami*, *Rhizoma Ligustici Chuanxiong*, and Radix *Paeoniae Rubra*). It is well-documented that Xuebijing confers protection to sepsis-induced animals. Efforts have been made to investigate the efficacy of XBJ in Acinetobacter baumannii-mediated sepsis. A partial reversal was observed in the pulmonary histopathology and associated inflammatory responses in the sepsis rat models (A. baumannii-infected male Wistar rats) treated simultaneously with XBJ through the intravenous route. Notably, an XBJ-induced enhanced annexin A1 (anti-inflammatory mediator) expression was observed in these septic rat models. Further, there occurred a decline in the IL-8 and TNF- α levels. These results are suggestive of the fact that XBJ exerts its anti-inflammatory anti-septic activity by regulating the host cytokine levels. Conclusively, an early intravenous administration of XBJ might abrogate the progression of A. baumannii-mediated sepsis [95].

The therapeutic effects of XBJ were also studied in the sepsis-induced inflammation and pulmonary injuries in the CLP-induced septic mice models. There was an improvement in the mortality of CLP-induced septic mice models regarding pulmonary injury upon intravenous XBJ administration. The pulmonary morphology was restored along with a reduction in the alveolar infiltration of neutrophils and the ratio of pulmonary wet to dry weight. Other important XBJ-mediated immunological alterations were HMGB1 and RAGE (receptor for advanced glycation end-products) down-regulation. Furthermore, there was a decrement in the bronchoalveolar lavage (BAL) neutrophil count and pro-inflammatory cytokine (IL-1β, IL-6, and TNF-α) expression. Hence, the improved mortality effects of XBJ are possibly due to its HMGB1/RAGE-mediated pro-inflammatory cytokine inhibitor activity [96–98]. Conclusively, benzoylpaeoniflorin (18), danshensu (19), gallic acid (20), hydroxysafflor yellow A (21), oxypaeoniflorin (22), paeoniflorin (23), protocatechualdehyde (24), safflor yellow A (25), and senkyunolide I (26) were identified to be the key anti-inflammatory or NF-κB inhibitory molecules in XBJ (Figure 3.2 and 3.3) [96, 97].

3.2.26 *Bai- Hu-Tang* (BHT or White Tiger Decoction)

This traditional Chinese medicinal formula comprises *Gypsum fibrosum*, *Anemarrhenae rhizoma*, radix *Glycyrrhizae preparata*, and non-glutinous rice (powdered extract in the ratio of 3:8:1:4). Lin et al. demonstrated *in vivo* anti-septic potential of BHT [99]. The survival rate of CLP-induced, BHT-treated (BHT-H6 group, a high BHT dose of 0.9 g/kg at 6 hours post-operation) septic rats (male, Sprague-Dawley rats) was significantly higher (80%) than the CLP-induced, sepsis-control group ($P<0.005$). Additionally, there were a significant decrease in the plasma IL-6 and IL-10 cytokine levels after 12 hours of operation in comparison to the CLP-induced, sepsis-control group. These observations are suggestive of the anti-septic properties of BHT as a promising treatment alternative for sepsis [99].

3.2.27 *Huang-Lian-Jie-Du-Tang* (HLJDT)

This traditional Chinese formula comprises four herbs, namely, Rhizoma *Coptidis* (*Coptis chinensis*, family Ranunculaceae), Radix *Scutellariae* (*Scutellaria baicalensis*, family Lamiaceae), Cortex *Phellodendri* (*Phellodendron chinense*, family Rutaceae), and Fructus *Gardeniae* (*Gardenia jasminoides*, family Rubiaceae). HLJDT (in the ratio 3:2:2:3) conferred protection to the CLP-induced rats against lethal sepsis (male Sprague-Dawley rats) prophylactically. Notably, the liver and lung injuries in rats induced by CLP improved together with the suppression of pro-inflammatory cytokine release. Specifically, these included interleukins (IL-1, IL-6, and IL-17A) and TNF-α indicative of the fact that HLJDT has the potential to regulate inflammatory responses in sepsis. Bacterial phagocytosis by the peritoneal macrophages was also facilitated by HLJDT. CLP-induced IFN-γ was de-repressed, while IL-4 was blocked in the rat spleens at 24 h post-CLP-induction. Further, HLJDT caused a shift in the immunological response from Th1 (Type 1 helper T-lymphocytes) to Th2 (Type 2 helper T cells) in sepsis rat models. Finally, IL-17A and ROR-γt (retinoic acid receptor-related orphan receptor gamma in the thymus) splenic expression was inhibited by HLJDT in these septic rats. Under septic, non-inhibitory conditions, IL-17A and ROR-γt cause activation of Th17. The authors concluded that HLJDT has anti-septic activity and is a potent medication candidate for treating septic patients [100].

3.3 CONCLUSION

Sepsis is among the major causes of morbidity and mortality in the population of all age groups worldwide. The current treatment strategy largely relies on empirical administration of antibiotics or a combination of specific antimicrobials depending upon the diagnostic accuracy and sepsis severity. Unfortunately, this chemically synthesized antibiotics-based treatment regimen for sepsis is challenged by multidrug-resistant pathogens. With this view, the plant-based, cost-effective natural herbal formulations are expected to alleviate the problem of microbial resistance, which, in turn, might revolutionize the present treatment modalities for sepsis. Howbeit, this would depend upon the timely and accurate diagnosis of sepsis using advanced diagnostics, such as RNA sequencing, and its severity. Further, a barcode-based compilation of the herbal plants with a prior authentication of their medicinal attributes is indispensable for preparing pathogen-specific antimicrobials for sepsis treatment. Furthermore, the phyto-active compounds with antimicrobial properties must be appropriately documented for ease of accessibility and utilization. Moreover, structural elucidation is pertinent for these bioactive compounds to deduce structure-function relationships, explore their pharmacological relevance in sepsis treatment, and in designing novel anti-septic drug molecules. Also, it is suggested that advanced purification methods must be exploited for extracting the identified active phytoconstituents in their purest form followed by recombination into pre-determined (from clinical data) dose formulations. In conclusion, the anti-septic properties of some extensively studied important medicinal plants have been discussed to disseminate the desired information for developing potent anti-septic plant-based formulations.

REFERENCES

1. Singer M, Deutschman CS, Seymour CW et al. The third international consensus definitions for sepsis and septic shock (Sepsis-3). JAMA 2016; 315(8): 801–10.
2. Gotts JE, Matthay MA. Sepsis: Pathophysiology and clinical management. BMJ 2016; 353: i1585.
3. United States Centers for Disease Control and Prevention. Healthcare Professional (HCP) Resources: Sepsis. Updated August 27, 2019. T06:23:15Z. www.cdc.gov/sepsis/get-ahead-of-sepsis/hcp-resources.html (accessed February 25, 2020).
4. Global Sepsis Alliance. Toolkits. www.world-sepsis-day.org/toolkits/ (accessed February 25 2020).
5. UK Sepsis Trust. Education. 2018. https://sepsistrust.org/education/ (accessed February 25, 2020).
6. Kumar A, Roberts D, Wood KE et al. Duration of hypotension before initiation of effective antimicrobial therapy is the critical determinant of survival in human septic shock. Crit Care Med 2006; 34: 1589–96.
7. Ulldemolins M, Nuvials X, Palomar M et al. Appropriateness is critical. Crit Care Clin 2011; 27: 35–51.
8. Sherwin R, Winters ME, Vilke GM et al. Does early and appropriate antibiotic administration improve mortality in emergency department patients with severe sepsis or septic shock? J Emerg Med 2017; 53: 588–95.
9. Rhodes A, Evans LE, Alhazzani W et al. Surviving Sepsis Campaign: International guidelines for management of sepsis and septic shock: 2016. Intensive Care Med 2017; 43: 304–77.
   Garnacho-Montero J, Gutierrez-Pizarraya A, Escoresca-Ortega A et al. Adequate antibiotic therapy prior to ICU admission in patients with severe sepsis and septic shock reduces hospital mortality. Crit Care 2015; 19: 302.
10. Garnacho-Montero J, Gutierrez-Pizarraya A, Escoresca-Ortega A et al. Adequate antibiotic therapy prior to ICU admission in patients with severe sepsis and septic shock reduces hospital mortality. Crit Care 2015; 19: 302.
11. Vallés J, Rello J, Ochagavía A et al. Community-acquired bloodstream infection in critically ill adult patients: impact of shock and inappropriate antibiotic therapy on survival. Chest 2003; 123: 1615–24.
12. Suberviola Cañas B, Jáuregui R, Ballesteros MÁ et al. Effects of antibiotic administration delay and inadequacy upon the survival of septic shock patients. Med Intensiva 2015; 39: 459–66.
13. Zilberberg MD, Shorr AF, Micek ST et al. Multi-drug resistance, inappropriate initial antibiotic therapy and mortality in Gram-negative severe sepsis and septic shock: a retrospective cohort study. Crit Care 2014; 18: 596.
14. Vazquez-Guillamet C, Scolari M, Zilberberg MD et al. Using the number needed to treat to assess appropriate antimicrobial therapy as a determinant of outcome in severe sepsis and septic shock. Crit Care Med 2014; 42: 2342–9.
15. Shehab N, Patel P, Srinivasan A et al. Emergency department visits for antibiotic-associated adverse events. Clin Infect Dis 2008; 47: 735–43.
16. Wang H, Xu T, Lewin MR. Future possibilities for the treatment of septic shock with herbal components. Am J Emerg Med 2009; 27(1): 107–12.
17. Zhao KJ, Dong TT, Tu PF, Song ZH, Lo CK, Tsim KW. Molecular Genetic and Chemical Assessment of Radix Angelica (Danggui) in China. J Agric Food Chem 2003; 51(9): 2576–83.
18. Jia M, Yang TH, Yao XJ, Meng J, Meng JR, Mei QB. Anti-oxidative effect of Angelica polysaccharide sulphate. Zhong Yao Cai 2007; 30(2): 185–8.

19. Wang H, Li W, Li J et al. The aqueous extract of a popular herbal nutrient supplement, angelica sinensis, protects mice against lethal endotoxemia and sepsis. J Nutr 2006; 136(2): 360–5.
20. Wang H, Bloom O, Zhang M et al. HMG-1 as a late mediator of endotoxin lethality in mice. Science 1999; 285(5425): 248–51.
21. Rendon-Mitchell B, Ochani M, Li J et al. IFN-gamma induces high mobility group box 1 protein release partly through a TNF-dependent mechanism. J Immunol 2003; 170(7): 3890–7.
22. Chen G, Li J, Ochani M et al. Bacterial endotoxin stimulates macrophages to release HMGB1 partly through CD14- and TNF-dependent mechanisms. J Leukoc Biol 2004; 76(5): 994–1001.
23. Yang H, Ochani M, Li J et al. Reversing established sepsis with antagonists of endogenous high-mobility group box 1. Proc Natl Acad Sci U S A 2004; 101: 296–301.
24. Wang H, Liao H, Ochani M et al. Cholinergic agonists inhibit HMGB1 release and improve survival in experimental sepsis. Nat Med 2004a; 10(11): 1216–21.
25. Wang H, Yang H, Tracey KJ. Extracellular role of HMGB1 in inflammation and sepsis. J Intern Med 2004a; 255(3): 320–31.
26. Wang QL, Tao YY, Yuan JL, Shen L, Liu CH. Salvianolic acid B prevents epithelial-to-mesenchymal transition through the TGF-beta1 signal transduction pathway in vivo and in vitro. BMC Cell Biol 2010; 11: 31.
27. Liu CL, Xie LX, Li M, Durairajan SS, Goto S, Huang JD. Salvianolic acid B inhibits hydrogen peroxide-induced endothelial cell apoptosis through regulating PI3K/Akt signaling. PloS One 2007; 2(12): e1321.
28. Lee WY, Cheung CC, Liu KW et al. Cytotoxic effects of tanshinones from Salvia miltiorrhiza on doxorubicin-resistant human liver cancer cells. J Nat Prod 2010; 73(5): 854–9.
29. Kim SY, Moon TC, Chang HW, Son KH, Kang SS, Kim HP. Effects of tanshinone I isolated from Salvia miltiorrhiza bunge on arachidonic acid metabolism and in vivo inflammatory responses. Phytother Res 2002; 16(7): 616–20.
30. Kang BY, Chung SW, Kim SH, Ryu SY, Kim TS. Inhibition of interleukin-12 and interferon-γ production in immune cells by tanshinones from Salvia miltiorrhiza. Immunopharmocology 2000; 49(3): 355–361.
31. Jang SI, Jeong SI, Kim KJ et al. Tanshinone IIA from Salvia miltiorrhiza inhibits inducible nitric oxide synthase expression and production of TNF-α, IL-1β and IL-6 in activated RAW 264.7 cells. Planta Med 2003; 69(11): 1057–9.
32. Li W, Li J, Ashok M et al. A cardiovascular drug rescues mice from lethal sepsis by selectively attenuating a late-acting proinflammatory mediator, high mobility group box 11. J Immunol 2007; 178(6): 3856–64.
33. Graham HN. Green tea composition, consumption, and polyphenol chemistry. Prev Med 1992; 21(3): 334–50.
34. Chacko SM, Thambi PT, Kuttan R, Nishigaki I. Beneficial effects of green tea: A literature review. Chin Med 2010; 5: 1–9.
35. Yang CS, Wang H. Cancer Preventive Activities of Tea Catechins. Molecules 2016; 21(12): 1679.
36. Li W, Ashok M, Li J, Yang H, Sama AE, Wang H. A Major Ingredient of Green Tea Rescues Mice from Lethal Sepsis Partly by Inhibiting HMGB1. PLoS One 2007; 2(11): e1153.
37. Chen X, Li W, Wang H. More tea for septic patients? Green tea may reduce endotoxin-induced release of high mobility group box 1 and other pro-inflammatory cytokines. Med Hypotheses 2005; 66(3): 660–3.
38. Tang W, Eisenbrand G. Panax ginseng C.A. Mey. In: Chinese Drugs of Plant Origin. Springer, Berlin, Heidelberg, 1992.
39. Ganesan, P, Ko HM, Kim IS, Choi DK. Recent trends of nano bioactive compounds from ginseng for its possible preventive role in chronic disease models. RSC Adv 2015;5: 98634–42.
40. Lee, SH., Oh, M., Park, J. et al. Antioxidant and anti-inflammatory activities of the ethanolic extract of fermented red ginseng marc. Food Sci Biotechnol 2015; 24: 651–7.
41. Shergis JL, Zhang AL, Zhou W, Xue CC. Panax ginseng in randomised controlled trials: A systematic review. Phytother Res 2013; 27(7): 949–65.
42. Lee SY, Jeong JJ, Eun SH, Kim DH. Anti-inflammatory effects of ginsenoside Rg1 and itsmetabolites ginsenoside Rh1 and 20(S)-protopanaxatriol in mice with TNBS-induced colitis. Eur J Pharmacol 2015; 762: 333–43.
43. Bak M, Jeong W, Kim K. Detoxifying effect of fermented black ginseng on H_2O_2-induced oxidative stress in HepG2 cells. Int J Mol Med 2014; 34(6): 1516–22.
44. Oh HA, Kim DE, Choi HJ, Kim NJ, Kim DH. Anti-stress effects of 20(S)-protopanaxadiol and 20(S)-protopanaxatriol in immobilized mice. Biol Pharm Bull 2015; 38(2): 331–5.

45. Lee W, Cho SH, Kim JE et al. Suppressive Effects of Ginsenoside Rh1 on HMGB1-Mediated Septic Responses. Am J Chin Med 2019; 47(1): 119–33.
46. Zhang J, Wang J, Wu X, Wei Y. Ginsenoside Rb1 inhibits proliferation and promotes apoptosis by regulating HMGB1 in uterine fibroid cells. Artif Cells Nanomed Biotechnol 2019; 47(1): 2967–71.
47. Izawa B. Illustrated Cyclopedia of Medicinal Plants of Japan. Seibundo-Shinkosha Publishing Corporation, Tokyo, Japan 1980.
48. Peake PW, Pussell BA, Martyn P, Timmermans V, Charlesworth JA. The inhibitory effect of rosmarinic acid on complement involves the C5 convertase. Int J Immunopharmacol 1991; 13(7): 853–7.
49. Rimando A, Inoshiri S, Otsuka H, Kohda H, Yamasaki K, Padolina WG. Screening of mast cell histamine release inhibitory activity of Philippine medicinal plants. Active constituent of Ehretia microphylla. Shoyakugaku Zasshi. 1987; 41: 242–7.
50. Yamamoto H, Sakakibara J, Nagatsu A, Sekiya K. Inhibitors of arachidonate lipoxygenase from defatted perilla seed. J Agric Food Chem 1998; 46(3): 862–5.
51. Abedini A, Roumy V, Mahieux S et al. Rosmarinic acid and its methyl ester as antimicrobial components of the hydromethanolic extracts of Hyptis atrorubensPoit. (Lamiaceae). Evid Based Complement Alternat Med 2013; 2013: 604536.
52. Adomako-Bonsu AG, Chan SL, Pratten M, Fry JR. Antioxidant activity of rosmarinic acid and its principal metabolites in chemical and cellular systems: Importance of physico-chemical characteristics. Toxicol In Vitro 2017; 40: 248–55.
53. Al-Dhabi NA, Arasu MV, Park CH, Park SU. Recent studies on rosmarinic acid and its biological and pharmacological activities. EXCLI J 2014; 13: 1192–5.
54. Luan H, Kan Z, Xu Y, Lv C, Jiang W. Rosmarinic acid protects against experimental diabetes with cerebral ischemia: relation to inflammation response. J Neuroinflammation 2013; 10: 28.
55. Swarup V, Ghosh J, Ghosh S, Saxena A, Basu A. Antiviral and anti-inflammatory effects of rosmarinic acid in an experimental murine model of Japanese encephalitis. Antimicrob Agents Chemother 2007; 51(9): 3367–70.
56. Yang EJ, Ku SK, Lee W et al. Barrier protective effects of rosmarinic acid on HMGB1-induced inflammatory responses in vitro and in vivo. J Cell Physiol 2013; 228(5): 975–82.
57. Duke JA. Handbook of Edible Weeds. 1st ed. London: CRC Press 2001; pp 158.
58. Tabba HD, Chang R, Smith KM. Isolation, purification, and partial characterization of prunellin, an anti-HIV component from aqueous extracts of Prunella vulgaris. Antiviral Res 1989; 11(5-6): 263–73.
59. Psotová J, Kolár M, Sousek J, Svagera Z, Vicar J, Ulrichová J. Biological activities of Prunella vulgaris extract. Phytother Res 2003; 17(9): 1082–7.
60. Ryu SY, Oak MH, Yoon SK et al. Anti-allergic and anti-inflammatory triterpenes from the herb of Prunella vulgaris. Planta Med 2000; 66(4): 358–60.
61. Lee IK, Kim DH, Lee SY et al. Triterpenoic acids of Prunella vulgaris var. lilacina and their cytotoxic activities In Vitro. Arch Pharm Res 2008; 31(12): 1578–83.
62. Oh C, Price J, Brindley MA et al. Inhibition of HIV-1 infection by aqueous extracts of Prunella vulgaris L. Virol J 2011; 8: 188.
63. Kwon G, Kim HJ, Park SJ, et al. Anxiolytic-like effect of danshensu [(3-(3,4-dihydroxyphenyl)-lactic acid)] in mice. Life Sci 2014; 101(1-2): 73–78.
64. Jun MS, Kim HS, Kim YM et al. Ethanol Extract of Prunella vulgaris var. lilacina Inhibits HMGB1 Release by Induction of Heme Oxygenase-1 in LPS-activated RAW264.7 Cells and CLP-induced Septic Mice. Phytother Res 2011; 26(4): 605–12.
65. Krafczyk N, Glomb MA. Characterization of phenolic compounds in rooibos tea. J Agric Food Chem 2008; 14;56(9): 3368–76.
66. Lee W, Kim KM, Bae JS. Ameliorative effect of aspalathin and nothofagin from Rooibos (Aspalathus linearis) on HMGB1-induced septic responses in vitro and in vivo. Am J Chin Med 2015; 43(5): 991–1012.
67. Kazuno S, Yanagida, M, Shindo N, Murayama K. Mass spectrometric identification and quantification of glycosyl flavonoids, including dihydrochalcones with neutral loss scan mode. Anal Biochem 2005; 347(2): 182–92.
68. Kokotkiewicz A, Luczkiewicz M, Sowinski P, Glod D, Gorynski K, Bucinski A. Isolation and structure elucidation of phenolic compounds from Cyclopia subternata Vogel (honeybush) intact plant and in vitro cultures. Food Chem 2012; 133(4): 1373–82.

69. de Beer D, Schulze AE, Joubert E, de Villiers A, Malherbe CJ, Stander MA. Stander. Food ingredient extracts of cyclopia subternata (honeybush): Variation in phenolic composition and antioxidant capacity. Molecules 2012; 17(12): 14602–24.
70. Lee W, Yoon EK, Kim KM, Park DH, Bae JS. Antiseptic effect of vicenin-2 and scolymoside from Cyclopia subternata (honeybush) in response to HMGB1 as a late sepsis mediator in vitro and in vivo. Can J Physiol Pharmacol 2015; 93(8): 709–20.
71. Liu MW, Su MX, Zhang W, Zhang LM, Wang YH, Qian CY. Rhodiola rosea suppresses thymus T-lymphocyte apoptosis by downregulating tumor necrosis factor-α- induced protein 8-like-2 in septic rats. Int J Mol Med 2015; 36(2): 386–98.
72. Qi Z, Zhang Y, Qi S et al. Salidroside Inhibits HMGB1 Acetylation and Release through Upregulation of SirT1 during Inflammation. Oxid Med Cell Longev 2017; 2017: 9821543.
73. Lan KC, Chao SC, Wu HY et al. Salidroside ameliorates sepsis-induced acute lung injury and mortality via downregulating NF-κB and HMGB1 pathways through the upregulation of SIRT1. Sci Rep 2017; 7(1): 12026.
74. Yang EJ, Lee W, Song KS, Bae JS. Ameliorative effect of a rarely occurring C-methylrotenoid on HMGB1-induced septic responses in vitro and in vivo. Biochem Pharmacol 2016; 110–111: 58–70.
75. Yang YI, Woo JH, Seo YJ, Lee KT, Lim Y, Choi JH. Protective effect of brown alga phlorotannins against hyper-inflammatory responses in lipopolysaccharide-induced sepsis models. J Agric Food Chem 2016; 64(3): 570–578.
76. Park EJ, Kim YM, Park SW et al. Induction of HO-1 through p38 MAPK/Nrf2 signaling pathway by ethanol extract of Inula helenium L. reduces inflammation in LPS-activated RAW 264.7 cells and CLP-induced septic mice. Food Chem Toxicol 2013; 55: 386–95.
77. Yang J, Dai Y, Xia YF, Huang WZ, Wang ZT. Alpinia katsumadai Hayata prevents mouse sepsis induced by cecal ligation and puncture through promoting bacterial clearance and downregulating systemic inflammation. Phytother Res 2009; 23(2): 267–73.
78. Maciel MCG, Farias JC, Maluf MJ, Gomes EA, Pereira PVS, Aragão-Filho WC. Syzygium jambolanum treatment improves survival in lethal sepsis induced in mice. BMC Complement Altern Med 2008; 8(57): 1–7.
79. Rattmann YD, Souza LM, Malquevicz-Paiva SM et al. Analysis of flavonoids from eugenia uniflora leaves and its protective effect against murine sepsis. Evid Based Complement Alternat Med 2012; 2012: 1–9.
80. Fatemi F, Allameh A, Khalafi H, Ashrafihelan J. Hepatoprotective effects of gamma-irradiated caraway essential oils in experimental sepsis. Appl Radiat Isot 2010; 68(2): 280–5.
81. Dadkhah A, Fatemi F. Heart and kidney oxidative stress status in septic rats treated with caraway extracts. Pharm Biol 2011; 49(7): 679–86.
82. Bayir Y, Albayrak A, Can I, et al. Nigella sativa as a potential therapy for the treatment of lung injury caused by cecal ligation and puncture-induced sepsis model in rats. Cell Mol Biol (Noisy-le-grand) 2012; 58(2): 1680–7.
83. Liu MW, Su MX, Wang YH, Wei W, Qin LF, Liu X, Tian ML, Qian CY. Effect of melilotus extract on lung injury by upregulating the expression of cannabinoid CB2 receptors in septic rats. BMC Complement Altern Med 2014; 14: 94.
84. Liu MW, Su MX, Wang YH, Qian CY. Effect of Melilotus extract on lung injury via the upregulation of tumor necrosis factor-α-induced protein-8-like 2 in septic mice. Mol Med Rep 2015; 11(3): 1675–84.
85. Auyeung KK, Han QB, Ko JK. Astragalus membranaceus: A review of its protection against inflammation and gastrointestinal cancers. Am J Chin Med 2016; 44(1): 1–22.
86. Gao XH, Xu XX, Pan R et al. Saponin fraction from Astragalus membranaceus roots protects mice against polymicrobial sepsis induced by cecal ligation and puncture by inhibiting inflammation and upregulating protein C pathway. J Nat Med 2009; 63(4): 421–9.
87. Park SH, Roh E, Kim HS et al. Inhibition of IRAK-4 activity for rescuing endotoxin LPS-induced septic mortality in mice by lonicerae flos extract. Biochem Biophys Res Commun 2013; 442(3-4): 183–8.
88. Kim SJ, Yoon SJ, Kim YM et al. HS-23, Lonicera japonica extract, attenuates septic injury by suppressing toll-like receptor 4 signaling. J Ethnopharmacol 2014; 155(1): 256–66.
89. Kim SJ, Kim JS, Choi HS et al. HS-23, a Lonicera japonica extract, reverses sepsis-induced immunosuppression by inhibiting lymphocyte apoptosis. J Ethnopharmacol 2015; 171: 231–9.

90. Yun N, Lee CH, Lee SM. Protective effect of Aloe vera on polymicrobial sepsis in mice. Food Chem Toxicol 2009; 47(6): 1341–8.
91. Chen J, Raj N, Andrejko KM, Deutschman CS. Intrahepatic nuclear factor-kappa B activity and alpha 1-acid glycoprotein transcription do not predict outcome after cecal ligation and puncture in the rat. Crit Care Med 2001; 29(3): 589–96.
92. Rios CEP, Abreu AG, Braga Filho JAF et al. Chenopodium ambrosioides L. improves phagocytic activity and decreases bacterial growth and the systemic inflammatory response in sepsis induced by cecal ligation and puncture. Front Microbiol 2017; 8.
93. Yang CJ, Chen YC, Tsai YJ, Huang MS, Wang CC. Toona sinensis leaf aqueous extract displays activity against sepsis in both in vitro and in vivo models. Kaohsiung J Med Sci 2014; 30(6): 279–85.
94. Barroqueiro ESB, Prado DS, Barcellos PS et al. Immunomodulatory and antimicrobial activity of babassu mesocarp improves the survival in lethal sepsis. Evid Based Complement Alternat Med 2016; 2016: 1–7.
95. He XD, Wang Y, Wu Q et al. Xuebijing Protects Rats from Sepsis Challenged with Acinetobacter baumannii by Promoting Annexin A1 Expression and Inhibiting Proinflammatory Cytokines Secretion. Evid Based Complement Alternat Med 2013; 2013: 1–11.
96. Jiang M, Zhou M, Han Y et al. Identification of NF-kB inhibitors in Xuebijing injection for sepsis treatment based on bioactivity-integrated UPLC-Q/TOF. J Ethnopharmacol 2013; 147(2): 426–33.
97. Liu MW, Wang YH, Qian CY, Li H. Xuebijing exerts protective effects on lung permeability leakage and lung injury by upregulating Toll-interacting protein expression in rats with sepsis. Int J Mol Med 2014; 34(6): 1492–504.
98. Wang Q, Wu X, Tong X, Zhang Z, Xu B, Zhou W. Xuebijing ameliorates sepsis-induced lung injury by downregulating HMGB1 and RAGE expressions in mice. Evid Based Complement Alternat Med 2015; 2015: 1–9.
99. Lin CJ, Su YC, Lee CH, Li TC, Chen YA, Lin SJS. Bai-Hu-Tang, ancient Chinese medicine formula, may provide a new complementary treatment option for sepsis. Evid Based Complement Alternat Med 2013; 2013: 1–8.
100. Wei Y, Shan L, Qiao L, Liu R, Hu Z, Zhang W. Protective Effects of Huang-Lian-Jie-Du-Tang against Polymicrobial Sepsis Induced by Cecal Ligation and Puncture in Rats. Evid Based Complement Alternat Med 2013; 2013: 909624.

4 Antimicrobials

Advances in Pharmaceutical and Medicinal Research

Rahul Sharma,[1] Himani Meena,[2] Lipsa Panda[1] and Ajit Ramesh Sawant[1]*
[1]Department of Biotechnology, School of Life Sciences, Pondicherry University, Puducherry, India
[2]Department of Microbiology, School of Life Sciences, Pondicherry University, Puducherry, India
*Corresponding Author: Rahul Sharma rsbiot77@gmail.com

CONTENTS

DOI: 10.1201/9781003268932-4

4.1 INTRODUCTION

Reminiscence about maladies traces us back to the pre-antibiotic era when it was arduous to decipher the nature of prevailing ailments. Anxious-ridden human population across the border relied on natural products that ranged from using moulds by the ancient Egyptians to the barks of cinchona plant and other plant-derived compounds, such as artemisinin and ginseng, as an antimicrobial agent (Aminov, 2009). However, during the early 1900s, most of the global population suffered from common ailments, namely cold and flu, owing to the paucity of antimicrobial agents. The substances that inhibit the growth and development of microorganisms such as bacteria, viruses, protozoa and fungi are called antimicrobials (Burnett-Boothroyd and McCarthy 2011). The term antibiotic, on the other hand, originates from antibiosis which means 'against life'. Earlier, the use of this term was confined to the natural chemical ingredients produced by microbes that were either lethal (bactericidal) or inhibitory (bacteriostatic) to other related microorganisms (such as algae, bacteria and fungi) but now they also include chemically-synthesized compounds (Aminov and Mackie, 2007; Amnirov, 2009).

Till now, ~250 types of antibiotics have been noted for use in humans and animals where they have been shown to affect the biosynthesis of components of cell wall, protein synthesis machinery, folic acid synthesis, metabolism and genetic element of the organisms (Figure 4.1; Kummerer, 2003; Singh et al., 2021). The prime source for antimicrobials/antibiotics includes algae, bacteria, fungi and plants (Hayashi et al., 2014; Inamuddin et al. 2021). With advances in therapeutics, the multitude of antibiotics usage and lack of surveillance gathered crucial setbacks, of which acquired resistance is the prominent one. Bacterial resistance can be categorized into multidrug-resistant (MDR), pan drug-resistant (PDR) and extreme drug-resistant (XDR) (Magiorakos et al., 2012). MDR bacteria have become an alarming concern owing to their steadily growing resistance to last drug resorts such as vancomycin and colistin. Acquiring resistance to the in-use antibiotics pose significant threats that can be accounted to the following grounds (Reygaert, 2018).

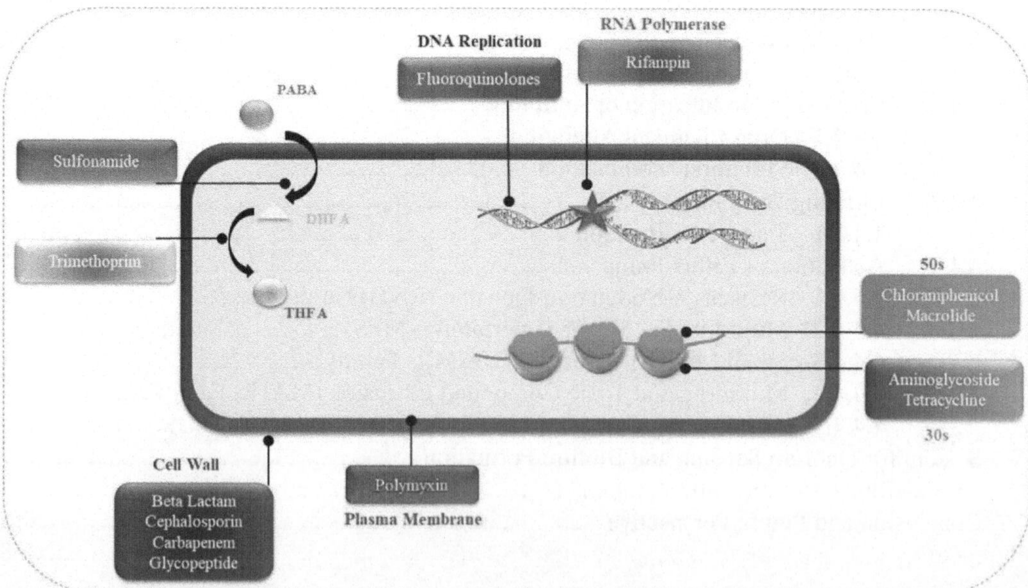

FIGURE 4.1 Mode of action and molecular targets for selected antibiotics.

- Genetic mutations in target gene(s)
- Acquiring resistant genes such as HGT
- Antibiotic modification
- Decreased accumulation of drug due to active efflux pump
- Alteration in target site(s)

Antimicrobial resistance (AMR) is a continuous evolutionary process that makes the drugs ineffective due to their continuous exposure and use in a short period of time. This leads to decline in private investment and lack of new technological innovations in discovery and development of antibiotics (Magiorakos et al., 2012). The major challenge before the medical community is to overcome the increasing AMR by finding new strategies or antimicrobials. The increasing incidences of AMR have also underlined the importance of antimicrobial stewardship which aims for the selection of an optimal dose and duration of the drug resulting in an improved effectiveness with negligible toxicity to the patients (Doron and Davidson, 2011). The World Health Organization (WHO) is the main regulatory body that oversees the functionality of stewardship in hospitals to scrutinize the implementation of antibiotics (Doron and Davidson, 2011). In 2017, WHO published a list of priority pathogens against new antibiotics to address the increasing AMR world-wide. Over 60% of antibiotics are revived from natural sources but their production remains insufficient owing to the arduous process of drug discovery. Therefore, to overcome such challenges, the production of antibiotics could be facilitated through industrial microbiology using recombinant DNA technology (Pham et al., 2019). Based on these scientific backgrounds, a growing interest of research on antimicrobial peptides and aptamers is observed in recent times (Lee et al., 2017; Colombo et al., 2020). The chapter is focussed on chemical nature and evolution of AMR against existing antimicrobials.

4.2 CHRONOLOGY OF ANTIMICROBIALS/ANTIBIOTICS

With the boom in industrialization, mankind took giant strides towards modernization which paved the path to more lethal diseases, syphilis was one of them. In 1910, scientists deployed a cure for syphilis (drug marketed as *salvarsan*) that suppressed *Treponema pallidium* infection (Hutchings et al., 2019). The drug exhibited bacteriostatic nature and inhibited bacterial infection by affecting essential cellular biomolecules/pathways responsible for bacterial multiplication and growth. Later, bacteriologists Florey and Fleming, in 1928, put forward an elixir by the breakthrough discovery of penicillin isolated from a mould, *Penicillium crysogenum*, was more of a miracle (Ligon, 2004). Penicillin, the first beta-lactam antibiotic, upon arrival into the clinical setting (in 1948) attained new heights as it was highly useful to treat wounded soldiers during the World War II (Ligon, 2004; Hutchings et al., 2019). However, in 1944, the AMR in *S. aureus* was attributed to the overuse of β-lactamase-type antibiotics. The milestone discoveries that led to the generation of different kinds and classes of antibiotics are presented in Figure 4.2.

During 1932, pharmaceutical company Bayer Laboratory took ample interest in treating wound infections caused in the aftermath of the First World War (Wainwright and Kristiansen, 2011). Later, Gerhard and colleagues from the same laboratory screened a library of synthetic compounds for their potential antibacterial activity against the haemolytic streptococcal infections and discovered a red diazo dye *streptozon* which became popular with the trade name prontosil. This first ever discovered sulpha-drug elicited its activity upon conversion to its apo-form to the active sulphanilamide counterpart (Gaynes, 2017). Gerhard studied the clinical efficacy of prontosil on animal models as well as on his daughter who then suffered from streptococcal infection on her amputated arm. In contrast, Dr. Walksman and his students, in 1943, observed that certain soil actinomycetes inhibit the growth of other co-occurring microbes (Woodruff, 2014). These milestone discoveries laid the red carpet for scientists to discover or develop new antibiotics. These sparks led to the

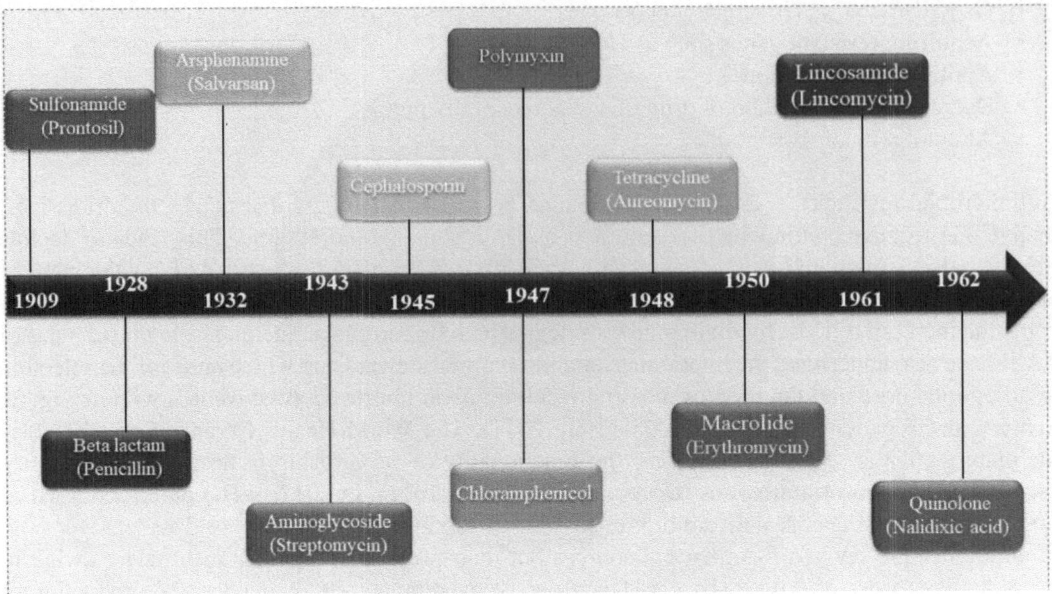

FIGURE 4.2 Timeline for the evolution of antibiotics.

discovery of the first ever class of aminoglycoside antibiotic, streptomycin. This bactericidal antibiotic exerted its effect by binding to the 30s ribosomal subunit and thus inhibiting translation. Giuseppe Brotzuin, an Italian bacteriologist in the year 1948, discovered cephalosporins from a mould *Acremonium strictum* which was the first ever antibiotic against *Salmonella typhi*, the causative agent of typhoid fever (Bo, 2000). These expeditions in the search for new antibiotics led to the discovery of polymyxin antibiotics, the last line of the drug against MDR Gram-negative infections (Velkov and Roberts, 2019). It has been noted that the majority of the antimicrobial/antibiotic agents have been revived from actinomycetes (45%), followed by fungi (38%) and unicellular eubacterial (17%) species (Mullis et al. 2019).

4.3 CLASSIFICATION AND PHYSICOCHEMICAL PROPERTIES OF ANTIMICROBIAL AGENTS

The antimicrobial agents can be divided into specific groups based on their mode of action that enhances the effectiveness of the drug molecules via inhibition of cell wall synthesis, blocking metabolic pathways, interfering with nucleic acid and protein synthesis (Reygaert 2018; Hotinger et al., 2021). Based on the structural similarity, the drugs can exhibit similar antagonistic activity against specific targets in a microbial cell, and one can be as effective as another drug with a correlative absorption, distribution, metabolism, excretion and toxicology (ADMET) profile. Structural-activity relationship (SAR) studies focus on providing insightful information about the drug specificity index and exploring the opportunity to discover new antimicrobial compounds closely associated with lead molecules (Elkamhawy et al., 2021; Guo et al., 2021; Zhou and Huang 2021). For example, a chemical analog of AHL molecule can mimic the similar structural conformation and bind with the cognate receptor (LasR/RhlR in *Pseudomonas aeruginosa*; AbaR in *Acinetobacter baumannii*) and possess a competitive inhibitory effect. Computer-aided drug design strategies contribute significantly to drug discovery, where structural similarity is crucial for future drug development towards multidrug-resistant strains (Elkamhawy et al., 2021; Mardirossian et al., 2021; Zhang et al., 2021; Pena-Diaz and Ventura, 2022).

4.3.1 B-LACTAMS

Penicillin and cephalosporin are bactericidal drugs containing four-membered lactone rings attached to the molecular structure known as β-lactam drugs. These drugs interfere with peptidoglycan biosynthesis by exhibiting transpeptidase and transglycosylase activity (Gontijo et al., 2021; Dutta and Phull, 2021; Bouza, 2021; Melenotte et al., 2021; Hussain et al., 2021; Carcione et al., 2021).

4.3.2 SULFONAMIDES

Sulphonamides are amphoteric, non-antibiotic compounds used to treat allergic reactions. They are composed of a sulphonamide group with an amine group that inhibits folate biosynthesis, an essential element for bacterial growth and multiplication *via* blocking dihydropteroate synthase (DHPS). These synthetic compounds are used as anti-allergen, while being used as antibiotics for treating inflammatory bowel disease (Christensen et al., 2021; Mphahlele 2021; Bora et al., 2021)

4.3.3 AMINOGLYCOSIDES

Aminoglycosides are bactericidal compounds that hinder bacterial translation process by binding to the 30s ribosomal binding site at the penicillin-binding domain. These drugs are composed of more than two amine groups associated with aminocyclitol structure by glycoside bonds. Tobramycin and streptomycin are well-reported aminoglycosides as antimycobacterial drugs and can treat urinary tract infections in females (Diepstraten et al., 2021; Obasi et al., 2021; Singh et al., 2021; Chaves and Tadi, 2022). Gentamicin, kanamycin, tobramycin and amikacin are used in general to treat Gram-negative infections alone or in combination.

4.3.4 TETRACYCLINES

Tetracyclines are bacteriostatic polyketides produced by *Streptomyces* species and are effective against a wide range of human pathogens. Based on the structural diverisity, they can be classified into four groups called tetracycline, oxytetracycline, doxycycline and chlortetracycline. They are composed of four cyclic hydrocarbon rings that diffuse through the porin channels and interfere with the protein synthesis or translation. They exhibit antagonistic activity *via* binding with 30s ribosomal subunit and hinder the binding of tRNA with mRNA-ribosomal unit leading to an inhibitory effect on the bacterial growth (Matlock et al., 2021; Ahn et al., 2021; Zhuang et al., 2021; Bunduki et al., 2021; Del Rosso et al., 2021).

4.3.5 MACROLIDES

Macrolides are high molecular weight compounds with a large macrocyclic lactone ring that incorporates more than one deoxy sugar that contributes to their lipophilic nature. Erythromycin and azithromycin treat bacterial infection by inhibiting their protein biosynthesis *via* binding with the P site of 50s ribosomal subunit in a reversible manner. They are derived from S*treptomyces erythreus* and act as antimicrobial agents against Gram-positive bacteria (Pollock and Chalmers, 2021; De Lemos Neto et al., 2021; Huang et al., 2021; Miklasińska-Majdanik, 2021; Hema-Ouangraoua et al., 2021; Yang et al., 2021a).

4.3.6 QUINOLONES

Quinolones are broad-spectrum antimicrobials that have dual nature for microorganisms as they display bactericidal and bacteriostatic properties. These antibiotics restrict DNA replication in bacteria *via* interrupting enzymes such as DNA gyrase and topoisomerase IV involved in the DNA replication. Quinolones are comprised of a central ring system attached with a functional group at C-6 or C-7 carbon position. Similarly, fluoroquinolone is developed *via* fluorine atoms at C-6 position. The first generation of quinolones namely, nalidixic acid, ciprofloxacin and levofloxacin were helpful in treating mycobacterial and urinary tract infections (Albuquerque et al., 2021; Doña et al., 2021; Yan and Bryant, 2021; Mohamed et al., 2021; Terreni et al., 2021; Romo et al., 2021).

4.3.7 Polyether Ionophores

Salinomycin is one of the well-known polyether ionophores responsible for the transportation of K^+ across the lipid-bilayer due to lipophilic nature produced by *Streptomyces* spp. Polyether ionophores consist of a cyclic ether chain with a terminal alcohol group and a carboxylic group at both ends (Huczyński, 2012; Naujokat and Steinhart, 2012; Zhou et al., 2013; Abdelrahman et al., 2014; Chapman and Jeffers, 2015).

4.4 ANTIMICROBIAL RESISTANCE (AMR)

Antimicrobial compounds have been a significant part of the scientific community and the general public fighting microbial infections for decades. The discovery of substantial antimicrobials has been a great challenge due to emerging MDR microbial species against conventional antimicrobials. Due to the evolution of MDR strains, WHO estimated considerable increase in mortality because of untreatable infections. The organization has identified ESKAPE pathogens (*Escherichia coli*, *Staphylococcus aureus*, *Klebshiella pneumoniae*, *Acinetobacter baumannii*, *Pseudomnonas aeruginosa* and *Enterococcus faecium*) as clinically important due to the complexity of their pathogenicity (Fortier and Sekulovic, 2013; Pacios et al., 2020; Maryam et al., 2021). Based on the survey, considering AMR for global threat, the scientific community has been assigned to find novel drug molecules without increasing the risk of AMR in microbial agents. Selective pressure of antibiotics leads to the evolution of microbial machinery on molecular and cellular levels that have compromised drug efficacy (Gupta et al., 2020). To understand AMR in microbes, we must learn the ethics of antimicrobial targets to specify the anti-infective agents for the specific infection and mitigate the threat *via* targeting the molecular pathways associated with the pathogenicity (Wright, 2010; Erken et al., 2013; Amaro et al., 2021). By focusing on the particular targets, the scientific community can provide an ultimate solution for the global health issues raised by AMR.

4.4.1 Evolution and Mechanisms of AMR

This section emphasizes a brief account of the defence mechanisms contributing to antibiotic resistance against existing antimicrobial agents. The microbial communities can be intrinsic (overexpression of resistance genes or under stimulating conditions) or acquire resistance *via* self-medication, overdose of the drug or genetic mutation and gene transfer (conjugation, translocation or transduction). The acquired resistance is most often induced by drug overdose, unprescribed drugs, or self-medication, leading to the magnification of acquired resistance. These resistance mechanisms protect microbial cells by enzymatic degradation of the antibiotics, such as β-lactamase and aminoglycosides, limiting drug influx, target site modification, activation of efflux pump and thick biofilm formation (Aminov et al., 2007, 2009; Davidovich et al., 2020).

4.4.2 Enzymatic Modification of Antibiotics

An enzymatic modification of drug molecules can be achieved using two mechanisms: alteration of the functional group of the specific drug molecule and disintegration of structure of the molecules leading to deactivation of the antimicrobial agents. A schematic representation of mechanisms of AMR against antimicrobials is provided in Figure 4.3.

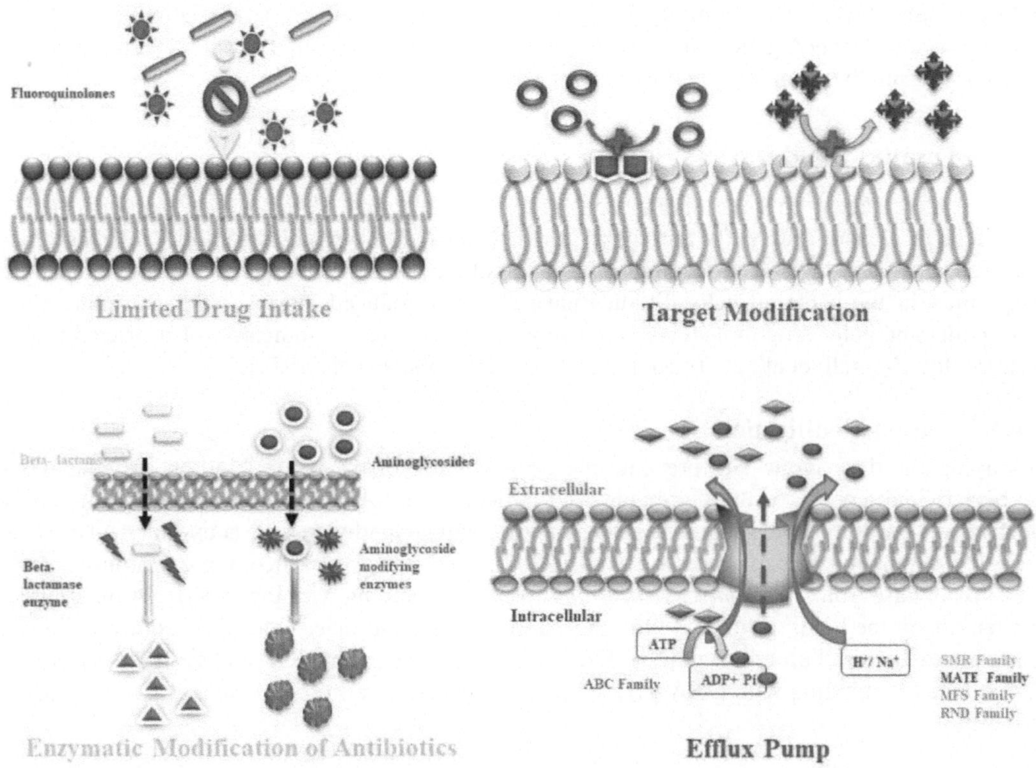

FIGURE 4.3 Schematic representation of microbial resistance mechanism against antimicrobial agents.

4.4.2.1 Drug Chemical Alteration

Bacterial enzymes can alter the chemical structure of drug molecules through chemical group transfer mediated by acetylation, phosphorylation and adenylation (Chen et al., 2014). Aminoglycosides such as gentamycin and streptomycin inhibit bacterial translation *via* binding with 16S ribosomal unit of the 30S rRNA. Aminoglycosides promote the misreading of codons to prevent protein synthesis in bacteria. On the other hand, chloramphenicol crosses the bacterial membrane and binds with 50S ribosomal unit facilitating inhibition of peptide bond formation. The process blocks amino acid transfer suppressing peptidyl transferase activity. Similarly, fluoroquinolone blocks bacterial synthesis *via* impeding DNA gyrase and topoisomerase and halting normal cell division. Aminoglycoside N-acetyltransferase uses acetyl-CoA to transfer acetyl group to aminoglycoside drugs and is classified into four specific groups based on their particular site for acetylation. AAC-2′ (aminoglycoside N-acetyltransferase) from *M. tuberculosis* has been implicated for N-acetylation of drugs containing 2'-hydroxyl or amino group substituent (Hegde et al., 2001; Prabu et al., 2015). *Vibrio fluvialis* secretes aminoglycoside 6-N-acetyltransferase (AAC-6′) and alters the substrate specificity of ciprofloxacin (Chowdhury et al., 2011), while AAC-6′ from *Pseudomonas fluorescens* grants resistance against gentamycin (Lambert et al., 1994).

4.4.2.2 Structural Modification

Cephalosporin and penicillin belong to β-lactam class of antimicrobial compounds, containing a central β-lactam ring (Yang et al., 2021b). Penicillin inhibits cell wall formation *via* binding with penicillin-binding protein specifically to DD-transpeptidase, an essential enzyme for facilitating cell wall peptidoglycan formation. The β-lactam antibiotics cause blockage of peptidoglycan synthases and hydrolases, further restricting the biosynthesis of the bacterial cell wall (Fatima et al., 2021). *S. aureus* acquired *mecA* gene confers resistance against β-lactam antibiotics *via* cleavage of the β-lactam ring from the complex structure (Jabila Mary et al., 2021).

4.4.3 Limiting Drug Influx

Gram-negative bacteria possess an extra peptidoglycan layer that creates a barrier around the bacterial cells to prevent drug entry and make it less permeable for certain antibiotics such as vancomycin. A decreased level of intracellular fluoroquinolones in *E. coli* is a prime example of limited drug influx in the bacterial cells. Studies have shown a reduced intrabacterial accumulation of hydrophilic and polar drug molecules (tetracyclines and β-lactams), indicative of restricted cellular permeability (Vergalli et al., 2020; Richter et al., 2020; Sabnis et al., 2021).

4.4.3.1 Target Modification

Modifying the drug target site prevents the drug binding to the active binding site and blocks the bactericidal activity of the particular drug molecules. RNA ribosylation is one of the known mechanisms for drug target modifications where the enzyme methylates the active site and alters the bactericidal function (Vazquez-Laslop et al., 2008). The *ermC* gene encoding erythromycin ribosomal methylase from *Streptomyces erythraeus* promotes binding with the 16S rRNA unit, leading to alteration at the binding site and blocking streptogramin and macrolides that require 16S rRNA to inhibit the bacterial protein synthesis. Fluoroquinolone inhibits nucleic acid synthesis, which can be prevented by binding with DNA gyrase and topoisomerase V in *S. aureus* and *E. coli* (Seefeldt et al., 2021).

4.4.4 Activation of Efflux Pump

Efflux pumps are membrane-spanning proteins that mediate the transportation of toxic compounds from an intracellular region to the outside of the bacterial cells. These efflux pumps are divided into five specific classes based on their substrate specificity, energy sources, membrane-anchored structure and transmembrane spanning region, enlisted as RND family, MFS family, SMR family, MATE family and ABC family (Soto, 2013; Thakur et al., 2021).

4.4.4.1 Resistance-Nodulation-Division (RND) Family

The RND family is one of the largest efflux pumps present in Gram-negative bacteria with considerably large periplasmic domains that assemble outer membrane channels with periplasmic adaptor proteins to translocate cationic drugs across the cell membrane. The efflux pump is a tripartite structure that contains three specific transmembrane domains (TMs), the pore channel and a drug-binding pocket (Nikaido and Takatsuka, 2009). The TM is made up of 12 α-helices that help in drug movement using protons as an energy source. The TM domain coupled with the pore domains provides an energy-driven channel for the lipophilic drugs. The pore channel is comprised of four sub-domains; PN1 and PN2 (proximal domain) and PC1 & PC2 (distal domain) for substrate specificity due to specified polar and charged residues at binding pocket (Blair and Piddock, 2009). For example, *A. baumanni* & *Acinetobacter nosocomialis* displays resistance towards tigecycline by elevated expression of *adeB* (adeABC efflux pump), *adeG* (adeFGH efflux pump), and *adeJ* (adeIJK efflux pump) (Yang et al., 2019; Foong et al., 2020). In *E. coli*, the RND efflux pump illustrates antibiotic resistance against β-lactams, tetracyclines, and chloramphenicols using acriflavine (AcrAB, AcrAD, AcrEF) and the multidrug transporter (MdtABC, and MdtEF) efflux pump system (Pan et al., 2020). *P. aeruginosa* provides evidence of RND-type efflux pump involvement in reduced antibiotic for ciprofloxacin, carbepenem, and aminoglycosides correlated with MexD, MexA, and MexY efflux pumps (Pourakbari et al., 2016).

4.4.4.2 Multidrug Facilitator Superfamily (MFS)

MFS utilizes the proton motive force to transport the drug molecules outside the bacterial cell. The efflux pump comprises 12 α-helices transmembrane proteins that confer resistance *via* three distinct modes, namely uniport, symport and antiport, using "Rocket-switch" strategy. The model requires binding of the proton followed by the substrate or *vice versa* depending on the type of MFS efflux pumps (Hinchliffe et al., 2014; Li et al., 2019). *S. aureus* reported overexpression of *norA* gene-regulating extrusion of fluoroquinolones that leads to the lowering the susceptibility towards certain biocides and dyes such as ethidium bromide and acridine orange (Utegenova et al., 2018; Elhidar et al., 2021). Overexpression of proteins involved in Nor A efflux pump in *E. coli* and *A. baumannii* is accounted for the resistance against methylene blue (Rineh et al., 2018).

4.4.4.3 Small Multidrug Resistance (SMR) Family

The SMR family is the smallest efflux pump comprising of four TM α-helices that span cell membrane and small amino acyl residues that are responsible for translocation of cationic/lipophilic molecules through the hydrophobic region. The mechanism involves four α-helices dichotomized with dimer interface providing substrate binding with three TM regions during the fourth spanner control homodimer dimerization. The fourth TM helix forms a β-stand structure nearby the substrate-binding domain responsible for an allosteric movement of the efflux pump (Jack et al., 2000; Ranaweera et al., 2015; Ebbensgaard et al., 2020). *A. baumannii* has shown resistance towards amikacin, ciplofloxacin and imipenem owing to the overexpression of a putative *adeS* gene encoding AdeS efflux pump belonging to the SMR family (Srinivasan et al., 2009; Lin et al., 2017; Alav et al., 2021). *E. coli* illustrated similar efflux pump EmrE which is known for decreasing bacterial cell susceptibility towards quaternary cation compounds (Bay et al., 2012; Ovchinnikov et al., 2018; Shcherbakov et al., 2021). Meanwhile, *E. coli* was reported for the overexpression of ynfA gene-regulating efflux pump mediated resistance towards tetracyclines, carbepenem and aminoglycoside (Sarkar et al., 2015).

4.4.4.4 Multidrug and Toxic Compound Extrusion (MATE) Family

The extrusion of polyaromatic and cationic drugs can be mediated through a proton driven force or a Na^+ gradient based motion forming a TM channel called the multidrug and toxic compound extrusion or MATE efflux pump. The MATE family has an open-facing position at membrane leaflet, forming a centrally located V-shaped cavity involving amino acid terminal and a carboxyl terminal associated with pseudo-two-fold symmetry and spanning bilayer membrane in cell membrane (Kuroda and Tsuchiya, 2009; Wang et al., 2016). The MATE efflux pump mediates drug transportation with binding of proton or Na^+ to the channel followed by conformational changes in the structure and opening of the V-shaped cavity. This step enables drug assimilation at the channel and initiates deprotonation of the membrane to transfer drug molecules. The mechanism involves four main steps: binding of protons, conformational changes in the channel protein, restoring the open face state and relocating the drugs (Du et al., 2021). MepA protein belongs to MepRAB efflux of the MATE family, restoring the resistance mechanism in *S. aureus* against tigecycline (Opperman et al., 2010; López et al., 2020; Herrera et al., 2021; Rocha et al., 2021).

4.4.4.5 ATP-binding Cassette (ABC)Transporter Family

The ABC family facilitates the transportation of drug molecules initiated by ATP hydrolysis, providing enough energy to drive the influx and efflux process. The mechanism involves harbouring the substrate-binding pocket and nucleotide-binding pocket that mediate ATP hydrolysis *via* the TM spanning protein. This process can be summarized as the binding of drug molecules to the nucleotide-binding site, followed by the dimerization of the nucleotide-binding domain leading to conformational changes and dissociation of the dimerized complex controlled *via* ATP hydrolysis

(Pohl et al., 2011; Koh-Tan et al., 2016; Sunkara et al., 2022; Mohanty et al., 2021). For example, *Candida albicans* have increased tolerance against antifungal compounds due to candida drug resistance (CDR) known as cdr1p and cdr2p belonging to the ABC transporter family (Shukla et al., 2006; Banerjee et al., 2020). *Riemerella anatipestifer* reduces antibiotic susceptibility for aminoglycosides, cephalosporins, quinolones and macrolides *via* overexpression of the RanB efflux pump (Guo et al., 2017; Li et al., 2020), and *Enterococcus faecalis* for the ABC transporter gene (*lsaE*) regulating antibiotic resistance against macrolides, clindamycin and aminoglycoside (Salah et al., 2021).

4.5 SYSTEM FOR QUORUM SENSING AND BIOFILM FORMATION

Quorum sensing is a bacterial communication system regulated by diffusible chemical signals that recruit other bacterial species to form biofilm in order to enhance the bacterial pathogenicity (Sonawane et al. 2022). A graphical presentation for the same in Gram-negative and Gram-positive bacteria is represented in Figure 4.4. Biofilm can be defined as a complex structure comprised of a heterogeneous bacterial population that produces an enormous amount of virulence factors in a synchronized manner that prevent antibiotic penetration in the community. The QS system is an enzymatic reaction of the signalling molecule production that binds with a membrane-anchored receptor protein to control numerous physiological parameters in the bacteria. These parameters are

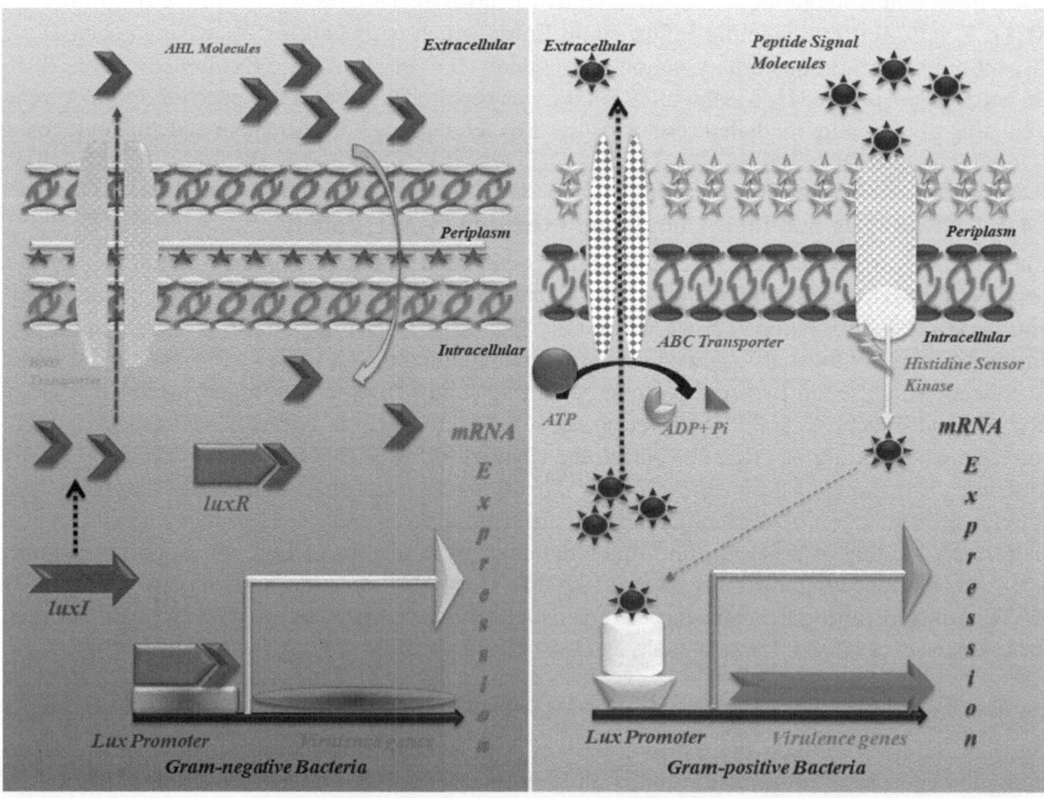

FIGURE 4.4 Quorum sensing system in Gram-negative and Gram-positive bacteria. Gram-negative utilizes AHL molecules as signalling molecule whereas Gram-positive depends on peptide signaling molecule for quorum sensing regulation. The QS system led to biofilm formation restricting antibiotic entry and conferring resistance towards antimicrobial agents.

enlisted as bacterial motility, horizontal gene transfer, the production of pathogenic determinants (toxins, cellulolytic enzymes) and biofilm associated factors such as alginate, rhamnolipid, exopolysaccharides and extracellular DNA (Thornton et al., 2021; Simonetti et al., 2021; Kreve and Reis, 2021; Tripathi et al., 2021). The Gram-negative bacteria utilize acyl-homoserine lactone as a signaling molecule. The Gram-positive bacteria, on the other hand, requires an autoinducer peptide to activate the bacterial QS system. AHL molecules play a significant role in intra-species bacterial communication regulated by LuxI/R type regulatory system, mainly found in Gram-negative bacteria such as *P. aeruginosa* (LasI/R, RhlI/R, PQS) (Reynolds and Kollef, 2021; Chadha et al., 2021) and *A. baumannii* (AbaI/R) (Niu et al., 2008; Tang et al., 2020; Sun et al., 2021). Autoinducer peptides require activation *via* phosphorylation of histidine sensor kinase in Gram-positive bacteria. Some of the peptide-based QS systems are Agr-RNAIII operon in *S. aureus* (Canovas et al., 2016; Xu et al., 2017) and comQXPA operon in *Bacillus subtilis* (Dogsa et al., 2014; Kalamara et al., 2018).

4.6 ANTIMICROBIAL STEWARDSHIP

Over the decades, microorganisms have modified their resistance mechanism to fight the antimicrobials and strengthen their survival skills by establishing a robust immune system against antimicrobials. International organizations have declared AMR a global menace and provided guidelines to deal specifically with the emerging health problems (Berhe et al., 2021; Aslam et al., 2021). John McGowan and Dale Gerding, in 1996, from Emory University alarmed the scientific community about factors associated with the emerging rate of microbial resistance suggesting the necessity of antimicrobial stewardship. In 1997, the Infectious Diseases Society of America (IDSA) and Society for Healthcare Epidemiology of America (SHEA) curated clinical guidelines for health professionals to prevent AMR among pathogenic strains. After a decade, the Centre for Disease Control and Prevention (CDC) was concerned about the rapid increase in the AMR in bacteria, virus and fungal strains. Later, IDSA associated with SHEA released a new clinical practice protocol to minimize the risk of AMR. Paediatric Infectious Diseases Society (PIDS) along with SHEA and IDSA declared a joint policy statement related to treatment, control and eradication of infectious diseases under antimicrobial stewardship (Ruiz-Ramos et al., 2021; Granata et al., 2021; Hamada et al., 2021).

In 2014, Mr. Barack Obama, the then U.S. president, formed a Presidential Advisory Council on Combating Antibiotic-Resistant Bacteria (PACCARB) that issued an executive order (13676) on *Combating Antibiotic-Resistant Bacteria* for five years to reduce the unauthorized drug prescription, control the infection rate, rise in antimicrobial resistance (AMR) and provide effective treatment for the patients. In 2017, the joint commission regulations were implied to comply with hospitals following clinical practice guidelines *via* antimicrobial stewardship by forming a three-member regulatory committee consisting of infection preventionists, pharmacists and practitioners or doctors (Wils et al., 2021; Weragama et al., 2021). These strategies acknowledge developing countries can execute proper ethical practices for consumption of antimicrobials *via* specific protocols for drug use. An antimicrobial stewardship program implies monitoring the optimal drug prescription, perpetuating patient records, minimizing antimicrobial stewardship evolution and tracking financial records (Thursky et al., 2021; Van den Bergh and Brink 2021). The World Health Assembly initiated five strategies to conquer the situation with self-medication and overuse of drugs without worsening the situation (Rizk et al., 2021; Sahitya et al., 2021; Mariappan et al., 2021; Gu et al., 2021). These five strategies are as follows:

1. Education and Training: Provide antimicrobial stewardship-related information to the general people and health workers *via* specialized workshops and educational training.
2. Surveillance and Research: Data collection based on knowledge and evidence through case studies.

3. Maintain Hygiene Conditions: Minimize the infection rate by maintaining hygienic conditions through regular sanitization and preventive protocol.
4. Drug Dose Optimization: Prescribe an adequate dose for human and animal consumption.
5. Economic Development: Encourage government and businesses to invest in medicinal research to enhance diagnostic tools and vaccines production.

4.7 CONCLUSION AND FUTURE PERSPECTIVE

Over the last few decades, antibiotic resistance has emerged as a global menace due to the overuse of drugs and environmental factors that contribute to horizontal gene transfer and mutations. The increasing rate of effective antimicrobials due to the emergence and spread of multidrug-resistant (MDR) pathogens has alarmed the world's healthcare system. Prior to the COVID-19 pandemic, the healthcare experts predicted that AMR will result in ~10 million deaths annually by 2050, and will have a major impact on the economy of the healthcare system. The awareness increased during this pandemic about personal hygiene, basic microbiology, infections and ways of preventing infectious diseases can be exploited to manage the infections related to antimicrobial resistance, prevent the spread of infection and ensure sensible use of antimicrobials. Computational approaches play an important role in modern drug discovery involving structure activity relationships using *in silico* approaches and ADMET profiling. Antimicrobial stewardship aims to instruct health professionals to utilize drugs and keep a record for future drug prescriptions. It is hoped that in-depth research in the health sector diagnoses the disease and provides an effective, affordable treatment to the patients and monitors the situation to avoid drug resistance. It is believed that the lack of contribution by the private sector to new antimicrobial development is the major barrier to the discovery of new antimicrobial agents. The obvious reasons for this are the cost and time required for new drug development. The accurate design of antimicrobial delivery systems specific to the type of infection is the need of the hour in order to avoid the spread of the AMR. The ideal antimicrobial agent should be cost-effective, and effective in treating specific infections at low concentrations with a simple route of administration. It is mandatory to implement the surveillance for use of antimicrobials on the ground with strict measures/control on its sale and use.

REFERENCES

Abdelrahman W, Mohnl M, Teichmann K, Doupovec B, Schatzmayr G, Lumpkins B, Mathis G. Comparative evaluation of probiotic and salinomycin effects on performance and coccidiosis control in broiler chickens. *Poult Sci.* 2014;93(12):3002–3008. doi:10.3382/ps.2014-04212.

Ahn JG, Cho HK, Li D, Choi M, Lee J, Eun BW, Jo DS, Park SE, Choi EH, Yang HJ, Kim KH. Efficacy of tetracyclines and fluoroquinolones for the treatment of macrolide-refractory *Mycoplasma pneumoniae* pneumonia in children: a systematic review and meta-analysis. *BMC Infect Dis.* 2021;21(1):1003. doi:10.1186/s12879-021-06508-7

Alav I, Kobylka J, Kuth MS, Pos KM, Picard M, Blair JMA, Bavro VN. Structure, assembly and function of tripartite efflux and type 1 secretion systems in Gram-negative bacteria. *Chem Rev.* 2021;121(9):5479–5596. doi:10.1021/acs.chemrev.1c00055

Albuquerque HMT, Pinto DCGA, Silva AMS. Microwave irradiation: Alternative heating process for the synthesis of biologically applicable chromones, quinolones, and their precursors. *Molecules.* 2021;26(20):6293. doi:10.3390/molecules26206293

Amaro F, Martín-González A. Microbial warfare in the wild-the impact of protists on the evolution and virulence of bacterial pathogens. *Int Microbiol.* 2021;24(4):559–571. doi:10.1007/s10123-021-00192-y

Aminov RI, Mackie RI. Evolution and ecology of antibiotic resistance genes. *FEMS Microbiol Lett.* 2007;271(2):147–161. doi:10.1111/j.1574-6968.2007.00757.x

Aminov RI. The role of antibiotics and antibiotic resistance in nature. *Environ Microbiol.* 2009;11(12): 2970–2988. doi:10.1111/j.1462-2920.2009.01972.x

Aslam B, Khurshid M, Arshad MI, Muzammil S, Rasool M, Yasmeen N, Shah T, Chaudhry TH, Rasool MH, Shahid A, Xueshan X, Baloch Z. Antibiotic resistance: One health one world outlook. *Front Cell Infect Microbiol.* 2021;11:771510. doi:10.3389/fcimb.2021.771510

Banerjee A, Moreno A, Khan MF et al. Cdr1p highlights the role of the non-hydrolytic ATP-binding site in driving drug translocation in asymmetric ABC pumps. *Biochim Biophys Acta Biomembr.* 2020;1862(2):183131. doi:10.1016/j.bbamem.2019.183131

Bay DC, Turner RJ. Small multidrug resistance protein EmrE reduces host pH and osmotic tolerance to metabolic quaternary cation osmoprotectants. *J Bacteriol.* 2012;194(21):5941–5948. doi:10.1128/JB.00666-12

Berhe DF, Beyene GT, Seyoum B, Gebre M, Haile K, Tsegaye M, Boltena MT, Tesema E, Kibret TC, Biru M, Siraj DS, Shirley D, Howe R, Abdissa A. Prevalence of antimicrobial resistance and its clinical implications in Ethiopia: A systematic review. *Antimicrob Resist Infect* Control. 2021;10(1):168. doi:10.1186/s13756-021-00965-0

Blair JM, Piddock LJ. Structure, function and inhibition of RND efflux pumps in Gram-negative bacteria: An update. *Curr Opin Microbiol.* 2009;12(5):512–519. doi:10.1016/j.mib.2009.07.003

Bo, G. (2000). Giuseppe Brotzu and the discovery of cephalosporins. *Clinical microbiology and infection*, 6. https://doi.org/10.1111/j.1469-0691.2000.tb02032.x

Bora ES, Karaali R, Akyol PY, Yurtsever G, Erbaş O. The effect of sulfasalazine in pentylenetetrazole-induced seizures in rats. *Braz J Med Biol Res.* 2021;54(12):e11541. doi:10.1590/1414-431X2021e11541

Bouza E. The role of new carbapenem combinations in the treatment of multidrug-resistant Gram-negative infections. *J Antimicrob Chemother.* 2021;76(Suppl 4):38–45. doi:10.1093/jac/dkab353

Bunduki GK, Heinz E, Phiri VS, Noah P, Feasey N, Musaya J. Virulence factors and antimicrobial resistance of uropathogenic *Escherichia coli* (UPEC) isolated from urinary tract infections: a systematic review and meta-analysis. *BMC Infect Dis.* 2021;21(1):753. doi:10.1186/s12879-021-06435-7

Burnett-Boothroyd, SC, McCarthy, BJ (2011). Antimicrobial treatments of textiles for hygiene and infection control applications: An industrial perspective. In *Textiles for Hygiene and Infection Control* (pp. 196–209). Woodhead Publishing.

Carcione D, Siracusa C, Sulejmani A, Leoni V, Intra J. Old and new beta-lactamase inhibitors: Molecular structure, mechanism of action, and clinical use. *Antibiotics* (Basel). 2021;10(8):995. doi:10.3390/antibiotics10080995

Chaves BJ, Tadi P. Gentamicin. In: Stat Pearls. Treasure Island (FL): *Stat Pearls* Publishing; September 29, 2021.

Canovas J, Baldry M, Bojer MS, Andersen PS, Grzeskowiak PK, Stegger M, Damborg P, Olsen CA, Ingmer H. Cross-Talk between *Staphylococcus aureus* and other staphylococcal species via the agr quorum sensing system. *Front Microbiol.* 2016;7:1733. doi:10.3389/fmicb.2016.01733

Chadha J, Harjai K, Chhibber S. Repurposing phytochemicals as anti-virulent agents to attenuate quorum sensing-regulated virulence factors and biofilm formation in *Pseudomonas aeruginosa. Microb Biotechnol.* 2021. doi:10.1111/1751-7915.13981

Chapman HD, Jeffers TK. Restoration of sensitivity to salinomycin in Eimeria following 5 flocks of broiler chickens reared in floor-pens using drug programs and vaccination to control coccidiosis. *Poult Sci.* 2015;94(5):943–946. doi:10.3382/ps/pev077

Chen D, Murchie AI. An aminoglycoside sensing riboswitch controls the expression of aminoglycoside resistance acetyltransferase and adenyltransferases. *Biochim Biophys Acta.* 2014;1839(10):951–958. doi:10.1016/j.bbagrm.2014.02.019

Chowdhury G, Pazhani GP, Nair GB, Ghosh A, Ramamurthy T. Transferable plasmid-mediated quinolone resistance in association with extended-spectrum β-lactamases and fluoroquinolone-acetylating aminoglycoside-6'-N-acetyltransferase in clinical isolates of *Vibrio fluvialis. Int J Antimicrob Agents.* 2011;38(2):169–173. doi:10.1016/j.ijantimicag.2011.04.013

Christensen SB. Drugs that changed society: history and current status of the early antibiotics: Salvarsan, sulfonamides, and β-lactams. *Molecules.* 2021;26(19):6057. doi:10.3390/molecules26196057

Colombo, M., Masiero, S., Rosa, S., Caporali, E., Toffolatti, S. L., Mizzotti, C. ... Pesaresi, P. NoPv1: A synthetic antimicrobial peptide aptamer targeting the causal agents of grapevine downy mildew and potato late blight. *Scientific reports* 2020;10(1):1–18. https://doi.org/10.1038/s41598-020-73027-x

Davidovich NV, Kukalevskaya NN, Bashilova EN, Bazhukova TA.. General principles of antibiotic resistance evolution in bacteria. Klin Lab Diagn 2020; 65(6):387–393. doi:10.18821/0869-2084-2020-65-6-387-393

De Lemos Neto M, Alexandre RCV, Morra ROG, da Paz JAS, Barroso SPC, Resende AC, de Medeiros-Lima DJM, Alexandre PCB. Use of glucocorticoids and azithromycin in the therapy of COVID-19. *Pharmacol Rep.* 2021;73(6):1513–1519. doi:10.1007/s43440-021-00286-4

Del Rosso JQ, Webster G, Weiss JS, Bhatia ND, Gold LS, Kircik L. Nonantibiotic properties of tetracyclines in rosacea and their clinical implications. *J Clin Aesthet Dermatol.* 2021;14(8):14–21.

Diepstraten FA, Hoetink AE, van Grotel M, Huitema ADR, Stokroos RJ, van den Heuvel- Eibrink MM, Meijer AJM. Aminoglycoside- and glycopeptide-induced ototoxicity in children: a systematic review. *JAC Antimicrob Resist.* 2021;3(4):dlab184. doi:10.1093/jacamr/dlab184

Doña I, Blanca-López N, Boteanu C, Cueva-Oliver B, Fernández-Sánchez FJ, Gajate P, García-Avilés MC, García-Núñez I, Lobera T, Moreno E, Rojas P, Rosado A. Clinical practice guidelines for diagnosis and management of hypersensitivity reactions to quinolones. *J Investig Allergol Clin Immunol.* 2021;31(4):292–307. doi:10.18176/jiaci.0669

Dogsa I, Choudhary KS, Marsetic Z, Hudaiberdiev S, Vera R, Pongor S, Mandic-Mulec I. ComQXPA quorum sensing systems may not be unique to *Bacillus subtilis*: A census in prokaryotic genomes. *PLoS One.* 2014;9(5):e96122. doi:10.1371/journal.pone.0096122

Doron S, Davidson LE. Antimicrobial stewardship. In *Mayo Clinic Proceedings* (2011, 86, 11, pp. 1113–1123). Elsevier.

Du Z, Su Q, Wu Z, Huang Z, Bao J, Li J, Tu H, Zeng C, Fu J, He H. Genome-wide characterization of MATE gene family and expression profiles in response to abiotic stresses in rice (*Oryza sativa*). *BMC Ecol Evo.* 2021:141 https://doi.org/10.1186/s12862-021-01873-y.

Dutta AK, Phull PS. Treatment of helicobacter pylori infection in the presence of penicillin allergy. *World J Gastroenterol.* 2021;27(44):7661–7668. doi:10.3748/wjg.v27.i44.7661

Ebbensgaard AE, Løbner-Olesen A, Frimodt-Møller J. The role of efflux pumps in the transition from low-level to clinical antibiotic resistance. *Antibiotics (Basel).* 2020;9(12):855. doi:10.3390/antibiotics9120855.

Elhidar N, Nafis A, Goehler A, Abbad A, Hassani L, Mezrioui NE, Bohnert JA. Novel DiOC3 96-well real-time efflux assay for discovery of NorA efflux pump inhibitors in *Staphylococcus aureus. J Microbiol Methods.* 2021;181:106128. doi:10.1016/j.mimet.2020.106128

Elkamhawy A, Ali EMH, Lee K. New horizons in drug discovery of lymphocyte-specific protein tyrosine kinase (Lck) inhibitors: a decade review (2011–2021) focusing on structure-activity relationship (SAR) and docking insights. *J Enzyme Inhib Med Chem.* 2021;36(1):1574–1602. doi:10.1080/14756366.2021.1937143

Erken M, Lutz C, McDougald D. The rise of pathogens: predation as a factor driving the evolution of human pathogens in the environment. *Microb Ecol.* 2013;65(4):860–868. doi:10.1007/s00248-013-0189-0

Fatima H, Goel N, Sinha R, Khare SK. Recent strategies for inhibiting multidrug-resistant and β-lactamase producing bacteria: A review. *Colloids SurfB Biointerfaces.* 2021;205:111901. doi:10.1016/j.colsurfb.2021.111901

Foong WE, Wilhelm J, Tam HK, Pos KM. Tigecycline efflux in *Acinetobacter baumannii* is mediated by TetA in synergy with RND-type efflux transporters. *J Antimicrob Chemother.* 2020;5(75):1135–1139. https://doi.org/10.1093/jac/dkaa015.

Fortier LC, Sekulovic O. Importance of prophages to evolution and virulence of bacterial pathogens. *Virulence.* 2013;4(5):354–365. doi:10.4161/viru.24498

Gaynes, R. (2017). The discovery of penicillin—new insights after more than 75 years of clinical use. *Emerging Infectious Diseases*, 23(5):849. http://dx.doi.org/10.3201/eid2305.161556

Geske GD, O'Neill JC, Blackwell HE. Expanding dialogues: From natural autoinducers to non-natural analogues that modulate quorum sensing in Gram-negative bacteria. *Chem Soc Rev.* 2008;37(7):1432–1447. doi:10.1039/b703021p

Gontijo AVL, Pereira SL, de Lacerda Bonfante H. Can Drug Repurposing be Effective Against Carbapenem-Resistant *Acinetobacter baumannii?*. *Curr Microbiol.* 2021;79(1):13. doi:10.1007/s00284-021-02693-5

Gu Y, Fujitomo Y, Ohmagari N. Outcomes and future prospect of Japan's National Action Plan on Antimicrobial Resistance (2016–2020). *Antibiotics* (Basel). 2021;10(11):1293. doi:10.3390/antibiotics10111293

Granata G, Petrosillo N, Al Moghazi S, Caraffa E, Puro V, Tillotson G, Cataldo MA. The burden of *Clostridioides difficile* infection in COVID-19 patients: A systematic review and meta-analysis. *Anaerobe.* 2021:102484. doi:10.1016/j.anaerobe.2021.102484

Guo HY, Chen ZA, Shen QK, Quan ZS. Application of triazoles in the structural modification of natural products. *J Enzyme Inhib Med Chem.* 2021;36(1):1115–1144. doi:10.1080/14756366.2021.1890066

Guo Y, Hu D, Guo J, Li X, Guo J, Wang X, Xiao Y, Jin H, Liu M, Li Z, Bi D, Zhou Z. The role of the regulator fur in gene regulation and virulence of *riemerella anatipestifer* assessed using an unmarked gene deletion system. *Front Cell Infect Microbiol.* 2017;7:382. doi:10.3389/fcimb.2017.00382

Gupta MN, Alam A, Hasnain SE. Protein promiscuity in drug discovery, drug-repurposing and antibiotic resistance. *Biochimie.* 2020;175:50–57. doi: 10.1016/j.biochi.2020.05.004

Hamada Y, Ebihara F, Kikuchi K. A Strategy for hospital pharmacists to control antimicrobial resistance (AMR) in Japan. *Antibiotics* (Basel). 2021;10(11):1284. doi:10.3390/antibiotics10111284

Hayashi MA, Bizerra FC, Junior PIDS (Eds.). (2014). *Antimicrobial Compounds from Natural Sources.* Frontiers E-books. https://doi.org/10.3389/fmicb.2013.00195

Hegde Subray S, Javid-Majd Farah, Blanchard John S. Overexpression and mechanistic analysis of chromosomally encoded aminoglycoside 2′-N-acetyltransferase (AAC(2′)-Ic) from *Mycobacterium tuberculosis. Int. J. Biol. Chem.* 2001;276(49):45876–45881, https://doi.org/10.1074/jbc.M108810200.

Hema-Ouangraoua S, Tranchot-Diallo J, Zongo I, Kabore NF, Nikièma F, Yerbanga RS, Tinto H, Chandramohan D, Ouedraogo GA, Greenwood B, Ouedraogo JB. Impact of mass administration of azithromycin as a preventive treatment on the prevalence and resistance of nasopharyngeal carriage of *Staphylococcus aureus. PLoS One.* 2021;16(10):e0257190. doi: 10.1371/journal.pone.0257190

Herrera M, Gregorio SD, Haim MS, Posse G, Mollerach M, Di Conza J. Genetic changes associated with tigecycline resistance in *Staphylococcus aureus* in vitro-selected mutants belonging to different lineages. *Int J Antimicrob Agents.* 2021;57(4):106304. doi:10.1016/j.ijantimicag.2021.106304

Hinchliffe P, Greene NP, Paterson NG, Crow A, Hughes C, Koronakis V. Structure of the periplasmic adaptor protein from a major facilitator superfamily (MFS) multidrug efflux pump. *FEBS Lett.* 2014, 25;588(17):3147–53. doi: 10.1016/j.febslet.2014.06.055

Hotinger JA, Morris ST, May AE. The case against antibiotics and for anti-virulence therapeutics. *Microorganisms.* 2021;9(10):2049. doi:10.3390/microorganisms9102049

Huang H, Liang Z, Zheng X, Qing Q, Du X, Tang Z, Wei M, Wang C, Zhong Q, Lin X. Tacrolimus versus cyclophosphamide for patients with idiopathic membranous nephropathy and treated with steroids: a systematic review and meta-analysis of randomized controlled trials. *Ren Fail.* 2021;43(1):840–850. doi:10.1080/0886022X.2021.1914655

Huczyński A. Polyether ionophores-promising bioactive molecules for cancer therapy. *Bioorg Med Chem Lett.* 2012;22(23):7002–7010. doi:10.1016/j.bmcl.2012.09.046

Hussain HI, Aqib AI, Seleem MN, Shabbir MA, Hao H, Iqbal Z, Kulyar MF, Zaheer T, Li K. Genetic basis of molecular mechanisms in β-lactam resistant gram-negative bacteria. *Microb Pathog.* 2021;158:105040. doi:10.1016/j.micpath.2021.105040

Hutchings MI, Truman AW, Wilkinson B. Antibiotics: Past, present and future. *Current Opinion in Microbiology* 2019;51:72–80. https://doi.org/10.1016/j.mib.2019.10.008

Inamuddin, Ahamed MI, and Prasad R (2021) *Advanced Antimicrobial Materials and Applications.* Springer Singapore (ISBN: 978-981-15-7098-8). https://www.springer.com/gp/book/9789811570971

Jabila Mary TR, Kannan RR, Muthamil Iniyan A et al. β-lactamase inhibitory potential of kalafungin from marine Streptomyces in *Staphylococcus aureus* infected zebrafish. *Microbiol Res.* 2021;244:126666. doi:10.1016/j.micres.2020.126666

Jack DL, Storms ML, Tchieu JH, Paulsen IT, Saier MH Jr. A broad-specificity multidrug efflux pump requiring a pair of homologous SMR-type proteins. *J Bacteriol.* 2000;182(8):2311–2313. doi:10.1128/JB.182.8.2311-2313.2000

Kalamara M, Spacapan M, Mandic-Mulec I, Stanley-Wall NR. Social behaviours by *Bacillus subtilis*: quorum sensing, kin discrimination and beyond. *Mol Microbiol.* 2018;110(6):863–878. doi:10.1111/mmi.14127

Koh-Tan HH, Strachan E, Cooper K, Bell-Sakyi L, Jonsson NN. Identification of a novel β-adrenergic octopamine receptor-like gene (βAOR-like) and increased ATP-binding cassette B10 (ABCB10) expression in a *Rhipicephalus microplus* cell line derived from acaricide-resistant ticks. *Parasit Vectors.* 2016;9(1):425. doi:10.1186/s13071-016-1708-x

Kreve S, Reis ACD. Bacterial adhesion to biomaterials: What regulates this attachment? A review. *Jpn Dent Sci Rev.* 2021;57:85–96. doi:10.1016/j.jdsr.2021.05.003

Kümmerer, K. (2003). Significance of antibiotics in the environment. *Journal of Antimicrobial Chemotherapy* 52(1):5–7. https://doi.org/10.1093/jac/dkg293.

Kuroda T, Tsuchiya T. Multidrug efflux transporters in the MATE family. *Biochim Biophys Acta.* 2009;1794(5):763–768. doi:10.1016/j.bbapap.2008.11.012

Lambert T, Ploy MC, Courvalin P. A spontaneous point mutation in the aac(6')-Ib' gene results in altered substrate specificity of aminoglycoside 6'-N-acetyltransferase of a *Pseudomonas fluorescens* strain. *FEMS Microbiol Lett.* 1994;115(2-3):297–304. doi:10.1111/j.1574-6968.1994.tb06654.x

Lee B, Park J, Ryu M, Kim S, Joo M, Yeom JH ... Bae J (2017). Antimicrobial peptide-loaded gold nanoparticle-DNA aptamer conjugates as highly effective antibacterial therapeutics against *Vibrio vulnificus*. *Scientific Reports*, 7(1):1–10. https://doi.org/10.1038/s41598-017-14127-z.

Li X, Li P, Ruan C, Xie LX, Gu Y, Li J, Yi Q, Lv X, Xie J. *Mycobacterium tuberculosis* Rv0191 is an efflux pump of major facilitator superfamily transporter regulated by Rv1353c. *Arch Biochem Biophys.* 2019;667:59–66. doi: 10.1016/j.abb.2019.04.010

Li S, Chen Q, Gong X, Liu Y, Zheng F. RanB, a putative ABC-type multidrug efflux transporter contributes to aminoglycosides resistance and organic solvents tolerance in *Riemerella anatipestifer*. *Vet Microbiol.* 2020;243:108641. doi: 10.1016/j.vetmic.2020.108641

Ligon BL. Penicillin: its discovery and early development. In *Seminars in Pediatric Infectious Diseases* (2004, 15, 1, pp. 52–57). WB Saunders. https://doi.org/10.1053/j.spid.2004.02.001.

Lin MF, Lin YY, Tu CC, Lan CY. Distribution of different efflux pump genes in clinical isolates of multidrug-resistant *Acinetobacter baumannii* and their correlation with antimicrobial resistance. *J Microbiol Immunol Infect.* 2017;50(2):224–231. doi: 10.1016/j.jmii.2015.04.004

López Y, Tato M, Gargallo-Viola D, Cantón R, Vila J, Zsolt I. Comparative activity of ozenoxacin and other quinolones in *Staphylococcus aureus* strains overexpressing the efflux pump-encoding genes mepA and norA. *Int J Antimicrob Agents.* 2020;56(3):106082. doi: 10.1016/j.ijantimicag.2020.106082

Mardirossian M, Rubini M, Adamo MFA, Scocchi M, Saviano M, Tossi A, Gennaro R, Caporale A. Natural and synthetic halogenated amino acids-structural and bioactive features in antimicrobial peptides and peptidomimetics. *Molecules.* 2021; 26(23):7401. doi:10.3390/molecules26237401

Magiorakos AP, Srinivasan A, Carey RB, Carmeli Y, Falagas ME, Giske CG. ... Monnet DL. (2012). Multidrug-resistant, extensively drug-resistant and pandrug-resistant bacteria: an international expert proposal for interim standard definitions for acquired resistance. *Clinical Microbiology and Infection* 18(3):268–281. https://doi.org/10.1111/j.1469-0691.2011.03570.x.

Mariappan V, Vellasamy KM, Mohamad NA, Subramaniam S, Vadivelu J. OneHealth approaches contribute towards antimicrobial resistance: Malaysian perspective. *Front Microbiol.* 2021;12:718774. doi:10.3389/fmicb.2021.718774

Maryam L, Usmani SS, Raghava GPS. Computational resources in the management of antibiotic resistance: Speeding up drug discovery. *Drug Discov Today.* 2021;26(9):2138–2151. doi: 10.1016/j.drudis.2021.04.016

Matlock A, Garcia JA, Moussavi K, Long B, Liang SY. Advances in novel antibiotics to treat multidrug-resistant gram-negative bacterial infections. *Intern Emerg Med.* 2021;16(8):2231–2241. doi:10.1007/s11739-021-02749-1

Melenotte C, Pontarotti P, Pinault L, Mège JL, Devaux C, Raoult D. Could β-Lactam Antibiotics Block Humoral Immunity?. *Front Immunol.* 2021;12:680146. Published 2021 Sep 15. doi:10.3389/fimmu.2021.680146

Miklasińska-Majdanik M. Mechanisms of resistance to macrolide antibiotics among *staphylococcus aureus*. *Antibiotics (Basel).* 2021;10(11):1406. doi:10.3390/antibiotics10111406

Mohamed S, Mvungi HC, Sariko M, Rao P, Mbelele P, Jongedijk EM, van Winkel CAJ, Touw DJ, Stroup S, Alffenaar JC, Mpagama S, Heysell SK. Levofloxacin pharmacokinetics in saliva as measured by a mobile microvolume UV spectrophotometer among people treated for rifampicin-resistant TB in Tanzania. *J Antimicrob Chemother.* 2021;76(6):1547–1552. doi:10.1093/jac/dkab057

Mohanty H, Pachpute S, Yadav RP. Mechanism of drug resistance in bacteria: efflux pump modulation for designing of new antibiotic enhancers. *Folia Microbiol (Praha).* 2021;66(5):727–739. doi:10.1007/s12223-021-00910-z

Mphahlele MJ. Synthesis, structural and biological properties of the ring-a sulfonamido substituted chalcones: A review. *Molecules.* 2021;26(19):5923. doi:10.3390/molecules26195923

Mullis MM, Rambo IM, Baker BJ, Reese BK. Diversity, ecology, and prevalence of antimicrobials in nature. *Frontiers in microbiology*, 2019:2518. https://doi.org/10.3389/fmicb.2019.02518

Naujokat C, Steinhart R. Salinomycin as a drug for targeting human cancer stem cells. *J Biomed Biotechnol.* 2012;950658. doi:10.1155/2012/950658

Nikaido H, Takatsuka Y. Mechanisms of RND multidrug efflux pumps. *Biochim Biophys Acta.* 2009;1794(5):769–781. doi:10.1016/j.bbapap.2008.10.004

Niu C, Clemmer KM, Bonomo RA, Rather PN. Isolation and characterization of an autoinducer synthase from *Acinetobacter baumannii. J Bacteriol.* 2008;190(9):3386–3392. doi:10.1128/JB.01929-07.

Obasi M, Abovich A, Vo JB, Gao Y, Papatheodorou SI, Nohria A, Asnani A, Partridge AH. Statins to mitigate cardiotoxicity in cancer patients treated with anthracyclines and/or trastuzumab: A systematic review and meta-analysis. *Cancer Causes Control.* 2021;32(12):1395–1405. doi:10.1007/s10552-021-01487-1

Opperman TJ, Williams JD, Houseweart C, Panchal RG, Bavari S, Peet NP, Moir DT, Bowlin TL. Efflux-mediated bis-indole resistance in *Staphylococcus aureus* reveals differential substrate specificities for MepA and MepR. *Bioorg Med Chem.* 2010;18(6):2123–2130. doi: 10.1016/j.bmc.2010.02.005.

Ovchinnikov V, Stone TA, Deber CM, Karplus M. Structure of the EmrE multidrug transporter and its use for inhibitor peptide design. *Proc Natl Acad Sci.* 2018;115(34):E7932–E7941. doi:10.1073/pnas.1802177115

Pacios O, Blasco L, Bleriot I, Fernandez-Garcia L, González Bardanca M, Ambroa A, López M, Bou G, Tomás M. Strategies to Combat Multidrug-Resistant and Persistent Infectious Diseases. *Antibiotics (Basel).* 2020;9(2):65. doi:10.3390/antibiotics9020065

Pan X, Cen Y, Kuang M, et al. Artesunate interrupts the self-transcriptional activation of MarA to inhibit RND family pumps of *Escherichia coli. IJMM.* 2020;310(8):151465. doi: 10.1016/j.ijmm.2020.151465

Pena-Diaz S, Ventura S. One ring is sufficient to inhibit α-synuclein aggregation. *Neural Regen Res.* 2022;17(3):508–511. doi:10.4103/1673-5374.320973

Pham JV, Yilma MA, Feliz A, Majid MT, Maffetone N, Walker JR ... Yoon YJ (2019). A review of the microbial production of bioactive natural products and biologics. *Frontiers in Microbiology* 2019;10:1404. https://doi.org/10.3389/fmicb.2019.01404.

Pohl PC, Klafke GM, Carvalho DD, Martins JR, Daffre S, da Silva Vaz I Jr, Masuda A. ABC transporter efflux pumps: a defense mechanism against ivermectin in *Rhipicephalus (Boophilus) microplus. Int J Parasitol.* 2011;41(13-14):1323–1333. doi: 10.1016/j.ijpara.2011.08.004

Pollock J, Chalmers JD. The immunomodulatory effects of macrolide antibiotics in respiratory disease. *Pulm Pharmacol Ther.* 2021;71:102095. doi: 10.1016/j.pupt.2021.102095

Pourakbari B, Yaslianifard S, Yaslianifard S, Mahmoudi S, Keshavarz-Valian S, Mamishi S. Evaluation of efflux pumps gene expression in resistant *Pseudomonas aeruginosa* isolates in an Iranian referral hospital. *Iran J Microbiol.* 2016;8(4):249–256.

Prabu A, Hassan S, Prabuseenivasan, Shainaba AS, Hanna LE, Kumar V. Andrographolide: A potent antituberculosis compound that targets Aminoglycoside 2'-N-acetyltransferase in *Mycobacterium tuberculosis. J Mol Graph Model.* 2015;61:133–140. doi: 10.1016/j.jmgm.2015.07.001

Ranaweera I, Shrestha U, Ranjana KC, Kakarla P, Willmon TM, Hernandez AJ, Mukherjee MM, Barr SR, Varela MF. Structural comparison of bacterial multidrug efflux pumps of the major facilitator superfamily. *Trends Cell Mol Biol.* 2015;10:131–140.

Reygaert WC. An overview of the antimicrobial resistance mechanisms of bacteria. *AIMS Microbiol.* 2018;4(3):482–501. doi:10.3934/microbiol.2018.3.482

Reynolds D, Kollef M. The epidemiology and pathogenesis and treatment of *pseudomonas aeruginosa* infections: An update. *Drugs* 2021;81(18):2117–2131. doi:10.1007/s40265-021-01635-6

Richter R, Kamal MAM, García-Rivera MA et al. A hydrogel-based in vitro assay for the fast prediction of antibiotic accumulation in Gram-negative bacteria. *Mater Today Bio.* 2020;8:100084. doi:10.1016/j.mtbio.2020.100084

Rineh A, Bremner JB, Hamblin MR, Ball AR, Tegos GP, Kelso MJ. Attaching NorA efflux pump inhibitors to methylene blue enhances antimicrobial photodynamic inactivation of *Escherichia coli* and *Acinetobacter baumannii in vitro* and *in vivo. Bioorg Med Chem Lett.* 2018;28(16):2736–2740. doi:10.1016/j.bmcl.2018.02.041

Rizk NA, Moghnieh R, Haddad N, Rebeiz MC, Zeenny RM, Hindy JR, Orlando G, Kanj SS. Challenges to Antimicrobial Stewardship in the Countries of the Arab League: Concerns of Worsening Resistance during the COVID-19 Pandemic and Proposed Solutions. *Antibiotics* (Basel). 2021;10(11):1320. doi:10.3390/antibiotics10111320

Rocha JE, de Freitas TS, da Cunha Xavier J, Pereira RLS, Junior FNP, Nogueira CES, Marinho MM, Bandeira PN, de Oliveira MR, Marinho ES, Teixeira AMR, Dos Santos HS, Coutinho HDM. Antibacterial and antibiotic modifying activity, ADMET study and molecular docking of synthetic chalcone (E)-1-(2-hydroxyphenyl)-3-(2,4-dimethoxy-3-methylphenyl)prop-2-en-1-one in strains of *Staphylococcus aureus* carrying NorA and MepA efflux pumps. *Biomed Pharmacother.* 2021;140:111768. doi:10.1016/j.biopha.2021.111768

Romo SA, Martinez-Morales F, Aragon-Martinez OH. Do fluoroquinolone agents produce therapeutic benefits or harmful effects in patients with periodontitis? A systematic review and meta-analysis. *Dent Med Probl.* 2021;58(2):253–266. doi:10.17219/dmp/133512.

Ruiz-Ramos J, Vallvé Alcón E, Moreno Ramos F, Santolaya-Perrín R, Guardiola Tey JM. Antimicrobial stewardship programs in emergency departments: How do we measure antimicrobial use? A systematic review. *Rev Esp Quimioter* 2021;34(6):610–617. doi:10.37201/req/028.2021

Sabnis A, Hagart KL, Klöckner A, Becce M, Evans LE, Furniss RCD, Mavridou DA, Murphy R, Stevens MM, Davies JC, Larrouy-Maumus GJ, Clarke TB, Edwards AM. Colistin kills bacteria by targeting lipopolysaccharide in the cytoplasmic membrane. *Elife* 2021;10:e65836. doi:10.7554/eLife.65836

Sahitya DSK, Jandiyal A, Jain A, Senapati J, Nanda S, Aggarwal M, Kumar P, Mohapatra S, Ray P, Malhotra P, Mahapatra M, Dhawan R. Prevention and management of carbapenem-resistant Enterobacteriaceae in haematopoietic cell transplantation. *Ther Adv Infect Dis.* 2021;8. doi:10.1177/20499361211053480

Salah AN, Elleboudy NS, El-Housseiny GS, Yassien MA. Cloning and sequencing of lsaE efflux pump gene from MDR *Enterococci* and its role in erythromycin resistance. *Infect. Genet. Evol.* 2021;94:105010. https://doi.org/10.1016/j.meegid.2021.105010.

Sarkar SK, Bhattacharyya A, Mandal SM. YnfA, a SMR family efflux pump is abundant in *Escherichia coli* isolates from urinary infection. *Indian J Med Microbiol.* 2015;33(1):139–142. doi:10.4103/0255-0857.148415

Seefeldt AC, Aguirre Rivera J, Johansson M. Direct measurements of erythromycin's effect on protein synthesis kinetics in living bacterial cells. *J. Mol. Biol.* 2021;433(10):166942. https://doi.org/10.1016/j.jmb.2021.166942.

Shcherbakov AA, Hisao G, Mandala VS, Thomas NE, Soltani M, Salter EA, Davis Jr. JH, Henzler-Wildman KA, Hong M. Structure and dynamics of the drug-bound bacterial transporter EmrE in lipid bilayers. *Nat Commun* 2021;12: 172. https://doi.org/10.1038/s41467-020-20468-7

Shukla S, Rai V, Banerjee D, Prasad R. Characterization of Cdr1p, a major multidrug efflux protein of *Candida albicans*: purified protein is amenable to intrinsic fluorescence analysis. *Biochemistry.* 2006;45(7):2425–2435. doi:10.1021/bi0519147

Simonetti O, Rizzetto G, Radi G, Molinelli E, Cirioni O, Giacometti A, Offidani A. New perspectives on old and new therapies of *staphylococcal* skin infections: The role of biofilm targeting in wound healing. *Antibiotics (Basel).* 2021;10(11):1377. doi:10.3390/antibiotics10111377

Singh L, Mathibe LJ, Bangalee V. The efficacy of bedaquiline versus kanamycin in multi-drug resistant tuberculosis: A systematic scoping review. *Health SA.* 2021;26:1708. doi:10.4102/hsag.v26i0.1708

Sonawane, J.M, Rai, A.K, Sharma M, Tripathi, M., Prasad, R. Microbial biofilms: Recent advances and progress in environmental bioremediation. *Science of the Total Environment 2022*; 824(153843). https://doi.org/10.1016/j.scitotenv.2022.153843

Soto SM. Role of efflux pumps in the antibiotic resistance of bacteria embedded in a biofilm. *Virulence* 2013;4(3):223–229. https://doi.org/10.4161/viru.23724.

Srinivasan VB, Rajamohan G, Gebreyes WA. Role of AbeS, a novel efflux pump of the SMR family of transporters, in resistance to antimicrobial agents in *Acinetobacter baumannii. Antimicrob Agents Chemother.* 2009;53(12):5312–5316. doi:10.1128/AAC.00748-09

Sun X, Ni Z, Tang J, Ding Y, Wang X, Li F. The abaI/abaR quorum sensing system effects on pathogenicity in *acinetobacter baumannii. Front Microbiol.* 2021;12:679241. doi:10.3389/fmicb.2021.679241

Sunkara P, Chennuru S, Krovvidi S, Chitichoti J. Cypermethrin and ivermectin resistance in field populations of *Rhipicephalus sanguineus* sensu lato (Latrielle, 1806) collected from dogs in south India. *Vet Res Commun.* 2021; 46(1):67–78. doi:10.1007/s11259-021-09830-2

Tang J, Chen Y, Wang X, Ding Y, Sun X, Ni Z. Contribution of the AbaI/AbaR Quorum Sensing System to Resistance and Virulence of *Acinetobacter baumannii* Clinical Strains. Infect Drug Resist. 2020;13:4273–4281. doi:10.2147/IDR.S276970

Terreni M, Taccani M, Pregnolato M. New antibiotics for multidrug-resistant bacterial strains: Latest research developments and future perspectives. *Molecules.* 2021;26(9):2671. doi:10.3390/molecules26092671

Thakur V, Uniyal A, Tiwari V. A comprehensive review on pharmacology of efflux pumps and their inhibitors in antibiotic resistance. *Eur J Pharmacol.* 2021;903:174151. doi:10.1016/j.ejphar.2021.174151

Thornton JM, Yin K. Role of specialized pro-resolving mediators in modifying host defense and decreasing bacterial virulence. *Molecules* 2021;26(22):6970. doi:10.3390/molecules26226970

Thursky KA et al. Antimicrobial stewardship in Australia: The role of qualitative research in programme development. JAC Antimicrob Resist. 2021;3(4):dlab166. doi:10.1093/jacamr/dlab166

Tripathi AK, Thakur P, Saxena P, Rauniyar S, Gopalakrishnan V, Singh RN, Gadhamshetty V, Gnimpieba EZ, Jasthi BK, Sani RK. Gene sets and mechanisms of sulfate-reducing bacteria biofilm formation and quorum sensing with impact on corrosion. *Front Microbiol.* 2021;12:754140. doi:10.3389/fmicb.2021.754140

Utegenova GA, Pallister KB, Kushnarenko SV, Özek G, Özek T, Abidkulova KT, Kirpotina LN, Schepetkin IA, Quinn MT, Voyich JM. Chemical composition and antibacterial activity of essential oils from *ferula* l. species against methicillin-resistant *staphylococcus aureus. Molecules.* 2018;23(7):1679. doi:10.3390/molecules23071679

Van den Bergh D, Brink A. A commitment and call to strengthen and expand qualitative research efforts to improve the impact of antimicrobial stewardship. *JAC Antimicrob Resist.* 2021;3(4):dlab151. doi:10.1093/jacamr/dlab151

Vazquez-Laslop N, Thum C, Mankin AS. Molecular mechanism of drug-dependent ribosome stalling. *Mol Cell.* 2008;30(2): 190–202. doi:10.1016/j.molcel.2008.02.026

Velkov T, Roberts KD. Discovery of novel polymyxin-like antibiotics. *Polymyxin Antibiotics: From Laboratory Bench to Bedside* 2019:343–362. https://doi.org/10.1007/978-3-030-16373-0_20.

Vergalli J, Atzori A, Pajovic J. et al. The challenge of intracellular antibiotic accumulation, a function of fluoroquinolone influx versus bacterial efflux. *Commun Biol.* 2020;3:198. https://doi.org/10.1038/s42003-020-0929-x.

Wainwright M, Kristiansen JE. On the 75th anniversary of Prontosil. *Dyes and Pigments* 2011;88(3):231–234. https://doi.org/10.1016/j.dyepig.2010.08.012

Wang L, Bei X, Gao J, Li Y, Yan Y, Hu Y. The similar and different evolutionary trends of MATE family occurred between rice and *Arabidopsis thaliana. BMC Plant Biol.* 2016;16:207. https://doi.org/10.1186/s12870-016-0895-0.

Weragama K, Mudgil P, Whitehall J. Paediatric antimicrobial stewardship for respiratory infections in the emergency setting: A systematic review. *Antibiotics* (Basel). 2021;10(11):1366. doi:10.3390/antibiotics10111366

Wils J, Saegeman V, Schuermans A. Impact of multiplexed respiratory viral panels on infection control measures and antimicrobial stewardship: a review of the literature. *Eur J Clin Microbiol Infect Dis.* 2021;1–16. doi:10.1007/s10096-021-04375-3

Woodruff HB. Selman A. Waksman, winner of the 1952 Nobel Prize for physiology or medicine. *Applied and environmental microbiology* 2014;80(1):2–8. https://doi.org/10.1128/AEM.01143-13

Wright GD. Antibiotic resistance in the environment: A link to the clinic? *Curr Opin Microbiol.* 2010;13(5):589–594. doi:10.1016/j.mib.2010.08.005

Xu T, Wang XY, Cui P, Zhang YM, Zhang WH, Zhang Y. The agr quorum sensing system represses persister formation through regulation of phenol soluble modulins in *staphylococcus aureus. Front Microbiol.* 2017;8:2189. doi:10.3389/fmicb.2017.02189

Yan A, Bryant EE. Quinolones. In: StatPearls. Treasure Island (FL): StatPearls Publishing; July 26, 2021.

Yang H, Wang J, Liu H, Zhao Y, Lakshmi S, Li X, Nie R, Li C, Wang H, Cao Y, Menezes L, Cui L. Efficacy and safety of a naphthoquine-azithromycin coformulation for malaria prophylaxis in Southeast Asia: A phase 3, double-blind, randomized, placebo-controlled trial. *Clin Infect Dis.* 2021a;73(7):e2470–e2476. doi:10.1093/cid/ciaa1018

Yang L, Hu D, Liu H, Wang X, Liu Y, Xia Q, Deng S, Hao Y, Jin Y, Xie M. Biodegradation pathway of penicillins by β-lactamase encapsulated in metal-organic frameworks. *J. Hazard. Mater.* 2021b;414:125549. https://doi.org/10.1016/j.jhazmat.2021.125549.

Yang YS, Chen HY, Hsu WJ, Chou YC, Perng CL, Shang HS, Hsiao YT, Sun JR, Chang YY, Liu YM, Kuo SC, Liu CP, Chen TL, Lee YT, Yang YS. Overexpression of AdeABC efflux pump associated with tigecycline resistance in clinical *Acinetobacter nosocomialis* isolates. *Clin Microbiol Infect.* 2019;25(4):512–516. https://doi.org/10.1016/j.cmi.2018.06.012

Zakrzewska S, Mehdipour AR, Malviya VN, Nonaka T, Koepke J, Muenke C, Hausner W, Hummer G, Safarian S, Michel H. Inward-facing conformation of a multidrug resistance MATE family transporter. *Proc Natl Acad Sci.* 2019;116 (25):12275–12284. doi: 10.1073/pnas.1904210116.

Zhang RH, Guo HY, Deng H, Li J, Quan ZS. Piperazine skeleton in the structural modification of natural products: A review. *J Enzyme Inhib Med Chem.* 2021;36(1):1165–1197. doi:10.1080/14756366.2021.1931861

Zhuang M, Achmon Y, Cao Y, Liang X, Chen L, Wang H, Siame BA, Leung KY. Distribution of antibiotic resistance genes in the environment. *Environ Pollut.* 2021;285:117402. doi:10.1016/j.envpol.2021.117402

Zhou S, Huang G. Preparation, structure and activity of polysaccharide phosphate esters. *Biomed Pharmacother.* 2021;144:112332. doi:10.1016/j.biopha.2021.112332

Zhou S, Wang F, Wong ET, Fonkem E, Hsieh TC, Wu JM, Wu E. Salinomycin: a novel anti-cancer agent with known anti-coccidial activities. *Curr Med Chem.* 2013;20(33):4095–4101. doi:10.2174/15672050113109990199

5 Phyto-Nanoconjugates in Combating Multidrug Resistance in Medical Research

Charu Gupta[*1], *Worapong Kitdamrongtham*[2] *and*
M.F. Abu Bakar[3]
[1] Amity Institute of Herbal Research & Studies, Amity University-UP, Sector-125, NOIDA (UP, India)
[2] Pharmaceutical Chemistry & Division, Division of Cosmetic Technology, Faculty of Engineering and Technology, North-Chiang Mai University, Thailand
[3] Faculty of Applied Sciences & Technology, Universiti Tun Hussein Onn Malaysia, Muar, Johor, Malaysia
*Corresponding author email: charumicro@gmail.com

CONTENTS

DOI: 10.1201/9781003268932-5

5.1 INTRODUCTION

The unprecedented use of antibiotics has led to the development of resistance affects human health worldwide. The prescription of antibiotics imprudently and irrationally in different diseases progressed with the acquisition and development of antibiotic-resistant microbes that led to the resurgence of pathogenic strains harboring enhanced armors against existing therapeutics. The emergence of resistance among microbes has limited the effectiveness of antibiotics, shifting the lifesaving paradigm built around them. This prominent trend is regarded as a matter of concern by agencies like the World Health Organization and has become a challenging situation for the medical fraternity worldwide (WHO, 2017).

Antimicrobial resistance (AMR) is increasing globally and the latest projection showed drug-resistant diseases could cause 10 million deaths each year and impose a cumulative \$100 trillion burden on the global economy and a GDP loss of \$100.2 trillion by 2050 (Prestinaci et al. 2015). By 2030, antimicrobial resistance could force up to 24 million people into extreme poverty. The European Centre for Disease Prevention and Control (2015) estimated about 6,71,689 infections in the EU and a huge loss in the economy. More and more common diseases, including respiratory tract infections, sexually transmitted infections and urinary tract infections, are untreatable, and risky. Therefore, the WHO has declared AMR as the top global public health threat against humanity and immediate action to tackle it (CDC, 2019).

India carries one of the largest burdens of drug-resistant pathogens worldwide, including the highest burden of multidrug-resistant tuberculosis, and alarmingly high resistance among Gram-negative and Gram-positive bacteria even to newer antimicrobials such as carbapenems and faropenem since 2010. Reporting high AMR among pathogens such as *Salmonella typhi*, *Shigella*, *Pseudomonas* and *Acinetobacter* annually, more than 50,000 newborns are estimated to die from sepsis due to pathogens resistant to first-line antibiotics (Ventola, 2015). Two million deaths are projected in India due to AMR by 2050 (Centre for Disease Dynamics, 2015). The emergence of New Delhi metallo-β-lactamase (NDM-1), named after the national capital of India in 2008, rapidly spread to other countries (Nordmann et al. 2011). Available data indicate rising rates of AMR across multiple pathogens of clinical importance in the country. In 2008, about 29% of isolates of *Staphylococcus aureus* were methicillin resistant, and by 2014 this had risen to 47% in countries with effective control of antibiotics (Chokshi et al. 2019).

Superbug is a term used to describe those strains of bacteria that are resistant to the majority of antibiotics commonly used today. It is a naturally occurring phenomenon that can be slowed, but not stopped. Over the time, bacteria adapt to the drugs that are designed to kill them and change to ensure their survival. This makes previously standard treatments for bacterial infections less effective, and in some cases, ineffective. The main bacteria described as antibiotic-resistant threats and referred to as "superbugs" are Carbapenem-resistant Enterobacteriaceae (CRE), Methicillin-resistant *Staphylococcus aureus* (MRSA), ESBL-producing Enterobacteriaceae (extended-spectrum β-lactamases), Vancomycin-resistant *Enterococcus* (VRE), Multidrug-resistant *Pseudomonas aeruginosa*, Multidrug-resistant *Acinetobacter* and *E. coli* H30-Rx (Dadgostar, 2019).

5.2 FACTORS ACCELERATING THE RATE OF ANTIMICROBIAL RESISTANCE (AMR)

The high burden of AMR in India is driven by multiple factors. The main factor in developing antimicrobial resistance is due to the uncontrolled sale of antibiotics without a prescription, in many low- or middle-income countries. Besides, releasing large quantities of antibiotics into the environment during pharmaceutical manufacturing without adequate waste water treatment also increases the risk. Antibiotic resistance increases with duration of treatment. Therefore, as long as an effective minimum is kept, shorter courses of antibiotics are likely to decrease rates of resistance, reduce cost and have better outcomes with fewer complications. Within the hospitals, lack of monitoring of antibiotic use is one of the major factors driving the spread of resistance (Ventola, 2015).

In India a study found the median cost of treatment of a resistant bacterial infection to be more than a year's wages of a rural worker (Chandy et al. 2014). The lack of newer antibiotics is a worldwide problem and certainly a challenge for India to tackle AMR challenge. To fulfil these gaps, coordinated new policies and research focus is urgently needed to manage the AMR crisis.

A compromised treatment regime of a broad range of antibiotics and the rise in resistance has threatened human health and increased the treatment cost of diseases. Despite incredible advancement in scientific and modern medicine, the hope in antibiotics has faded away by the appearance of different resistance mechanisms that present a grave concern against frontline antibiotics. Of them, antibiotic inactivation through the production of a range of enzymes, change in cell permeability, alteration of drug targets, intrinsic expression of efflux pumps and biofilm formation all act as a defense against drugs and contribute to the sustained persistence of resistance bacteria (Dunphy et al. 2019). In addition to mutations, mobile genetic elements (plasmids, insertion sequences, transposons and integrative conjugative elements) play a crucial role in the expansion of resistance among diverse groups of bacteria (Beceiro et al. 2013). The four main mechanisms by which bacteria exhibit resistance to antibiotics are drug inactivation or modification, alteration of target or binding site, alteration of metabolic pathway and reduced drug accumulation: by decreasing drug permeability or increasing active efflux (pumping out) of the drugs across the cell surface (Pang et al. 2019).

5.3 NATURAL PRODUCTS FOR ANTIMICROBIAL RESISTANCE

Researchers have encouraged considering plant-derived substances (PDS) as an alternative or in complementing antibiotics against diseases due to the lack of a strategic management plan on diverse on metabolic, genetic, physiological fronts, rapid progression of resistant microbes.. Considering the quantitative characteristics of plant constituents that attribute beneficial health effects, analytical procedures for their isolation, characterization and phytochemical testing for elucidating ethno pharmacological effects has been worked out for employment in the treatment of different diseases. With an immense potential to combat bacterial infections, PDSs such as polyphenols, alkaloids, and tannins present a great potential for use, either as antimicrobials or as antibiotic resistance modifiers. Natural products from medicinal plants may play an important role for the design of novel combination therapy protocols. With an immense potential to combat bacterial infections, plant-derived substances such as polyphenols, alkaloids and tannins present a great potential for use, either as antimicrobials or as antibiotic resistance modifiers (AlSheikh et al. 2020).

Conventional curative systems have counted upon traditional herbs that are rich in compounds, such as alkaloids, terpenoids, tannins, steroids, coumarins and flavonoids, and do not normally cause resistance (Yother, 2011; Lewis and Ausubel, 2006). Essential oils from parsley, lovage, basil and thyme disrupt the physiological status of the bacterial cell by causing an increase in cell permeability, leakage of cell constituents, alterations in bacterial cell wall and cell membrane, ATP loss, inhibition of protein synthesis, pH disturbance, intracytoplasmic damage, DNA damage and inhibition of quorum sensing among bacteria like *Bacillus cereus*, *Staphylococcus aureus*, *Pseudomonas aeruginosa*, *Escherichia. coli* and *Salmonella enterica serovar typhimurium* (Kolli et al. 2016; Gemeda et al. 2018). Table 5.1 describes some of the important plant-derived substances and their action against different microbes.

5.4 PARADIGM SHIFT IN CHEMOTHERAPY

Meanwhile, a paradigm shift in chemotherapy involves a gradual transition, away from the mono-drug therapies that have been long and forcefully advocated, toward multidrug regimens.

TABLE 5.1
Plant-derived Substances and Their Action Against Different Microbes

Plant name	Plant derivatives	Bacterial sp.
Anogeissusa cuminata	terpenoids, flavonoids, saponins, tannins, alkaloids	*S. aureus* *A. baumannii* *C. freundii* *E. coli* *K. oxytoca* *K. pneumoniae* *P. aeruginosa*
Azadirachta indica	beta-sitosterol, flavonoids	*S. aureus* *A. baumannii* *C. freundii* *E. coli* *K. oxytoca* *K. pneumoniae* *P. aeruginosa* *H. pylori*
Bauhinia variegata	terpenoids, flavonoids, tannins, saponins, glucoside	*S. aureus* *A. baumannii* *C. freundii* *E. coli* *K. oxytoca* *K. pneumoniae* *P. aeruginosa*
Punica granatum	flavonoids, ellagitannin, punicalagin, ellagic acid	*S. aureus* *A. baumannii* *C. freundii* *E. coli* *K. oxytoca* *K. pneumoniae* *P. aeruginosa*
Terminalia chebula	flavonoids and flavins, terpenoids, steroids, alkaloids, tannins and their derivatives, glycosides	*S. aureus* *A. baumannii* *C. freundii* *E. coli* *K. oxytoca* *K. pneumoniae* *P. aeruginosa*
Occimomum basilicum EO	rosmarinic acid, phenol and terpenoid	*B. cereus* *S. aureus* *P. aeruginosa* *E. coli* *S. typhimurium*
Eucalyptus globulus	1,8-cineole, alpha-pinene, p-cymene	*S. aureus*
Alpinia galanga	alpha-pinene, myrcene, limonene	*Mycobacterium smegmatis*
Salvia fruticosa	flavonoids, phenolics and rosemarinic acid	*S. epidermidis*
Hypericum olympicum	essential oils, beta-pinene, beta-ocimene, beta-caryophyllene, germacrene-D	*S. aureus*

Combinations of synthetic chemicals or antibiotics with phytotherapeutics were categorically rejected. This decision implied a strict separation of chemotherapy from phytotherapy. It is self-evident that this therapeutic practice was never officially recognized by the medical authorities, because most of the phytotherapeutics were not standardized and no independent medical control with double-blind studies could be established.

5.5 ISOLATION AND CHARACTERIZATION OF BIOACTIVE COMPOUNDS

Different approaches are employed for screening the bioactive constituents of plant-derived substances. The direct method involves total chemical characterization of a plant species using approaches of de-replication (identification of already known bioactive constituents) via liquid chromatography–mass spectrometry (LC–MS). The objective is to isolate novel plant compounds, which are then sent directly for biological testing (Deng et al. 2014; Guo et al. 2013). Isolated plant compounds are collected for libraries, such as the National Cancer Institute's Natural Product Repository (Liu et al. 2013). Libraries can be screened to test the bioactivities of various compounds in order to examine fully characterized, new natural products (Ramasamy et al. 2014; Wiedenfeld et al. 2007). The bioactivity-guided fractionation method uses bioassays as fractionation monitors; uninterrupted fractionation cycles are used together for testing bioactive extracts so that pure and active principle compounds are isolated. This is the most commonly used way of determining bio-actives. Despite significant increase in the number of studies reporting bio-actives (Agarwal et al. 2014) through this process, a large number of botanicals and their preparations are still in use without any knowledge of their bio-actives (Fakhrudin et al. 2014; Mabona et al. 2013).

The synergy directed fractionation method uses bioactivity-guided fractionation combined with bioactivity testing in order to understand synergistic connections between the compounds present in a mixture. The method uses mass spectrometry (MS)-profiling for a guided isolation of natural products, targeting the synergistic interactions among obtained extracts which could not have been taken into consideration through conventional guided fractionation (Junio et al. 2011). Metabolism directed method is focused on the identification of bioactive metabolites not present at the beginning of the test, but produced with time due to changes in the metabolism of the plant or plant organ under study (Goretta et al. 2012). Additionally, metabolic profiling (metabolomics) method correlates chemical profiling of plant extracts with isolation and identification of new or already known bioactive constituents at an early stage (de-replication) (Wolfender et al. 2015); the goal of metabolomics in general is to analyze all secondary metabolites in a sample, both qualitatively and quantitatively. Comparison of bioactivity data permits early stage de-replication. This method also gives information about possible synergistic effects between molecules (Keerthi et al. 2014). Table 5.2 describes the mode of action of some important PDSs against broad range of microbes.

5.6 HYBRID COMBINATIONS

There is a paradigm shift in chemotherapy from mono-drug therapy toward multidrug combination regimens. It was noticed that only a combination of synthetic chemical drugs or antibiotics with defined plant-derived constituents (e.g., low molecular weight phenolic or terpenoid compounds) increased the pharmacological activity of the hybrid combination and simultaneously minimized the toxic side effects of synthetic drugs. These properties were found only with phenolic and terpenoid compounds. The aforementioned natural products originate from the secondary metabolism of plants that produce these constituents themselves as defence, signal or storage substances. The most frequent constituents produced in significant amounts are preferred phenolic and terpenoids compounds (in concentrations up to 60–80% as compared to all other constituents). Within the structural class of phenolic compounds, phenol-carboxylic acids (e.g. caffeic acids, ferulic acids or shikimic acids), flavonoids, procyanidines, anthocyanins, catechins and lignan derivatives

TABLE 5.2
Mechanism of Action of Plant-Derived Substances (Pdss) Against Broad Range of Microbes

Phytochemicals	Extract	Mode of action	Antimicrobial resistant microbes
Flavonoids	*Vaccinium macrocarpon* Alt (cranberry)	Modifies biofilm formation	*Enterococcus faecalis, E. coli, Pseudomonas aeruginosa*
	Myricetin, robinetin, epigallocatechin	Blocks bacterial DNA synthesis	*E. coli*
	Quercetin	Inhibit ATPase activity, GrYB protein, elevates extracellular phosphatase and beta-galactosidase	*E. coli, S. aureus*
Plant-derived peptides	*Moringa oleifera*	Membrane disruption	*E. coli, S. aureus, P. aeruginosa, S. typhimurium*
Essential oils (EOs)	*Petroselinum crispum EO, Levisticum officinale EO, Ocimum basilicum EO, Thymus vulgaris EO* *Cannabis sativa* EO	Increase cell permeability, leakage of cell constituents, alteration of bacterial cell wall and membrane disturbance, ATP loss, inhibit protein synthesis, lead to pH disturbance, intracytoplasmic damage, DNA damage, inhibit quorum sensing inhibits biofilm formation	*Bacillus cereus, Staphylococcus aureus, P. aeruginosa, E. coli, S. typhimurium S. aureus*
Tea tree oil (TTO)	terpenes, monoterpenes, sesquiterpenes	disrupts membrane permeability, damages cell membrane, obstructs cell growth, cause cell death	*E. coli, S. aureus, C. albicans*
Natural Efflux pump inhibitors (EPIs)	reserpine, gallotannin, piperine, curcumin, berberine, chalcones, carnosic acid	Inhibit various efflux pump in bacteria (EtBr EP, MexAB-OprM)	MDR Uropathogenic *E. coli*, MDR, *P. aeruginosa* (clinical isolates)

predominate, while among the terpenoids mono-, di-, sesqui- and triterpenoids (the latter also in glycosidic form) represent the majority. Biosynthetically, C5-isopentenyl diphosphate (IPP) and dimethylallyl diphosphate (DMAPP) serve as C5-building blocks (Ndhlala et al. 2013).

Phenolic and terpenoid compounds are of low molecular weight, which means that they can be easily isolated or synthesized or are commercially available from companies that specialize in the production of such compounds. Some of these low molecular weight plant constituents were marketed for several years following pharmacological testing, but their production was discontinued in the meantime. The argument frequently was that their therapeutic efficacies were too weak for the treatment of severe diseases. Therefore, their production and marketing as single plant drugs was estimated as not being sufficiently profitable for the pharmaceutical industry (Ruddaraju et al. 2020).

5.7 INTERACTIONS OF PHENOLIC AND TERPENOIDS COMPOUNDS WITH SYNTHETIC AND ANTIBIOTIC DRUGS

Low molecular phenolic compounds with one or several OH-groups are water-soluble and are able to form ionic bonds or H-bridges to enzymes, receptors or several other biomarkers. Many phenolic compounds possess anti-oxidative properties, for example they are able to maintain the redox homeostasis in a definite state. This might be necessary to keep the balance between reactive oxygen species (ROS) and antioxidant enzymes such as superoxide dismutase (SOD), catalase, glutathione (GSH) peroxidase and GSH reductase. Therefore, it is of interest to determine the degree of the scavenging activity of the selected phenolic compound as a possible adjuvant with stable 2,2-diphenyl-1-picrylhydrazyl (DPPH). Some phenolic compounds such as salicylic acid possess anti-inflammatory activity, and a few also induce apoptosis of tumor cells. Due to their lipophilic structural features (e.g. OCH_3-groups, conjugated aliphatic chains or ester groups), the majority of terpenoids is able to penetrate cell walls or enter tumor cells. The triterpenoids with additional sugar molecules (glycosides) or carboxylic acids possess amphoteric (amphiphilic) properties and therefore reveal far-reaching penetration capabilities (Wagner 2011).

The most practicable and illustrative method for selecting the best suitable phenolic and terpenoids from several dozens of low molecular weight compounds occurring in plants is the "isobol" method of Berenbaum (1977), which has been previously discussed regarding its application in phytotherapy (Berenbaum 1977).

Based on specific isobolograms, the following effects can be recorded, if a phytochemical compound is combined with a synthetic chemical drug or an antibiotic compound: additive effects; over-additive or potentiated effects and zero or antagonistic effects. The over-additive or potentiated effects are unique effects that demonstrate suitability for use as adjuvants for hybrid combination therapy. All single drugs administered orally or per injection cause multi-target effects, which could be not clarified in all their detail, in spite of the great progress in molecular-biological research. Predictions made in scientific publications that modern medicine, based on progress, would shortly lead to the development of new drugs remained unfulfilled (Wagner and Ulrich-Merzenich, 2009).

5.8 NANOPARTICLES FOR ANTIMICROBIAL RESISTANCE

The conventional antibiotic therapy proved to be ineffective due to low absorption and low bioavailability. Therefore, novel drug delivery to specific target organs using metallic nanoparticles (NPs), of 1–100 nm seems promising for antibiotic resistance due to their tuned physicochemical properties and ability to modulate oxidative stress. Since NPs can provide specifically targeted or localized therapy, and their use seems a promising therapeutic option as compared to conventional therapy for multidrug resistance. The most commonly used nanoparticles which have wide applications are silver, gold, zinc, copper and iron. Recently, scientists were increasingly focused on the activity of silver, gold, selenium, zinc, titanium (TiO2) nanoparticles, or a mixture of Ag/ZnO nanoparticles that exhibit antibacterial, antiviral and antifungal effects (Rai et al. 2014).

Nanoparticles in particular have demonstrated broad-spectrum antibacterial properties against both Gram-positive and Gram-negative bacteria. For example, ZnO nanoparticles were found to inhibit *Staphylococcus aureus*, and Ag nanoparticles exhibit concentration-dependent antimicrobial activity against *Escherichia coli* and *Pseudomonas aeruginosa* (Ramalingam et al. 2016).

Due to the size of the nanoparticles, their properties are accustomed to giving a larger surface area compared to the bulk material. Thus, materials made up in such a way will have atoms that have more contact with the external environment; whereas those which are considered as bulk hold the atoms closer to the center (Mohanraj and Chen, 2006).

As discussed in the above paragraphs, natural molecules (phytochemicals) could serve as a potential alternate source to combat antibiotic resistance antimicrobials. The antibacterial activities of phytochemicals are a well-known phenomenon and represent a possible hope of effective, cheap and safe antimicrobial agents. They have the ability to bind to protein domains leading to modification or inhibition protein-protein interactions. It has been observed experimentally that some phytochemicals exert synergistic effect in combination of antibiotics (Stefanovic and Comic, 2012; Olajuyigbe and Afolayan, 2012). Indeed, this combination therefore allowed reducing bacterial resistance to drugs. The phytochemicals are found to be effective modulators of host-related cellular processes viz. immune response, mitosis, apoptosis and signal transduction. Thus they may have antimicrobial activity with reduced ability to develop resistance to botanicals. Nutritional and botanical approaches, combined, may provide powerful tools for controlling an array of infections (Abascal and Yarnell, 2002).

But usually, phytochemicals have low bioavailability, low hydro-solubility, low stability in gastrointestinal tract and rapid elimination.

A number of strategies have been used to increase their chemical stability or permeability such as encapsulation in nano-conjugate has emerged as a novel and effective drug delivery strategy to improve delivery, distribution and bioactivity (Wagnera and Efferth 2017). These drawbacks can be overcome by using different nano conjugates that perform a wide range of additional functions like reduction of oxidative stress, antimicrobial, antifungal, immuno-modulatory, chemo-preventive and antioxidant with low toxicity (Kim et al. 2018). Due to their high bioavailability, low toxicity and affordability, nano forms can be the most appropriate mode of drug delivery. Further, it has been indicated that nanoparticles are 60 times more effective in fighting infections caused by *S. aureus*, *E. coli* and *P. aeruginosa* than conventional treatments (Ananth et al. 2019).

Different types of nanoparticles with a potential to inhibit drug-resistant microorganisms are discussed below.

5.8.1 Selenium Nanoparticles

This is a metalloid and an essential micronutrient. It plays vital functions in the human body by improving the action of enzymes such as glutathione peroxidase and seleno-enzymes which defend the body against immunity related diseases (Romero-Perez et al. 2010).

Selenium nanoparticles can be used as antibacterial agent as they have a unique structure and larger surface area that can be more in contact with the external environment. Selenium is also an antioxidant and immunity booster. The antibacterial effect is due to the fact that at a particular concentration nano-selenium interacts with the bacterial cell surface and penetrates into the cell, thus causing damage. SeNPs are more stable in nature; possess excellent biocompatibility, bio-efficacy and low toxicity as compared to various organic and inorganic forms of selenium compounds (Chiou and Hsu 2011). The biosynthesis of SeNPs has gained much interest due to its excellent therapeutical and nutraceutical potential among the other nanoparticles. Further, it has been indicated that nano-selenium conjugates are 60 times more effective in fighting infections caused by *S. aureus*, *E. coli* and *P. aeruginosa* than conventional therapy. Due to their unique morphological and chemical nature, Se nanoparticles are more reactive and lethal to bacteria. Therefore, Nano-selenium can be a viable alternative to antibiotics to prevent and treat a number of multi-drug-resistant bacterial infections in humans (Conte et al. 2016).

The utilization of "conjugate combination" for the therapeutic application of polyphenolic nano conjugates along with synthetic antibiotic drugs may play a key and unique role by synergistically enhanced pharmacological activity and reduced toxicity (Conte et al. 2016). There is convincing evidence for such effects by hybrid/conjugate combinations and support for more intensive and scientific efforts should focus in this direction.

5.8.2 Silver Nanoparticles (AgNPs)

For a long time, silver has been used as an antimicrobial agent for wound healing, both in its solid state and with salt solutions to clean wounds. Nowadays, dressings impregnated with silver nitrate ($AgNO_3$) can be found (Chen et al. 2008). Silver exhibits very interesting properties due to its chemical stability, good conductivity, catalytic and antibacterial activity. Moreover, nanoparticles made of silver (silver nanoparticles, AgNPs) are one of the most widely studied nowadays (Aderibigbe, 2017). AgNPs have been applied in different fields such as textile, cosmetics, food industry and biomedicine. In the biomedical field, they are gaining strength especially due to their applications as antimicrobial agents, as coating for medical devices and as carrier for chemotherapeutic drugs (Zhang et al. 2016; Aziz et al. 2014, 2015, 2016, 2019).

Bactericidal properties of silver nanoparticles are the most widely studied, but a wide variety of other biomedical properties such as antifungal, antiviral, anti-amoebial, anti-cancer, anti-angiogenic and anti-inflammatory activity are also being exploited (Zhang et al. 2016). New antifungal agents are also a demand, in particular, for immunosuppressed patients. AgNPs have a high antifungal potential, for example in *Candida albicans* infections, AgNPs stabilized with dodecyl sulphate show a better activity than conventional treatment (Panáˇcek et al. 2009). AgNPs also show antiviral properties. In the case of activity against HIV1, AgNPs have demonstrated anti-retroviral ability in addition to a potent virus inhibition effect. They have also proven to be efficient inhibitors against hepatitis B virus (HBV) (Lara et al. 2010).

5.8.3 Zinc Oxide Nanoparticles (ZnONPs)

Zinc is an essential mineral involved in the catalytic activity of numerous enzymes present in the organism and is widely distributed throughout the body tissue (Król et al. 2017). In the pharmaceutical field it is recognized as one of the safest materials by the Food and Drug Administration (FDA) (Vijayakumar et al. 2018). ZnONPs exhibit significant optical properties that offer NPs the ability to be used as a drug delivery system, as antitumor, antibacterial, antidiabetic and as a theragnostic tool.

ZnONPs possess antimicrobial activity against Gram-positive (*S. aureus, S. epidermis, B. subtilis, B. cereus, L. monocytogenes, E. faecium*) and Gram-negative (*P. aeruginosa, E. coli, K. pneumoniae, Salmonella sp.*) bacteria (Vijayakumar et al. 2018). Moreover, Singh et al. compared antimicrobial and antifungal potential of zinc oxide nanoparticles. Pathogenic microorganisms selected included different bacteria, *Escherichia coli* (MTCC443), *Staphylococcus aureus* (MTCC 3160), *Bacillus subtilis* (MTCC 441) and two fungi, *Aspergillus niger* (MTCC 281) and *Candida albicans* (MTCC 227). They show a 50% efficacy when nanoparticles are used instead of particles (Singh and Nanda, 2013). Antimicrobial tests carried out with different bacterial strains showed that minimum inhibitory concentration (MIC) values are 50–85% lower in Gram-positive than Gram-negative bacteria. The difference in antibacterial activity of ZnONPs toward Gram-positive and Gram-negative bacteria could be clarified due to the interaction with the cell wall. In Gram-negative bacteria, the structure of lipopolysaccharide opposes the attachment of ZnO and restrains the ions passing across the outer membrane (Yu et al. 2015). The toxicity induced by antimicrobial drugs is due to modifications in the membrane potential through the blockage of K^+ ion channel present in the bacteria cell membrane. ZnONPs in aqueous medium are dissolved with consequent release of Zn^{2+}. Zn^{2+} ions, that are attracted to the bacterial surface causing this phenomenon. These changes increased permeability of the membranes, leading to destabilization. Moreover, ROS production by nanoparticles contributes to this result. Due to their charge, Zn^{2+} ions can easily penetrate the bacterial cell wall and interact with different molecules—such as lipids, proteins, and nucleic acids—disrupting important metabolic pathways (Bhuyan et al. 2015).

Several authors have used ZnONPs as antimicrobial agents either as ZnO alone or in combination with drugs (Sirelkhatim et al. 2015). Recently, ZnONPs have been developed using a plant extract (*Punica granatum*) and obtaining spherical and hexagonal shapes of 32.98 nm diameter with antibacterial activity against *E. coli* and *E. faecalis* (Nur et al. 2019). Moreover, Jayabalan et al. in their last work have demonstrated that the ZnONPs synthetized using a biological method based on the use of *Pseudomona putida*, obtaining nanoparticles with a spherical shape and an average diameter of 44.5 nm. This ZnONPs presented antimicrobial activity against *Pseudomonas otitidis*,

Pseudomonas oleovorans, Acinetobacter baumannii, Bacillus cereus and *Enterococcus faecalis* using micro-titer plate method and disk diffusion assay (Jayabalan et al. 2019).

5.8.4 Gold Nanoparticles (AuNPs)

AuNPs are colloidal or clustered particles composed of a gold core, an inert and biocompatible compound (Gerber et al. 2013). One of the advantages of these particles are their synthetic versatility, which allows the control of their size, shape and surface properties. Furthermore, their coating can be modified to control particle solubility, stability and interaction with the environment. Also, the particle surface can bind thiols and amines, providing functional groups to the AuNPs for labeling, targeting and conjugating pharmacologic molecules (Her et al. 2017). Their unique characteristics make them a material of extreme interest in the medical field due to their optical and electronic properties of Au. Some of the major areas of application of AuNPs include biosensors and bio-imaging, drug delivery systems and also the treatment of some cancers. Meanwhile, some researchers have been interested in the potential antibacterial activity of AuNPs since this material is less toxic to mammalian cells compared to AgNPs, the most common nanoparticles employed as antimicrobial agents (Table 5.4) (Abdel-Kareem and Zohri, 2018).

AuNPs are active against Gram-negative and Gram-positive bacteria, namely *E. coli, P. aeruginosa, S. typhi, Serratia sp., K. pneumoniae, S. aureus, B. subtilis* and *E. faecalis*, among others (Table 5.3). The fact that AuNPs are relatively inert implies that they exhibit no apparent intrinsic antibacterial activity. Thus, it is understood that their main mechanism of bacterial toxicity is based on direct adherence of AuNPs onto the bacterial surface driven by electrostatic forces. This mechanism is highly dependent on nanoparticles size, typically with smaller nanoparticles showing lower MIC. From their adhesion results alterations of membrane potential, inhibition of adenosine triphosphatase (ATPase) activity (resulting in inhibition of ATP synthesis) and inhibition of tRNA binding in subunit of the ribosome. This phenomenon will block instrumental metabolic processes which result in the loss of cellular integrity (Abdel-Kareem and Zohri, 2018). Due to their low reactivity, AuNPs show ion release and ROS production as a minor mechanism of action. Therefore, they need to achieve higher concentrations to produce the same antibacterial effect as other metal-based nanoparticles (for example AgNPs). In this sense, Zhang et al. showed that to produce the small zone of inhibition for *S. aureus*, AuNPs need a concentration of 197 μg/mL, whereas AgNPs need small concentrations as 4.86 μg/mL (Zhang et al. 2015).

TABLE 5.3
Antibacterial Applications of Au-Based Green Nanoparticles, with Respective Minimum Inhibitory Concentration (MIC) and Minimum Bactericidal Concentration (MBC) Values

Au-nanoparticles efficacy against microorganisms	Production method	MIB and MIC values
P. aeruginosa	Biological method (extract of *A. comosus*)	MIC, MIB: 4 μg/mL
S. aureus	Biological method (extract of *M. piperita*)	MIC: 3.92 μg/mL
E. coli	Biological method (extract of *G. elongate*)	MIB: 12–16 μg/mL MIC: 4 μg/mL
K. pneumoniae	Biological method (extract of *S. brachiate*)	MIC: 3.3 μg/mL
S. typhimurium	Not available	MIC, MIB: 8 μg/mL
K. oxytoca, E. faecalis, V. cholera, S. paratyphii, V. parahaemolyticus, P. vulgaris	Biological method (extract of *S. marginatum*)	Data not available

TABLE 5.4
Antimicrobial Activity of Cu and Cuo-Based Nanoparticles, with Respective Zone of Inhibition (Z), Minimum Inhibitory Concentration (MIC) and Minimum Bactericidal Concentration (MBC) Values

Nanoparticles efficacy against MOs	Therapeutic efficacy	Zone of inhibition (Z)/ MIB/MIC values
Staphylococcus aureus, Pseudomonas aeruginosa	Ultrasound increased the antibacterial effect of CuO nanoparticles against *S. aureus* and *P. aeruginosa*	Data not available
E. coli, S. epiderdimis, methicillin 655 resistant S.s aureus (superbug MRSA) isolate, *Spore-forming Bacillus megatarium*	Reaction of copper nanoparticles of 100 nm with *B. subtilis* showed the highest susceptibility	Z = 0.0734 mL/μg
B. megatarium, *S. epidermidis, E. coli MRSA*	Cu1X and Cu10X kill *B. megatarium,* *S. epidermidis, E. coli* and MRSA	Data not available
E. coli, *S. aureus*	$CuCl_2·2H_2O$ derived nanoparticles showed more antibacterial activity than $CuSO_4$ $.5H_2O$ derived nanoparticles	*E. coli* MIC:3.75 mg/mL *S. aureus* MIC: 2.50 mg/mL

5.8.5 Copper and Copper Oxide Nanoparticles (CuNPs, Cu_2ONPs and CuONPs)

Cooper is a semiconductor material considered to be an excellent candidate for the synthesis of metal-based nanoparticles. Besides being highly resistant to heat, it is also robust, stable, cheap and easily synthesized (Abbaszadegan et al. 2015). The antimicrobial activity of CuONPs has been shown in Table 5.4.

The mechanism of antibacterial activity of CuONPs is not well explicated yet, but it is thought that it enhanced the attachment to bacterial cell wall which is initiated by electrostatic interactions. Dissociation of Cu^{2+} induces the generation of ROS that make contact with cellular membranes. These ions also have the capacity to enter the cell, causing membrane damage which is associated with disruption of cells' internal content and bacterial cell leakage (Slavin et al. 2017; Yadav et al. 2017). Moreover, their antibacterial activity has been studied particularly against microorganisms such as *E. coli, V. cholera, P. aeruginosa, S. typhus, S. aureus. E. faecalis, B. subtilis* and *S. faecalis* (Rajendran et al. 2017).

Mirhosseini confirmed *in vitro* that CuONPs had antibacterial properties, reducing significantly the growth of *S. aureus* and *P. aeruginosa* at the concentration of 500 μg/mL and also showing growth reduction percentages by 24% for *S. aureus* and 7.9% for *P. aeruginosa* (Mirhosseini, 2015).

The antimicrobial effect of copper nanoparticles is reduced at C90 *E. coli* and *B. subtilis* at the concentrations of 33.49 μg/mL and 28.20 μg/mL, respectively (Yoon et al. 2007). Kumar et al. also synthesized colloidal CuONPs from glucose, starch and $CuCl_2$ and its antibacterial properties were assessed against *E. coli* (Gram-negative), *S. epidermis* (Gram-positive), a methicillin-resistant *S. aureus* (superbug MRSA) isolate and the spore-forming *Bacillus megatherium*.

Results showed that treatments of CuONPs with elemental copper concentrations of 0.0113 and 0.00113 113 mol L^{-1} kill all the microorganisms assessed (Kumar et al. 2019). Moniri et al. synthesized ultra-small CuONPs obtaining a MIC (derived from both precursors) against *E. coli* and *S. aureus* of 3.75 and 2.50 mg/mL, respectively (Moniri Javadhesari et al. 2019). The biosynthesized CuO NPs showed higher antibacterial activity (zone of inhibition) against various human and fish bacterial pathogens like *B. cereus, P. mirabilis, A. caviae* (Nabila and Kannabiran 2018).

5.8.6 Bimetallic NPs

Ag and Au may be used in a single nanoparticle to enhance the effects of a drug and reduce the required dose. Alternatively, they can be used alone since they possess antimicrobial properties that are enhanced when combined in the form of bimetallic NPs (Arvizo et al. 2010; Singh et al. 2016).

Gold(Au)-silver(Ag) alloys are an optimal solution since they combine the antimicrobial effect of silver with the ease of functionalization and improved stability in complex biological media provided by gold (dos Santos et al. 2012). Fakhri and co-workers synthetized and functionalized AgAuNPs with a tetracycline and concluded that there exists a synergetic effect of the antibiotic with the bimetallic nanoparticle, with greater bactericidal activity of this form in detriment of its free forms. The mechanism of action was established as being the generation of ROS (Fakhri et al. 2017). Also recently, Baker and collaborators described the synthesis and antimicrobial activity of bimetallic AgAuNPs from the cell free supernatant of *Pseudomonas veronii* strain AS41G inhabiting *Annona squamosa L.* The authors showed their synergistic effect with standard antibiotics with 87.5, 18.5, 11.15, 10, 9.7, and 9.4% fold increased activity with bacitracin, kanamycin, gentamicin, streptomycin, erythromycin and chloramphenicol, respectively, against bacitracin resistant strains of *Bacillus subtilis, E. coli*, and *K. pneumoniae* (Baker et al. 2017). Zhao and collaborators have demonstrated the antibacterial activity of AuPtNPs bimetallic NPs against sensitive and drug-resistant bacteria *via* the dissipation of the bacterial membrane potential and the elevation of adenosine triphosphate (ATP) levels (Zhao et al. 2014).

5.9 PHYTO-NANO CONJUGATES AND ANTIMICROBIAL RESISTANCE

Plants are renewable and economical sources of antimicrobials with little toxicity and rich chemical diversity. Metabolites (alkaloids, polyphenols, terpenes, glycosides, etc.) in plant extracts are mainly phenolic derivatives that can stop bacterial growth by binding to bacterial proteins or through reducing pH, which alters the bacterial cellular process and kills bacteria (Mostafa et al. 2018; Prasad 2014). These plant-based antimicrobials (PBA) possess certain clinical value, as the bioactivity does not impose any resistance (Cheesman et al. 2017). To date, there has been no claim on identification of bacteria, which developed resistance to plant antimicrobials. There are also recent reviews on plant extracts' antibacterial activity describing their effective action and the defense mechanisms of plants to bacterial resistance (Gupta and Birdi, 2017). Nevertheless, traditional medicine (Ayurveda, Traditional Chinese medicine, Kampo, Unani, and Siddha) has not entered into the modern medicinal framework due to the chemical complexity of plant extracts, the requirement of higher concentrations for activity, poor water solubility and the lack of standardization (Harvey, 2008). Beyond these minute limitations, their true chemical diversity (extensive functional group and chirality) has encouraged phytochemicals for industrial biological applications (Cowan, 1999; Lewis and Ausubel, 2006).

To reduce toxicity and improve efficacy of chemically synthesized nanoparticles (NPs) as antibacterials, some researchers have employed plant metabolites as adjuvants. Green synthesized NPs are known to inhibit popular resistance mechanisms exhibited by bacteria such as excessive and poor influx, biofilm and so on (Musarrat et al. 2015). The strategy of using green NPs as an adjuvant to antibiotics is to acquire toxic-free antimicrobials as the minimum criteria (Slavin et al. 2017).

It is encouraging to know that research is focused recently on microbial and plant assisted NPs in combination with antibiotics, considering them as least toxic to mammalian cells. Formerly, many reports and reviews are available on green synthesized NPs using various plant and microbial sources (Abdelghany et al. 2018), but very few are available solely on combinational therapy of green synthesized NPs as an adjuvant to antibiotics. Some selected examples of synergism between phyto-nano conjugates and antibiotics for antimicrobial activity is given in Table 5.5.

Some other examples including combination effects of silver nanoparticles with cinnamaldehyde and eugenol, Se-NPs with quercetin, Au-NPs and Ag-NPs with resveratrol have been reported with

TABLE 5.5
Synergism Between Phyto-Nanoconjugates and Antibiotics for Antimicrobial Activity

Nanoparticle	Plant constituent	Antibiotic	Targeted bacteria
Ag NP	Corn leaf waste of *Zea mays* extract	Kanamycin and rifampicin	*Bacillus cereus*, *Escherichia coli*, *Staphylococcus aureus*, *Listeria monocytogenes*, and *Salmonella typhimurium*
Ag NP	Gum kondagogu	Ciprofloxacin, streptomycin, and gentamicin	Gram-positive (*Staphylococcus aureus*) and Gram-negative (*E. coli*, *Pseudomonas aeruginosa*)
Au NP	Aqueous extracts of outer oriental melon peel (OMP) and peach	Kanamycin and rifampicin	*E. coli*, *S. aureus*
Cu NP	Green tea (*Camellia sinensis*) and β-cyclodextrin	Ampicillin, amoxicillin, gentamicin and ciprofloxacin	*E. coli* (Gram-negative rods), *S. typhi* (Gram-negative rods), *Micrococcus luteus* (Gram-positive cocci) and *Streptococcus mutans*
Ag NP	Flower broth of *Tagetes erecta*	Commercial antibiotics	Gram-positive (*Staphylococcus aureus and Bacillus cereus*), Gram-negative (*E. coli* and *Pseudomonas aeruginosa*) bacteria
Ag NP	*Adiantum philippense* extract	Amoxicillin	MRSA
Ag NP	Leaf extracts of *Ficus virens*	streptomycin	Gram-positive (*Bacillus subtilis, Staphylococcus epidermidis, Enterococcus faecalis*) and three gram-negative (*Klebsiella pneumoniae, Vibrio cholera* and *Vibrio vulnificus*)

(Ruddaraju et al. 2020)

irreversible membrane damage against Gram-positive and Gram-negative bacteria (Huang et al. 2016; Park et al. 2016). When metallic NPs are administered at therapeutic dose, they are not toxic by themselves, as there is daily requirement of metals for our body (Akter et al. 2018). But toxicity may arise due to agents in the chemical synthesis. There is a recent review by Khan et al. focusing on the toxicity associated with chemical methods and properties of NPs. To overcome the toxicity issue, green nanotechnology can be an undeniable approach as it does not employ any synthetic reagents (Vasanth et al. 2017).

Nanoparticles (NPs) and plant-based antimicrobial therapies need a better perceptual view for a rational design of NP and plant-based antimicrobials. Expectantly, green nanotechnology and their combinational therapies with antibiotics have recently been employed as modern methodology in standardized protocols, to provide alternate mechanisms against MDR (Multiple Drug Resistance) bacteria (Mostafa et al. 2018). Green NPs with antibiotics carry numerous moieties in one scaffold; where the bacteria cannot develop multiple gene mutations simultaneously in the same microbial cell (Plank, 2009). Thus, this combinational approach can command and modify molecular structures at the nano-scale to attain smart, targeted and controlled delivery in an economical way through securing natural microbiome.

5.10 LIMITATIONS AND FORTHCOMING CONTESTS OF NANOPARTICLES (NPS)

Regardless of the anticipated potential of NPs for medical applications, there are still several limitations related with their acute and long-term exposure in humans. Several routes of exposure must be considered when evaluating NPs exposure, such as oral and gastrointestinal tract, dermal, respiratory system and endovenous administration directly to the bloodstream (De Matteis, 2017). It is well known also that the physicochemical properties of NPs (e.g., size, shape and surface chemistry) affect their interaction with biological systems, influencing cellular uptake, pharmacokinetics and bio-distribution, all with a direct impact on final biological effects (Warheit, 2018; Prasad et al. 2016; Srivastava et al. 2021). This poses a major concern to effectively draw critical conclusions on NPs' safety due to the vast number of different types/shapes/surface modified nanoparticles, the different methods used to evaluate their safety and environmental effects and also by the fact that most of these under *in vitro*/*in vivo* studies present acute studies rather than long-term exposure (Xia et al. 2016).

Several studies have shown that intravenously injected NPs accumulate in the colon, lung, bone-marrow, liver, spleen and lymphatics (Hagens et al. 2007). Inhalation has also been shown to cause cytotoxicity in lung, liver, heart and spleen through systemic circulation (Leucuta, 2013). This is of particular relevance for small NPs because of efficient cellular uptake and transcytosis across epithelial and endothelial cells into the blood and lymphatic circulation. Several NPs systems have demonstrated toxicity in multiple organs, such as free radical-mediated oxidative stress generated by the interaction of antimicrobial NPs with cell components that can result in hepatotoxicity and nephrotoxicity (De Jong and Borm, 2008).

Nevertheless, these *in vivo* and *in vitro* studies have been providing clues to the specific mechanisms by which NPs trigger an adverse effect enabling future surface modification of NPs to make them safer and less toxic (De Matteis, 2017). These concerns relating to nano-safety have been addressed and implemented *via* European Commission FP7 and H2020-sponsored programs followed by some relevant conclusions issued by the US National Academy of Science Committee on Research Progress of Environmental Health and Safety Aspects of Engineered Nanomaterials (Warheit, 2018).

Due to the 3Rs (Replacement, Reduction and Refinement) policies of *in vivo* studies, the future challenge of Regulatory Agencies is the standardization of the *in vitro* methodologies to establish the toxicology profile of NPs based on good laboratory practice (GLP) and the construction of flexible and reliable databases in which NPs are classified according to the data derived from these toxicological investigations. Together, these efforts might provide information on the dosage at which a particular NP can be considered safe and thus appropriate for medical use.

5.11 CONCLUSION

All renowned antibiotic classes have earned notable resistance and monotherapy approaches have become limited in the landscape of MDR pathogens. As an alternative, researches have focused on the resistance mechanisms of bacteria and developed adjuvant therapy with antibiotics. This combinational approach can bypass bacterial resistance mechanisms, but due to adjuvant-antibiotic interactions, resistance is spreading further. In addition, the broad-spectrum activity of these combinations is affecting natural microbiome. Therefore, scientists have shifted their focus on nanotechnology as an interdisciplinary approach to target MDR bacteria.

NPs have the ability to modulate their physicochemical properties that grant novel molecules for effective bactericidal activity with feasible administration routes. Several methods are employed (physical, chemical and biological) to prepare NPs. Each method has its own pros and cons with the fore-most merit of resistance-free and principal demerits of heterogeneous (physical) and toxicity (chemical). Of all methods, the biological method may satisfy the criteria of resistance-free and homogenous NPs with limited toxicity. Thus, it is a reverse brain drain of plants and metals in nano forms, as green NPs work against MDR bacteria through targeted delivery with no resistance and toxicity. In spite of NPs efficacy as antimicrobials with a targeted approach, our world cannot ignore antibiotics as they have proved their outstanding performance in antibacterial activity since the 20th century. Thus, a combination of antibiotic with greener NPs can be a paradigm shift to deliver the optimum concentration of drug at the target place, by bypassing resistance mechanisms of bacteria.

In addition, adopting greener NPs as drug delivery vehicles can be a toxic-free approach with fewer administration frequencies along with biocompatibility.

REFERENCES

Abascal, K.; Yarnell, E. Herbs and drug resistance: Part 2-clinical implications of research on microbial resistance to antibiotics. *Altern. Complement. Ther.* 2002, 8, 284–90.

Abbaszadegan, A.; Ghahramani, Y.; Gholami, A.; Hemmateenejad, B.; Dorostkar, S.; Nabavizadeh, M.; Sharghi, H. The effect of charge at the surface of silver nanoparticles on antimicrobial activity against Gram-positive and Gram-negative bacteria: A preliminary study. *J. Nanomater.* 2015, 2015, 720654.

Abdelghany, T.M.; Aisha, M.H.; Mohamed, A.R.; Al Abboud, A.; Alawlaqi, M.M.; Magdah, A.G. et al. Recent advances in green synthesis of silver nanoparticles and their applications: about future directions: A review. *Bionanosc.* 2018, 8(3), 5–16.

Abdel-Kareem, M.M.; Zohri, A.A. Extracellular mycosynthesis of gold nanoparticles using Trichoderma hamatum: Optimization, characterization and antimicrobial activity. *Lett. Appl. Microbiol.* 2018, 67, 465–475.

Aderibigbe, B.A. Metal-based nanoparticles for the treatment of infectious diseases. *Molecules.* 2017, 22, 1370.

Agarwal, A.; D'Souza, P.; Johnson, T.S.; Dethe, S.M.; Chandrasekaran, C. Use of in vitro bioassays for assessing botanicals. *Curr. Opin. Biotechnol.* 2014, 25, 39–44.

Akter, M.; Sikder, M.T.; Rahman, M.M.; et al. A systematic review on silver nanoparticles induced cytotoxicity: physicochemical properties and perspectives. *J Adv Res.* 2018, 9, 1–16.

AlSheikh, H.M.A.; Sultan, I.; Kumar, V.; Rather, I.A.; Al-Sheikh, H.; Jan, A.T.; Haq, Q.M.R. Plant-based phytochemicals as possible alternative to antibiotics in combating bacterial drug resistance. *Antibiotics* 2020, 9, 480.

Ananth, A.; Keerthika, V.; Rajan, M.R. Synthesis and characterization of nano-selenium and its antibacterial response on some important human pathogens. *Current Sc.* 2019, 116(2), 285–290.

Antibiotic resistance: a global threat I features I CDC. www. cdc.gov/features/antibiotic-resistance-global/index.html. Accessed September 15, 2019.

Arvizo, R.; Bhattacharya, R.; Mukherjee, P. Gold nanoparticles: Opportunities and challenges in nanomedicine. *Expert Opin. Drug Deliv.* 2010, 7, 753–763.

Aziz, N.; Faraz, M.; Pandey, R.; Sakir. M; Fatma, T; Varma, A.; Barman, I.; Prasad, R. Facile algae-derived route to biogenic silver nanoparticles: Synthesis, antibacterial and photocatalytic properties. *Langmuir* 2015, 3, 11605–11612. DOI: 10.1021/acs.langmuir.5b03081

Aziz, N; Fatma, T.; Varma, A.; Prasad, R. Biogenic synthesis of silver nanoparticles using Scenedesmus abundans and evaluation of their antibacterial activity. *Journal of Nanoparticles* 2014, Article ID 689419. http://dx.doi.org/10.1155/2014/689419

Aziz, N.; Pandey, R.; Barman, I.; Prasad, R. Leveraging the attributes of Mucor hiemalis-derived silver nanoparticles for a synergistic broad-spectrum antimicrobial platform. *Front Microbiol* 2016, 7, 1984. DOI: 10.3389/fmicb.2016.01984

Aziz, N.; Faraz, M.; Sherwani, MA; Fatma, T; Prasad, R. Illuminating the anticancerous efficacy of a new fungal chassis for silver nanoparticle synthesis. *Front Chem* 2019, 7, 65. doi: 10.3389/fchem.2019.00065

Baker, S.; Pasha, A.; Satish, S. Biogenic nanoparticles bearing antibacterial activity and their synergistic effect with broad spectrum antibiotics: Emerging strategy to combat drug resistant pathogens. *Saudi Pharm. J.* 2017, 25, 44–51.

Beceiro, A.; Tomas, M.; Bou, G. Antimicrobial resistance and virulence: A successful or deleterious association in the bacterial world? *Clin. Microbiol. Rev.* 2013, 26, 185–230.

Berenbaum, M.C. Synergy, additivism and antagonism in immunosuppression: A critical review. *Clin. Exp. Immunol.* 1977, 28(1), 1–18.

Bhuyan, T.; Mishra, K.; Khanuja, M.; Prasad, R.; Varma, A. Biosynthesis of zinc oxide nanoparticles from *Azadirachta indica* for antibacterial and photocatalytic applications. *Mater Sci Semicond Process* 2015, 32, 55–61

Center for Disease Dynamics Economics and Policy. Resistance Map: India. [Last accessed on 2018 Jan 19]. Available from: www.resistancemap.cddep.org/CountryPage.php?country=India

Chandy, S.J.; Naik, G.S.; Balaji, V.; Jeyaseelan, V.; Thomas, K.; Lundborg, C.S. High cost burden and health consequences of antibiotic resistance: The price to pay. *J Infect Dev Ctries.* 2014, 8(9), 1096–102.

Cheesman, M.J.; Ilanko, A.; Blonk, B.; Cock, I.E. Developing new antimicrobial therapies: Are synergistic combinations of plant extracts/compounds with conventional antibiotics the solution? *Pharmacogn Rev.* 2017, 11(22), 57–72.

Chen, X.; Schluesener, H.J. Nanosilver: A nanoproduct in medical application. *Toxicol. Lett.* 2008, 176, 1–12.

Chiou, Y.D.; Hsu, Y.J. Advances in nanotechnology research and application: 2012 Ed. *Appl. Catal. Environ B.* 2011, 105, 211–219.

Chokshi, A.; Sifri, Z.; Cennimo, D.; Horng, H. Global contributors to antibiotic resistance. *J Glob Infect Dis.* 2019, 11(1), 36–42.

Conte, R.; Calarco, A.; Napoletano, A.; Valentino, A.; Margarucci, S. et al. Polyphenols nanoencapsulation for therapeutic applications. *J. Biomol. Res. Ther.* 2016, 5, 139.

Cowan, M.M. Plant products as antimicrobial agents. *Clin Microbiol Rev.* 1999, 12(4), 564–82.

Dadgostar, P. Antimicrobial resistance: Implications and costs. *Infection and Drug Resistance.* 2019, 12, 3903–3910.

De Jong, W.H.; Borm, P.J. Drug delivery and nanoparticles: Applications and hazards. *Int. J. Nanomedicine.* 2008, 3, 133–149.

De Matteis, V. Exposure to inorganic nanoparticles: Routes of entry, immune response, bio-distribution and *in vitro/in vivo* toxicity evaluation. *Toxics.* 2017, 5, E29.

Deng, X.H.; Fang, F.F.; Zheng, C.J.; Wu, Y.; Qin, L.P. Monoterpenoids from the whole herb of *Veronicastrum axillare. Pharm. Biol.* 2014, 52, 661–663.

dos Santos, M.M.; Queiroz, M.J.; Baptista, P.V. Enhancement of antibiotic effect via gold: silver-alloy nanoparticles. *J. Nanoparticle Res.* 2012, 14, 859.

Dunphy, L.J.; Yen, P.; Papin, J.A. Integrated experimental and computational analyses reveal differential metabolic functionality in antibiotic-resistant *Pseudomonas aeruginosa. Cell Syst.* 2019, 8, 3–14.

Fakhri, A.; Tahami, S.; Naji, M. Synthesis and characterization of core-shell bimetallic nanoparticles for synergistic antimicrobial effect studies in combination with doxycycline on burn specific pathogens. *J. Photochem. Photobiol. B Biol.* 2017, 169, 21–26.

Fakhrudin, N.; Waltenberger, B.; Cabaravdic, M.; Atanasov, A.G.; Malainer, C.; Schachner, D.; Heiss, E.H.; Liu, R.; Noha, S.M.; Grzywacz, A.M. et al. Identification of plumericin as a potent new inhibitor of the NF-kappaB pathway with anti-inflammatory activity in vitro and in vivo. *Br. J. Pharm.* 2014, 171, 1676–1686.

Gemeda, N.; Tadele, A.; Lemma, H.; Girma, B.; Addis, G.; Tesfaye, B.; Abebe, A.; Gemechu, W.; Yirsaw, K.; Teka, F. et al. Development, characterization, and evaluation of novel broad-spectrum antimicrobial topical formulations from *Cymbopogon martini* (Roxb.) W. watson essential oil. *Evid. Based Complement. Altern. Med.* 2018, 9812093.

Gerber, A.; Bundschuh, M.; Klingelhofer, D.; Groneberg, D.A. Gold nanoparticles: Recent aspects for human toxicology. J. *Occup. Med. Toxicol.* 2013, 8, 32.

Goretta, L.A.; Leveques, A.; Giu_rida, F.; Michailidis, F.R.; Viton, F.; Barron, D.; Paton, M.D.; Manzano, S.G.; Buelga, C.S.; Williamson, G. et al. Elucidation of (-)-epicatechin metabolites after ingestion of chocolate by healthy humans. *Free. Radic. Biol. Med.* 2012, 53, 787–795.

Guo, Y.; Liu, Y.X.; Kang, L.P.; Zhang, T.; Yu, H.S.; Zhao, Y.; Xiong, C.Q.; Ma, B.P. Two novel furostanol saponins from the tubers of *Ophiopogon japonicus. J. Asian Nat. Prod. Res.* 2013, 15, 459–465.

Gupta, P.D.; Birdi, T.J. Development of botanicals to combat antibiotic resistance. *J Ayurveda Integr Med.* 2017, 8, 266–275.

Hagens, W.I.; Oomen, A.G.; de Jong, W.H.; Cassee, F.R.; Sips, A.J.A.M. What do we (need to) know about the kinetic properties of nanoparticles in the body? *Regul. Toxicol. Pharmacol.* 2007, 49, 217–229.

Happy, A.; Soumyaa, M.; Kumara, V.; Rajeshkumar, S.; Sheba, D.R.; Lakshmi, T.; Nallaswamyd, V.D. Phyto-assisted synthesis of zinc oxide nanoparticles using Cassia alata and its antibacterial activity against *Escherichia coli. Biochem. Biophys. Rep.* 2019, 17, 208–211.

Harvey, A.L. Natural Products in Drug Discovery. *Drug Discov Today* 2008, 13 (19–20), 894–901.

Her, S.; Jaffray, D.A.; Allen, C. Gold nanoparticles for applications in cancer radiotherapy: Mechanisms and recent advancements. *Adv. Drug Deliv. Rev.* 2017, 109, 84–101.

Huang, X.; Chen, X.; Chen, Q.; Yu, Q.; Sun, D.; Liu, J. Investigation of functional selenium nanoparticles as potent antimicrobial agents against superbugs. *Acta Biomater.* 2016, 30, 397–407.

Jayabalan, J.; Mani, G.; Krishnan, N.; Pernabas, J.; Milton, J.; Tae, H. Green biogenic synthesis of zinc oxide nanoparticles using Pseudomonas putida culture and it's *in vitro* antibacterial and anti-biofilm activity. *Biocatal. Agric. Biotechnol.* 2019, 21, 1–9.

Junio, H.A.; Cordero, A.A.S.; Ettefagh, K.A.; Burns, J.T.; Micko, K.T.; Graf, T.N.; Richter, S.J.; Cannon, R.E.; Oberlies, N.H.; Cech, N.B. Synergy-directed fractionation of botanical medicines: A case study with goldenseal (*Hydrastis canadensis*). *J. Nat. Prod.* 2011, 74, 1621–1629.

Keerthi, D.; Geethu, C.; Nair, R.A.; Pillai, P. Metabolic profiling of zingiber zerumbet following Pythium myriotylum infection: Investigations on the defensive role of the principal secondary metabolite, zerumbone. *Appl. Biochem. Biotechnol.* 2014, 172, 2593–2603.

Kim, D.; Touros, A.; Kim, W.R. Non-alcoholic fatty liver disease and metabolic syndrome. *Clin. Liver Dis.* 2018, 22, 133–140.

Kolli, M.E.; Laouer, H.; Kolli, H.E.; Akkal, S.; Sahli, F. Chemical analysis, antimicrobial and anti-oxidative properties of *Daucus gracilis* essential oil and its mechanism of action. *Asian Pac. J. Trop. Biomed.* 2016, 6, 8–15.

Król, A.; Pomastowski, P.; Rafi ́nska, K.; Railean-Plugaru, V.; Buszewski, B. Zinc oxide nanoparticles: Synthesis, antiseptic activity and toxicity mechanism. *Adv. Colloid Interface Sci.* 2017, 249, 37–52.

Kumar, S.V.; Bafana, A.P.; Pawar, P.; Faltane, M.; Rahman, A.; Dahoumane, S.A.; Kucknoor, A.; Jeffryes, C.S. Optimized production of antibacterial copper oxide nanoparticles in a microwave-assisted synthesis reaction using response surface methodology. *Colloids Surf. A* 2019, 573, 170–178.

Lara, H.H.; Ayala-Nuñez, N.V.; Ixtepan-Turrent, L.; Rodriguez-Padilla, C. Mode of antiviral action of silver nanoparticles against HIV-1. *J. Nanobiotechnol.* 2010, 8, 1–10.

Leucuta, S.E. Systemic and biophase bioavailability and pharmacokinetics of nano-particulate drug delivery systems. *Curr. Drug Deliv.* 2013, 10, 208–240.

Lewis, K.; Ausubel, F.M. Prospects for plant-derived anti-bacterials. *Nat. Biotechnol.* 2006, 24, 1504–1507.

Liu, S.; Hsieh, D.; Yang, Y.L.; Xu, Z.; Peto, C.; Jablons, D.M.; You, L. Coumestrol from the national cancer institute's natural product library is a novel inhibitor of protein kinase CK2. *BMC Pharm. Toxicol.* 2013, 14, 36.

Mabona, U.; Viljoen, A.; Shikanga, E.; Marston, A.; Vuuren, S.V. Antimicrobial activity of southern African medicinal plants with dermatological relevance: From an ethnopharmacological screening approach, to combination studies and the isolation of a bioactive compound. *J. Ethnopharmacol.* 2013, 148, 45–55.

Mirhosseini, M. Synergistic antibacterial effect of metal oxide nanoparticles and ultrasound stimulation. *J. Biol. Today's World.* 2015, 4, 138–144.

Mohanraj, V.J.; Chen, Y. Nanoparticles – a review. *Trop. J. Pharm. Res.*, 2006, 5(1), 561–573.

Moniri Javadhesari, S.; Alipour, S.; Mohammadnejad, S.; Akbarpour, M.R. Antibacterial activity of ultra-small copper oxide (II) nanoparticles synthesized by mechano-chemical processing against *S. aureus* and *E. coli*. *Mater. Sci. Eng. C.* 2019, 105, 110011.

Mostafa, A.A.; Al-Askar, A.A.; Almaary, K.S.; Dawoud, T.M.; Sholkamy, E.N.; Bakri, M.M. Antimicrobial activity of some plant extracts against bacterial strains causing food poisoning diseases. *Saudi J Biol Sci.* 2018, 25(2), 361–366.

Musarrat, J.; Ali, K.; Mohammad, A.; Quaiser, S.; Maqsood, S.; Shams, K. et al. Green synthesis of nanoparticles and their role as nano-antibiotics and anti-biofilm agents. *Planta Medica.* 2015, 81(5), OA44.

Nabila, M.I.; Kannabiran, K. Biosynthesis, characterization and antibacterial activity of copper oxide nanoparticles (CuO NPs) from actinomycetes. *Biocatal. Agric. Biotechnol.* 2018, 15, 56–62.

Ndhlala, A.R.; Aderogba, M.A.; Ncube, B.; Staden, J.V. Anti-oxidative and cholinesterase inhibitory effects of leaf extracts and their isolated compounds from two closely related Croton species. *Molecules.* 2013, 18, 1916–1932.

Nordmann, P.; Naas, T.; Poirel, L. Global spread of Carbapenemase-producing Enterobacteriaceae. *Emerg Infect Dis.* 2011, 17(10), 1791–1798.

Nur, S.; Shameli, K.; Mei-Theng, W.M.; Teow, S.-Y.; Chew, J.; Ismail, N.A. Cytotoxicity and antibacterial activities of plant-mediated synthesized zinc oxide (ZnO) nanoparticles using *Punica granatum* (pomegranate) fruit peels extract. *J. Mol. Struct. J.* 2019, 1189, 57–65.

Olajuyigbe, O.O.; Afolayan, A.J. Synergistic interactions of methanolic extract of *Acacia mearnsii* De Wild. with antibiotics against bacteria of clinical relevance. *Int. J. Mol. Sci.* 2012, 13, 8915–8932.

Panáˇcek, A.; Kolár, M.; Vecerová, R.; Prucek, R.; Soukupová, J.; Krystof, V.; Hamal, P.; Zboril, R.; Kvítek, L. Antifungal activity of silver nanoparticles against *Candida*. *Biomaterials.* 2009, 30, 6333–6340.

Pang, Z.; Raudonis, R.; Glick, B.R.; Lin, T.J.; Cheng, Z. Antibiotic resistance in *Pseudomonas aeruginosa:* Mechanisms and alternative therapeutic strategies. *Biotechnol. Adv.* 2019, 37, 177–192.

Park, S.; Cha, S.H.; Cho, I.; Park, S.; Park, Y.; Cho, S. et al. Antibacterial nanocarriers of resveratrol with gold and silver nanoparticles. *Mater Sci Eng C Mater Biol Appl.* 2016, 58, 1160–1169.

Plank, C. Nanomedicine: Silence the target. *Nat Nanotechnol* 2009, 4(9), 544–345.

Prasad, R. Synthesis of silver nanoparticles in photosynthetic plants. *Journal of Nanoparticles* 2014, Article ID 963961. http://dx.doi.org/10.1155/2014/963961

Prasad ,R.; Pandey, R.; Barman, I. Engineering tailored nanoparticles with microbes: quo vadis. *WIREs Nanomed Nanobiotechnol* 2016, 8, 316–330. DOI: 10.1002/wnan.1363

Prestinaci, F.; Pezzotti, P.; Pantosti, A. Antimicrobial resistance: A global multifaceted phenomenon. *Pathog Glob Health.* 2015, 109(7), 309.

Rai, M.; Kon, K.; Ingle, A.; Duran, N.; Galdiero, S.; Galdiero, M. Broad-spectrum bioactivities of silver nanoparticles: The emerging trends and future prospects. *Appl. Microbiol. Biotechnol.* 2014, 98, 1951–1961.

Rajendran, K.; Anwar, A.; Khan, N.A.; Siddiqui, R. Brain-eating amoebae: Silver nanoparticle conjugation enhanced efficacy of anti-amoebic drugs against naegleria fowleri. *ACS Chem. Neurosci.* 2017, 8, 2626–2630.

Ramalingam, B.; Parandhaman, T.; Das, S.K. Antibacterial effects of biosynthesized silver nanoparticles on surface ultrastructure and nano-mechanical properties of gram-negative bacteria viz. *Escherichia coli and Pseudomonas aeruginosa. ACS Appl. Mater Interfaces.* 2016, 8(7), 4963–4976.

Ramasamy, D.; Saraswathy, A. Vitiquinolone – A quinolone alkaloid from *Hibiscus vitifolius* linn. *Food Chem.* 2014, 145, 970–975.

Romero-Perez, A.; Garcia-Garcia, E.; Zavaleta-Manceraetal, A. Designing and evaluation of sodium selenite nanoparticles *in vitro* to improve selenium absorption in ruminants. *Vet. Res. Commun.*, 2010, 34, 71–79.

Ruddaraju, L.K., Pammi, S.V.N.; Guntuku, G.S.; Padavala, V.S.; Kolapalli, V.R.M. A review on anti-bacterials to combat resistance: From ancient era of plants and metals to present and future perspectives of green nano-technological combinations. *Asian J. Pharmaceut. Sc.* 2020, 15, 42–59.

Singh, P.; Nanda, A. Antimicrobial and antifungal potential of zinc oxide nanoparticles in comparison to conventional zinc oxide particles. *J. Chem. Pharm. Res.* 2013, 5, 457–463.

Singh, R.; Nawale, L.; Arkile, M.; Wadhwani, S.; Shedbalkar, U.; Chopade, S. et al. Phytogenic silver, gold, and bimetallic nanoparticles as novel anti-tubercular agents. *Int. J. Nanomed.* 2016, 11, 1889–1897.

Sirelkhatim, A.; Mahmud, S.; Seeni, A. Review on zinc oxide nanoparticles: Antibacterial activity and toxicity mechanism. *Nano-Micro Lett.* 2015, 7, 219–242.

Slavin, Y.N.; Asnis, J.; Häfeli, U.O.; Bach, H. Metal nanoparticles: Understanding the mechanisms behind antibacterial activity. *J Nanobiotechnol.* 2017, 15(65), 1–20.

Srivastava, S;, Usmani, Z.; Atanasov, A.G.; Singh, V.K.; Singh, N.P.; Abdel-Azeem, A.M.; Prasad, R.; Gupta, G.;, Sharma, M.; Bhargava, A. Biological nanofactories: Using living forms for metal nanoparticle synthesis. *Mini-Reviews in Medicinal Chemistry* 2021, 21(2), 245–265 .

Stefanovic, O.; Comic, L. Synergistic antibacterial interaction between *Melissa officinalis* extracts and antibiotics. *J. Appl. Pharm. Sci.* 2012, 2, 1–5.

Vasanth, S.B.; Kuria, G.A. Toxicity evaluation of silver nanoparticles synthesized by chemical and green route in different experimental models. *Artif. Cells Nanomed Biotechnol*, 2017, 45(8), 1721–1727.

Ventola, C.L. The antibiotic resistance crisis: part 1: Causes and threats. *P T.* 2015, 40(4), 277–283.

Vijayakumar, S.; Krishnakumar, C.; Arulmozhi, P.; Mahadevan, S.; Parameswari, N. Biosynthesis, characterization and antimicrobial activities of zinc oxide nanoparticles from leaf extract of *Glycosmis pentaphylla* (Retz.). *Microb. Pathog.* 2018, 116, 44–48.

Wagner, H. Synergy research: Approaching a new generation of phytopharmaceuticals. *Fitoterapia.* 2011, 82(1), 34–37.

Wagner, H.; Ulrich-Merzenich, G. Synergy research: Approaching a new generation of phytopharmaceuticals. *Phytomedicine.* 2009, 16, 97–110.

Wagnera, H.; Efferth, T. Introduction: Novel hybrid combinations containing synthetic or antibiotic drugs with plant-derived phenolic or terpenoid compounds. *Phytomedicine.* 2017, 37, 1–3.

Warheit, D. B. Hazard and risk assessment strategies for nanoparticle exposures: How far have we come in the past 10 years? *F1000Res.* 2018, 7, 376.

WHO. World Health Organization Model List of Essential Medicines; amended August 2017; 20th List; WHO: Geneva, Switzerland, March 2017.

Wiedenfeld, H.; Dumaa, M.; Malinowski, M.; Furmanowa, M.; Narantuya, S. Phytochemical and analytical studies of extracts from *Rhodiola rosea* and *Rhodiola quadrifida*. *Pharmazie* 2007, 62, 308–311.

Wolfender, J.L.; Marti, G.; Thomas, A.; Bertrand, S. Current approaches and challenges for the metabolite profiling of complex natural extracts. *J. Chromatogr. A.* 2015, 1382, 136–164.

Xia, Q.; Li, H.; Xiao, K. Factors affecting the pharmacokinetics, bio-distribution and toxicity of gold nanoparticles in drug delivery. *Curr. Drug Metab.* 2016, 17, 849–861.

Yadav, L.; Tripathi, R.M.; Prasad, R.; Pudake, R.N.; Mittal, J. Antibacterial activity of Cu nanoparticles against *E. coli, Staphylococcus aureus* and *Pseudomonas aeruginosa. Nano Biomed Eng*. 2017, 9(1): 9–14. DOI: 10.5101/nbe.v9i1.p9-14

Yoon, K.Y.; Hoon Byeon, J.; Park, J.H.; Hwang, J. Susceptibility constants of *Escherichia coli* and *Bacillus subtilis* to silver and copper nanoparticles. *Sci. Total Environ.* 2007, 373, 572–575.

Yother, J. Capsules of *Streptococcus pneumoniae* and other bacteria: Paradigms for polysaccharide biosynthesis and regulation. *Annu. Rev. Microbiol.* 2011, 65, 563–581.

Yu, J.; Zhang, W.; Li, Y.; Wang, G.; Yang, L.; Jin, J.; Chen, Q.; Huang, M. Synthesis, characterization, antimicrobial activity and mechanism of a novel hydroxyapatite whisker/nano zinc oxide biomaterial. *Biomed. Mater.* 2015, 10, 15001

Zhang, X.F.; Liu, Z.G.; Shen, W.; Gurunathan, S. Silver nanoparticles: Synthesis, characterization, properties, applications, and therapeutic approaches. *Int. J. Mol. Sci.* 2016, 17, 9.

Zhang, Y.; Shareena Dasari, T.P.; Deng, H.; Yu, H. Antimicrobial activity of gold nanoparticles and ionic gold. J. Environ. Sci. Heal. Part C Environ. *Carcinog. Ecotoxicol. Rev.* 2015, 33, 286–327.

Zhao, Y.; Ye, C.; Liu, W.; Chen, R.; Jiang, X. Tuning the composition of AuPt bimetallic nanoparticles for antibacterial application. *Angew Chem Int.* 2014, 53, 8127–8131.

6 Essential Oil Components
Anti-viral Properties

Ashwini Khanderao Jadhav[1] *and Sankunny Mohan Karuppayil*[*1]
[1]Department of Stem Cell and Regenerative Medicine and Medical Biotechnology, Centre for Interdisciplinary Research,
DY Patil Education Society, Deemed University, (NAAC Accredited with 'A' grade), Kolhapur, Maharashtra, India.
*Corresponding Author: prof.karuppayil@gmail.com

CONTENTS

6.1 INTRODUCTION

The essential oil market in the countries in the Asia Pacific region is expected to grow at a fast pace. Various factors may facilitate this like favorable weather, availability of labor, and the availability of sufficient quantities of raw materials. A change in shift in preferences among customers toward the use of healthier and more nutritious food may provide lucrative growth opportunities. Large consumer demand for the skincare market along with the enhanced demand for fragrances is expected to boost the demand for essential oils. In the year 2021 the essential oils market was US$ 11,768.6 million globally. The market is predicted to touch 24,731.60 million US$ by 2030, growing with an annual growth rate of 8.1% from 2022 to 2030 [1]. The prospect of essential oils and other phytochemicals against SARS-CoV-2 was reviewed by Hensel et al. 2020 [2, 3]. The possible mechanism of action of phytochemicals and their components is reviewed by Barkat et al. (2022) [4].

Essential oils (EO) are complex mixtures of low molecular compounds found in various plants. These oils are produced by more than 17,500 species of plants and 300 of them are commercialized. EOs may include hundreds of different metabolites belonging to various chemical classes. Terpenoids and phenylpropanoids form the major constituents of the essential oils. In addition to the

DOI: 10.1201/9781003268932-6

aromatic and aliphatic constituents, monoterpenes, sesquiterpenes and their oxygenated derivatives are the largest group of entities in essential oils. The overall activity may be due to a combination of molecules. Essential oils find various applications in health, pharmaceuticals and cosmetic and food industries. The biological properties exhibited by essential oils include antibacterial, antifungal, antiviral, anti-mutagenic, anticancer, antioxidant, anti-inflammatory, immunomodulatory and anti-protozoal properties. The emergence of drug resistant strains of pathogens, increase in the immune-compromised population and limitations of the available drugs have encouraged people to use the complementary and alternative therapies, including the use of essential oils. Secondary metabolites which are synthesized by plants in response to attack by bacteria, fungi, viruses, and insect pests constitute a complex mixture of essential oils. These small molecule metabolites alone and in combination possess significant medicinal properties; hence, they may be used for chemotherapy of infectious as well as non-infectious diseases [5–7]. The traditional systems of medicine like ayurveda use a number of plants against viral diseases in humans. Essential oils may contain hundreds of different metabolites belonging to various chemical classes like monoterpene hydrocarbons (*p*-cymene, limonene, α-pinene and -terpinene), oxygenated monoterpenes (camphor, carvacrol, eugenol and thymol), diterpenes (cembrene C, kaurene and camphorene), sesquiterpene hydrocarbons (germacrene D, caryophyllene and humulene), oxygenated sesquiterpenes (spathulenol, caryophyllene oxide), monoterpene alcohols (geraniol, linalool and nerol), sesquiterpene alcohol (patchoulol), aldehydes (citral, cuminal), acids (geranic acid, benzoic acid), ketones (acetophenone, benzophenone), lactones (bergapten), phenols (thymol, eugenol and catechol), esters (bornyl acetate, ethyl acetate) and coumarins (fumarin, benzofuran). The overall activity may be due to a combination of molecules [8–10].

6.2 ANTIVIRAL PROPERTIES OF ESSENTIAL OILS

Essential oils possess antiviral properties against herpes simplex virus type 1 (HSV-1) and type 2 (HSV-2), HIV, HPV, Hepatitis B, Hepatitis C, dengue virus type 2, influenza virus, adenovirus type 3, Zika virus, SARS-CoV-2 virus, poliovirus, Japanese encephalitis and coxsackievirus B1(Figure 6.1). The potential of essential oils as antivirals is reviewed by Swamy et al. (2016) [8], da Silva et al. (2020) [9], Sharifi-Rad et al. (2022) [10], Corrêa et al. (2022) [11], Jadhav and Karuppayil (2021) [12], Tomas et al. (2022) [13]. Reviews by Ma and Yao (2020) [5] and da Silva et al. (2020) [9] provide an overview on the antiviral efficacy of essential oils from various plants and their components, as well as their mechanisms of action. The roles of individual components relative to the overall antiviral efficacy of essential oils, together with the antiviral activity of essential oils in comparison with commercial drugs are discussed [5, 9] (Table 6.1 and Table 6.2).

6.2.1 SARS-CoV-2

Anuj Kumar et al. (2020) [14] reported metabolites, namely ursolic acid, carvacrol and oleanolic acid as the potential inhibitors against main protease (M^{pro}) of COVID-19. It was found that these three ligands bind to protease [14]. Furthermore, these molecules possessed stable and favorable energies causing strong binding with the binding site of M^{pro} protein. The three molecules have passed the ADME (Absorption, Distribution, Metabolism, and Excretion) tests as well as Lipinski's rule of five. The study provides a basic foundation and suggests that the three phytochemicals could serve as potential inhibitors in regulating the M^{pro} protein's function and controling viral replication. Using molecular docking, Thuy et al. (2020) [15] report the inhibitory effect of garlic oil components on the human receptor, angiotensin-converting enzyme 2 (ACE2) protein. Seventeen organic sulfur compounds of the garlic essential oil exhibited interactions with the amino acids of the ACE2 protein and the main protease of SARS-CoV-2. Strong anti-coronavirus activity was shown by allyl disulfide and allyl trisulfide, which account for 51.3% in the garlic essential oil.

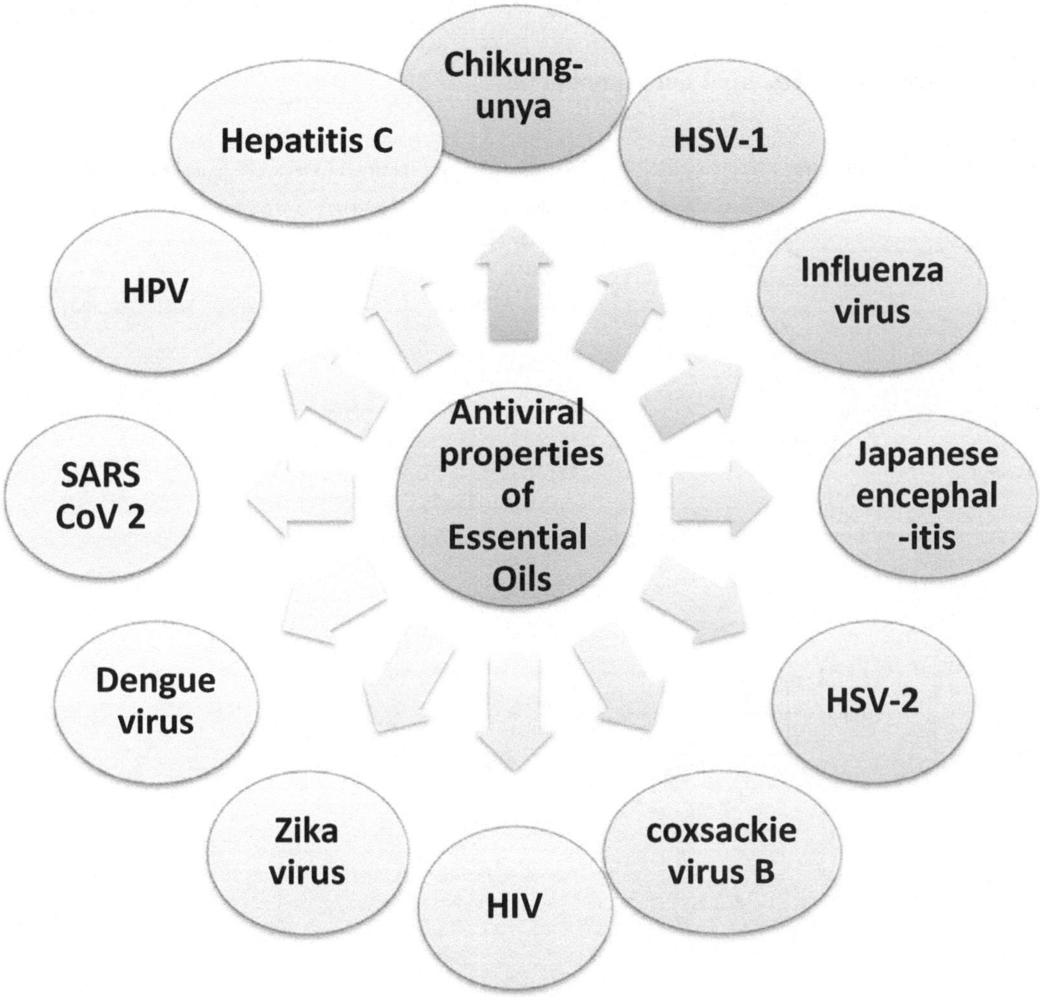

FIGURE 6.1 Antiviral activity of essential oil components.

These results suggest that the garlic essential oil may be a valuable source with antiviral compounds [1]. Domingues et al. (2022) developed the Essential Oils-Loaded Electrospun Fibrous Mat Mask to protect against Escherichia Virus MS2, Surrogate of SARS-CoV-2 [16].

The essential oil sage, *Salvia officinalis*, is reported to be active against SARS CoV-2 at IC 50, 870 mg/ml [17]. Based on *in silico* docking studies on main protesase of SARS-CoV-2, Bouchentouf and Noureddine report that Nigelledine and alpha Hedderin from *Nigella Sativa* as new Potential Inhibitors SARS-CoV-2 [18]. These molecules exhibited a better energy score than that exhibited by Hydroxychloroquine and Flavipiravir. Sharma and Kaur (2020) reported inhibition of SARS CoV2, main protease Mpro through a molecular docking study. They suggest that Eucalytol may represent a potential treatment option [19].

Da Silva *et al* (2020) (5) carried out a molecular docking analysis using 171 essential oil components with SARS-CoV-2 main protease (SARS-CoV-2 M^{pro}), SARS-CoV-2 endoribonuclease (SARS-CoV-2 Nsp15/NendoU), SARS-CoV-2 ADP-ribose-1″-phosphatase (SARS-CoV-2 ADRP), SARS-CoV-2 RNA-dependent RNA polymerase (SARS-CoV-2 RDRP), the binding domain of the SARS-CoV-2 spike protein (SARS-CoV-2 rS), and human angiotensin–converting enzyme

TABLE 6.1
Antiviral Activities of Essential Oil Components

Essential Oil Components	Struture	Assay	Name of Virus	IC_{50} (μg/mL)	Reference
(*E*)–Anethole	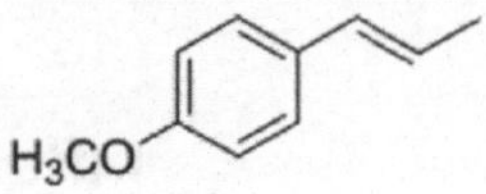	Plaque reduction assay (RC-37 cells),	HSV-1	20	[54]
Camphor	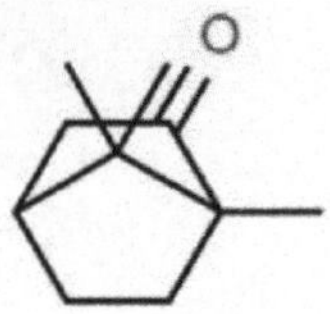	Plaque reduction assay (Vero cells),	HSV-1	2600	[55]
Carvacrol	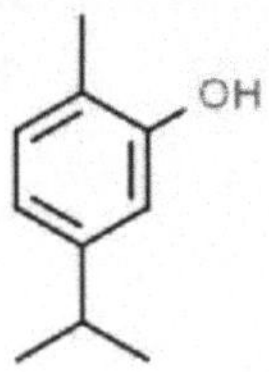	MTT assay (Mardin–Darby bovine kidney (MDBK) cells,	HSV-1	48.6	[56]
β-Caryophyllene	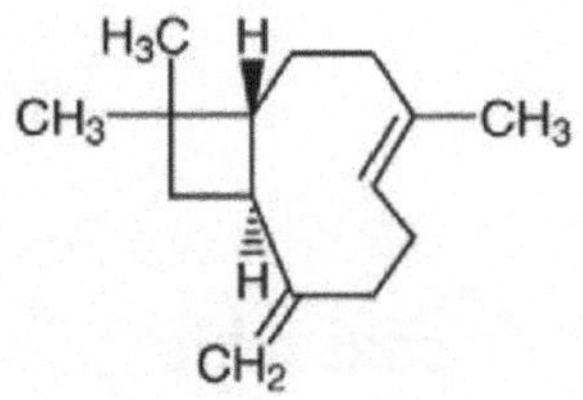	Plaque reduction assay (RC-37 cells),	HSV-1	0.25	[56]
Caryophyllene oxide	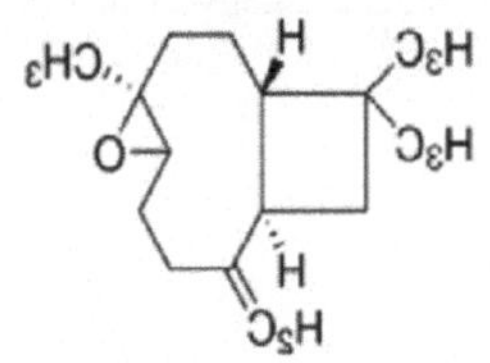	Plaque reduction assay (RC-37 cells),	HSV-1	0.7	[56]
1,8-Cineole	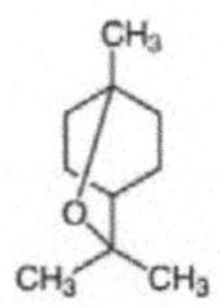	Plaque reduction assay (Vero cells),	HSV-1	1800	[55]
1,8-Cineole	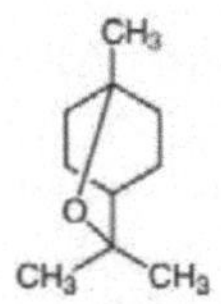	Plaque reduction assay (RC-37 cells),	HSV-1	1200	[54]
p-Cymene	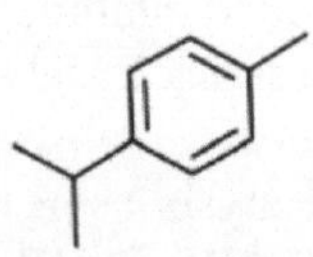	*In vitro* antiviral assay	Influenza, HSV-1, HSV-2, ECHO 9, Cox B1, Polio 1, Adeno 6	>500	[57]

TABLE 6.1 (Continued)
Antiviral Activities of Essential Oil Components

Essential Oil Components	Struture	Assay	Name of Virus	IC_{50} (μg/mL)	Reference
p-Cymene	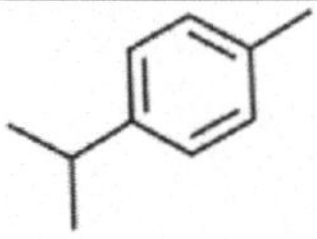	Plaque reduction assay (RC-37 cells),	HSV-1	16	[57]
α-Pinene	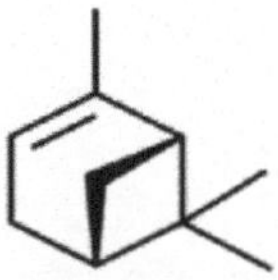	Plaque reduction assay (RC-37 cells),	HSV-1	4.5	[54]
β-Eudesmol	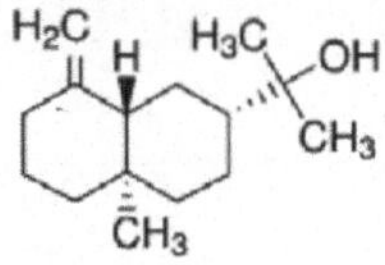	Plaque reduction assay (Vero cells),	HSV-1	6	[54]
Eugenol	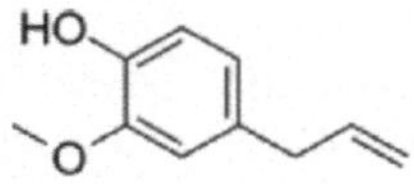	Plaque reduction assay (RC-37 cells),	HSV-1	35	[54]
Eugenol	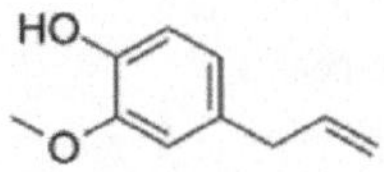	Plaque reduction assay (RC-37 cells),	HSV-1	25.6	[58]
Farnesol	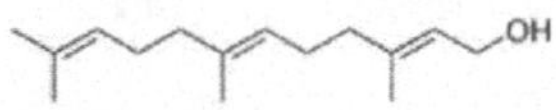	Plaque reduction assay (RC-37 cells),	HSV-1	3.5	[59]
Citral (Geranial + Neral)	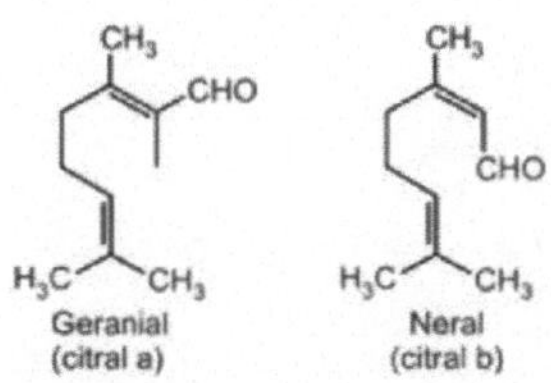	Plaque reduction assay (RC-37 cells),	HSV-1	3.50	[60]
Isoborneol	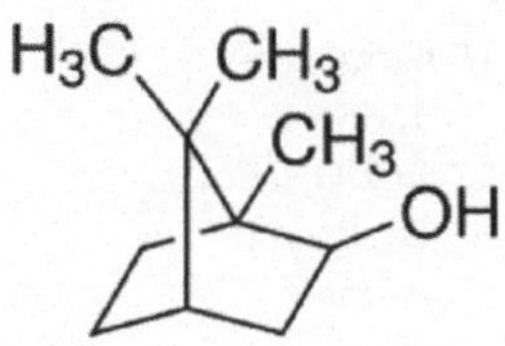	Plaque reduction assay (Vero cells),	HSV-1	<1000	[61]
Thymol		Plaque reduction assay (RC-37 cells),	HSV-1	30	**[54]**
Terpinen-4-ol	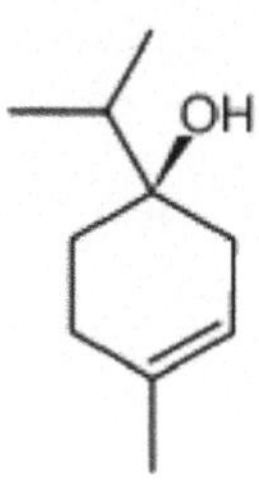	,	Influenza, HSV-1, HSV-2, ECHO 9, Cox B1, Polio 1, Adeno 2	25 (influenza) >50 (others)	[57]

(continued)

TABLE 6.1 (Continued)
Antiviral Activities of Essential Oil Components

Essential Oil Components	Struture	Assay	Name of Virus	IC_{50} (μg/mL)	Reference
Terpinen-4-ol	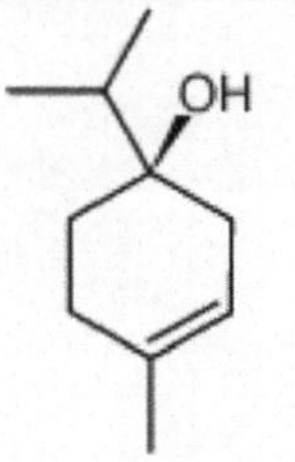	Plaque reduction assay (Vero cells),	HSV-1	60	**[54]**
α-Terpineol	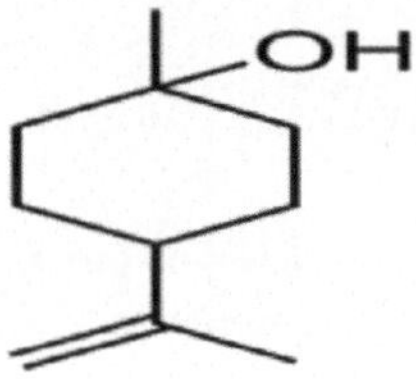	Plaque reduction assay (RC-37 cells),	HSV-1	22	**[54]**
Terpinolene	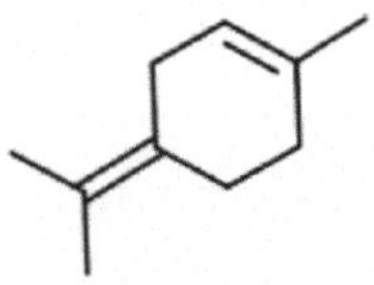	*In vitro* antiviral assay	Influenza, HSV-1, HSV-2, ECHO 9, Cox B1, Polio 1, Adeno 3	12 (influenza) >12 (others)	[57]
Thujones (α & β)	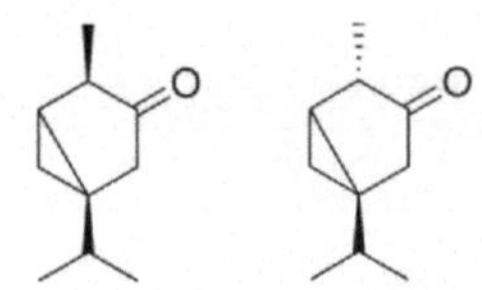 α-thujone β-thujone	Plaque reduction assay (RC-37 cells),	HSV-1	400	**[55]**
γ-Terpinene	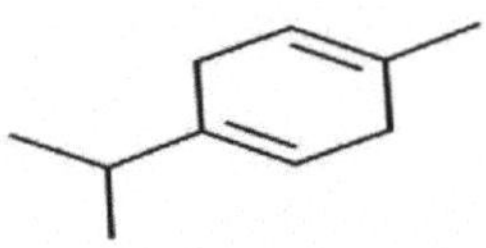	Plaque reduction assay (RC-37 cells),	HSV-1	7	**[54]**
γ-Terpinene	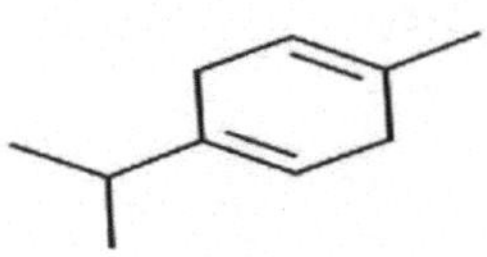	*In vitro* antiviral assay	Influenza, HSV-1, HSV-2, ECHO 9, Cox B1, Polio 1, Adeno 5	>120	[57]
α-Terpinene	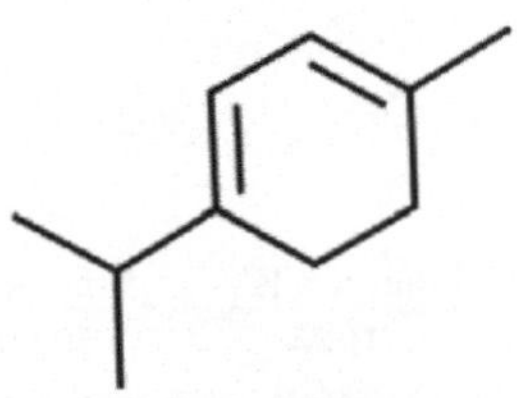	Plaque reduction assay (RC-37 cells),	HSV-1	8.5	[54]
Nerolidol (natural)	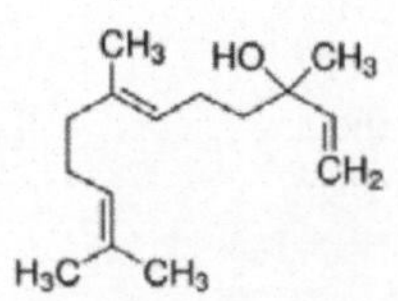	Plaque reduction assay (RC-37 cells),	HSV-1	4.2	**[60]**

TABLE 6.1 (Continued)
Antiviral Activities of Essential Oil Components

Essential Oil Components	Struture	Assay	Name of Virus	IC_{50} (μg/mL)	Reference
Nerolidol (synthetic)	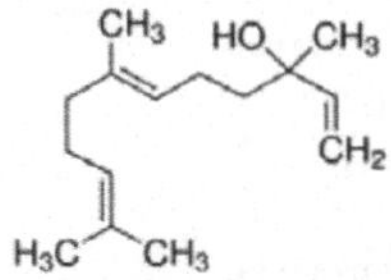	Plaque reduction assay (RC-37 cells),	HSV-1	1.5	[60]
Piperitenone oxide	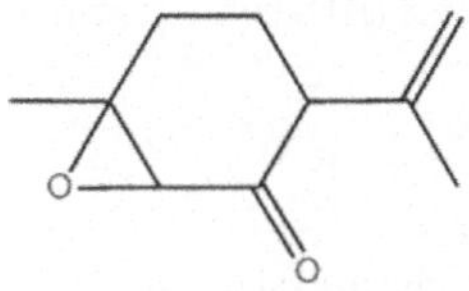	Plaque reduction assay (Vero cells)	HSV-1	1.4	[61]
(*E*)-Cinnamaldehyde	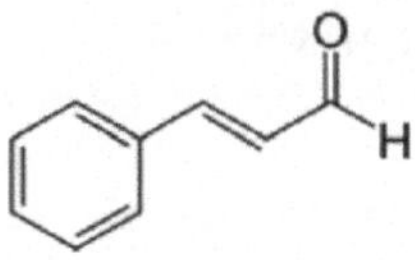	Virus growth assay *in vitro* and In vivo	Influenza type A (H1N1) virus	70% at a concentration of 0.53% after 3 h	[62]
(*E*)-Cinnamaldehyde	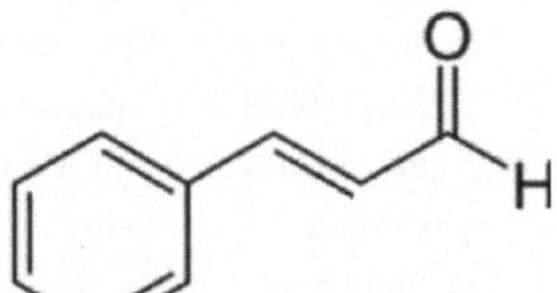	in vivo in a mouse model	H1N1	89% inhibition	[62]
Citral (Geranial + Neral)	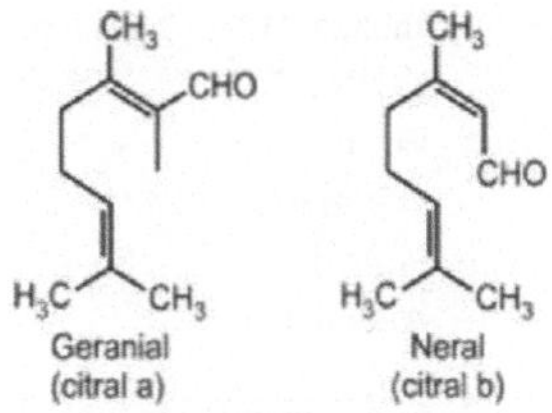	*In vitro* antiviral assay	YFV	17.6	[63]
p-Cymene	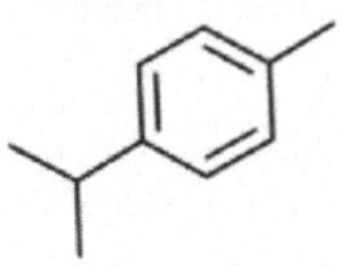	In Vivo assay in nice	H1N1	Inactive	[64]
Germacrone	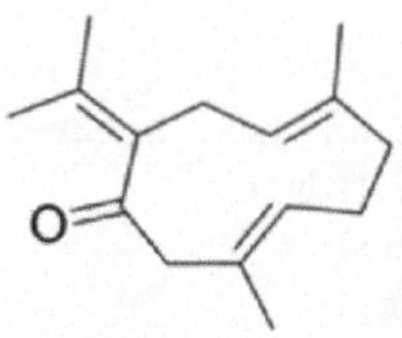	*In vitro* antiviral assay	Influenza type A (H1N1) virus	1.22–1.55	[65]
Germacrone	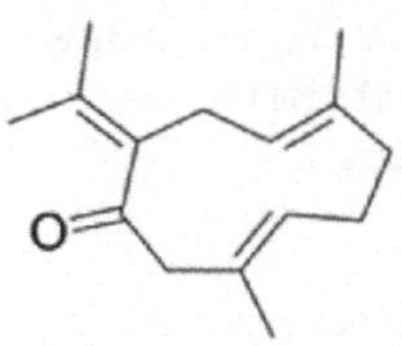	*In vitro* antiviral assay	Influenza type A (H3N2) virus	0.34	[65]
Germacrone	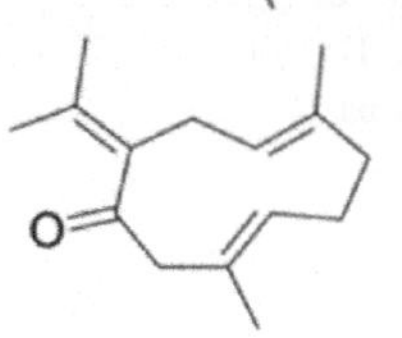	*In vitro* antiviral assay	Influenza type B virus	1.38	[65]

(continued)

TABLE 6.1 (Continued)
Antiviral Activities of Essential Oil Components

Essential Oil Components	Struture	Assay	Name of Virus	IC_{50} (μg/mL)	Reference
Octanal	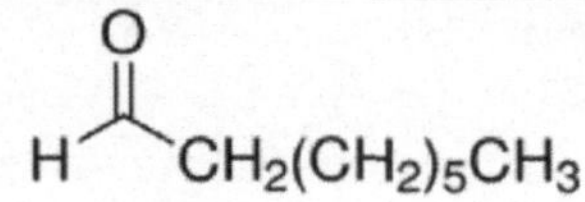	*In vitro* antiviral assay	Influenza type A (H1N1) virus	15	[65]
Patchouli alcohol	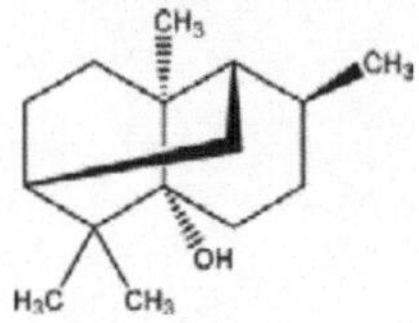	Plaque reduction assay	Influenza type A (H1N1)	89% inhibition at 10 μg/mL	[66]
Patchouli alcohol	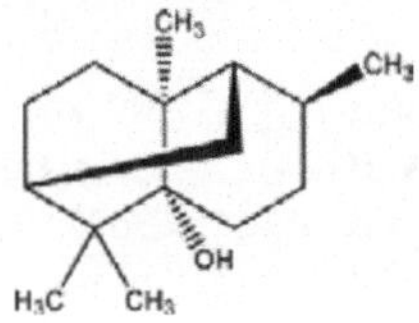	*in vitro, in vivo* and *in silico.*	Influenza type A (H2N2)	0.89	[66]
Patchouli alcohol	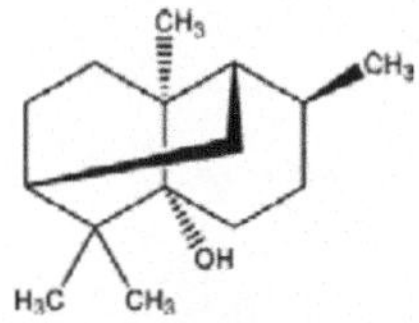	*in vitro, in vivo* and *in silico.*	Influenza type A (H2N2), in vivo test using a mouse model	70% survival rate at a dose of 5 mg/(kg day)	[67]
α-Terpinene	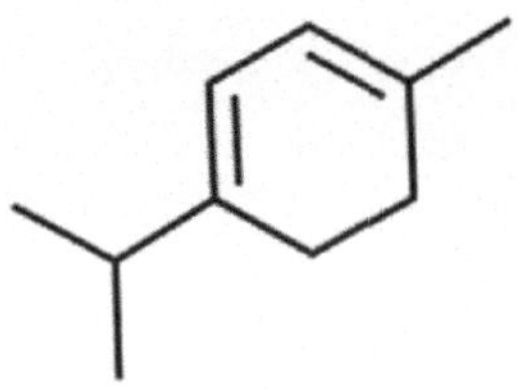	*In vitro* antiviral assay	Influenza, HSV-1, HSV-2, ECHO 9, Cox B1, Polio 1, Adeno 4	>12	[57]
α-Terpinene	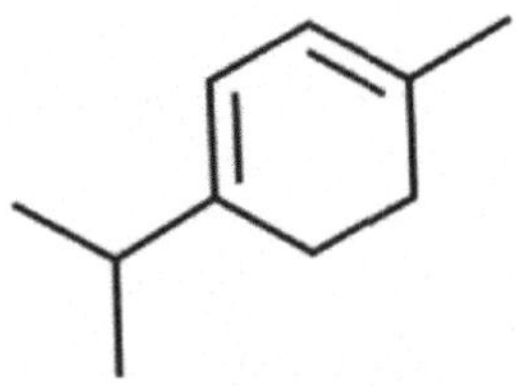	*In vitro* antiviral assay	Influenza type A (H1N1) virus	Inactive	[64]
γ-Terpinene	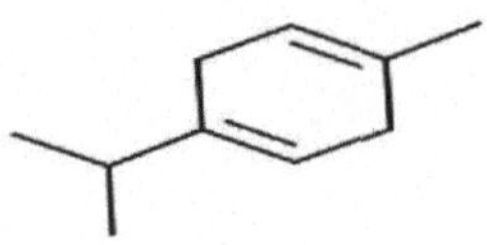	*In vitro* antiviral assay	influenza type A (H1N1) virus	inactive	[66]
Thymol	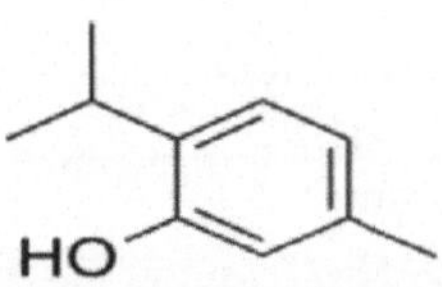	*In vitro* and in vivo assay	Influenza type A (H1N1) virus	Active	[64]
2-Undecanone	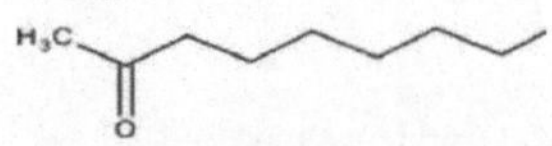	*In vitro* antiviral assay	influenza type A (H1N1) virus	62	[62]

TABLE 6.2
Molecules from Essential Oils Inhibiting Various Viruses [1, 5]

Name of Molecules	Antiviral property
Beta Tujone, linalool, Myrcene, Carvacrol, Beta caryophellene, bicyclogermacrene, alpha Phellandrene, p cymene, germacrene, D., alpha Pinene, estragalol, eugenol, Farnesol, Limonene, Trans-antheol, β-Eudesmol, Trans-anethole, Thymol, p-Cymene, (E)–Anethole Camphor Carvacrol, Caryophyllene oxide,1,8-Cineole,β-Pinene, β-Eugenol Isoborneol, Nerolidol (natural) Nerolidol (synthetic) α-Pinene,Piperitenone oxide, α-Terpinene, γ-Terpinene Terpinen-4-ol Terpinen-4-ol,Terpinolene, Thujones. Inhibitor of Acylovir resistant HSV 1: Carvacrol	Inhibitors of HSV 1
Myrcene, beta caryophellene, citral, Caryophellene oxide, Camphor, citranellal	Inhibitors of Avian influenza virus (AIV subtype H9N2)
Pachulol, alpha gurjunene, alpha guaine,aromadendrene,beta pachoulene. Patchouli alcohol	Inhibitors of Influena virus H2N2
Carvacrol Germacrone, Germacrene, Eugenol, β-Santalol I(E)-Cinnamaldehyde Eugenol Octanal, Patchouli alcohol	Inhibitors of IFV-A (H1N1)
Germacrene, Santalol	Inhibitors of IFV –A (H3N2)
Citral	Inhibitor of Yellow fever virus
B caryophellene	Inhibitor dengue virus:
Thymohydroquione dimethly ether	Inhibitor of Zikka Virus
Thymol and Carvacrol	Inhibitors of SARS-CoV-2 Main protease
Isothymol,Thymol, Limonene, P-cymene, gama terpinene	Inhibitors of ACE2

(hACE2). Sesquiterpene hydrocarbon (*E*)-β-farnesene has been shown to have binding affinity with SARS-CoV-2 M^{pro}. The best binding affinity for ligands includes (*E,E*)-α-farnesene, (*E*)-β-farnesene and (*E,E*)–farnesol with SARS-CoV Nsp15/NendoU. (*E, E*)–Farnesol also binds with SARS-CoV-2 RDRP efficiently [9].

6.2.2 HSV1

Minami et al. (2003) report the efficacy of lemongrass oil against HSV-1 infection. The topical use of essential oils, especially lemongrass, is recommended by these authors for the treatment of recurrent ocular and dermal infection with HSV-1 [20]. Alpha-pinene and alpha -terpineol revealed the highest selectivity index against HSV1 [21]. Peppermint oil exhibited high levels of virucidal activity against HSV-1 and HSV-2. At noncytotoxic concentrations of the oil, plaque formation was significantly reduced by 82% and 92% for HSV-1 and HSV-2, respectively. Peppermint oil reduced titers of herpes viruses by more than 90%. Herpes viruses were significantly inhibited when pretreated with the essential oil prior to adsorption. Peppermint oil affected the virus before adsorption, but not after penetration into the host cell. This oil was active against an acyclovir resistant strain of HSV-1 (HSV-1-ACV^{res}). Plaque formation was significantly reduced by 99%. These authors opined that peppermint oil might be suitable for topical therapeutic use as a virucidal agent in recurrent herpes infection [22]. The envelope of HSV-1 was disrupted when treated with oregano (*O. vulgare*) oil and clove (*S. aromaticum*) oil. Eugenol (4-ydroxyl-3-methoxy-allyl-benzene), the main component of clove oil, was shown to be a very effective against HSV-1 and HSV-2. Eucalyptus oil is reported to suppress HSV 1 virus multiplication by 96% [22].

Mieres-Castro et al. (2021)[23] reported that molecular mechanisms involved in the antiviral activity by the direct binding of monoterpenes with free viruses, particularly with viral proteins involved in the entry and penetration of the host cell, thus avoiding viral infection [23].

Tea tree oil has been shown to inhibit replication of the Herpes simplex virus by 96% [24]. B caryophyllene inhibited the propagation of HSV1 with IC50 and SI values of 0.25 μg/mL and 140 μg/mL, respectively. The plants which contain β caryophyllene include pepper, cannabis, oregano, cloves and cinnamon. A wide range of common components in essential oils have been screened for their anti-herpes activities and all exhibited high antiviral activities at the concentration range of 0.025-0.8 μg/mL [5]. Germacrone was shown to effectively inhibit multiple strains of feline Calici viruses and nonenveloped RNA viruses. These results suggested that Germacrone, a major component from the rhizome curcuma, could be used as a promising broad spectrum therapeutic agent. 1,8-cinole and Eugenol are effective against HSV 1 [12–13]. Several phenyl propanoids and sesquiterpenes including eugenol, trans-anethole, *β*-eudesmol, *β*-caryophyllene, and farnesol, which are present in essential oils, also have antiviral properties against HSV [25].

6.2.3 Influenza

Essential oils from Umbelliferae, Labiatae, Myrtaceae, *Thymus vulgaris*, *Cinnamomum zeylanicum*, *Citrus bergamia* are reported to show high antiviral efficacy [26]. An EO-derived molecule, Germacrone, is among the most effective against IFV which may be due to its broad spectrum of mechanism of action by inhibiting multiple steps of viral replication [27]. β-Santalol was investigated for its activity against influenza A/HK (H_3N_2) virus. β-Santalol exhibited anti-influenza activity of 86% with no cytotoxicity at a concentration of 100 μg/ml. Oseltamivir showed an antiviral activity of 83% against A/HK (H3N2) virus at 100 μg/ml. The presence of 100 μg/ml of β-santalol caused inhibition of late viral RNA synthesis compared with oseltamivir [28].

The *in vitro* and *in vivo* anti-influenza activities of Maqian essential oil and its major ingredient, d/L-limonene, were assessed against Influenza A Virus H1N1 [29]. Virucidal activity was also investigated of D-limonene, the major terpene in essential oils of *C. xaurantium*, and its enantiomer L-limonene [30]. A mixture of the of the essential oils of eucalyptus, *Cinnamom zyleanicum*, *Rosmarinus*, *Daucus carota* and *Camelina satiava* oil significantly reduced H_1N_1 titer after 60 minutes of exposure [31]. Sood et al. (2009) studied crude extracts of leaves and bark of *E. jambolana* for antiviral activity against a highly pathogenic strain of avian influenza virus (H5N1) [32]. The cold and hot aqueous extracts of bark and hot aqueous extract of leaves of *E. jambolana* showed 100% inhibition. Choi et al. (2011) evaluated the effect of quercetin 3-rhamnoside (Q3R) on influenza A/WS/33 virus infected mice [33]. Significant decreases in weight loss and decreased mortality in mice was observed by treatment of Q3R orally (6.25 mg/kg per dose) after influenza virus infection was exhibited. Lung virus titers of mice killed after infection were about 2000 times lower than that of control mice and about two times lower than that for the oseltamivir treated mice. Development and progression of pulmonary lesions were also delayed by the administration of Q3R. This suggests that Q3R could be an attractive lead for the development of agents against influenza virus. Choi et al. (2018) found that 11 of the 62 essential oils (100 μg/mL) tested possessed anti-influenza activity > 30%. Marjoram, clary sage and anise oils exhibited anti A/WS/33 virus activity of > 52.8%. Clary sage and anise essential oils were the most effective. All the three oils contained linalool, suggesting that this may have anti-influenza activity [33].

6.2.4 HIV

Salehi et al. (2018) [34] has reviewed the use of 717 species of plants against human immunodeficiency virus [34]. Of these, several plant species have shown remarkable anti-HIV activity, especially *Artemisia annua*, *Garcinia edulis*, *Justicia gendarussa*, *Phyllanthus pulcher*, *Rhus chinensis*, *Smilax corbularia*, *Terminalia paniculata* and *Tuberaria lignose*. Activities against HIV-RT, HIV-PR, and HIV-IN are reported for 206, 254 and 43 species, respectively. About 390 species of plants are studied for enzyme inhibition activity against HIV [34].

Antiviral activity of ethanol extracts of the *Alpinia eremochlamys* leaves, pseudostems, and rhizomes parts on HIV-infected MT-4 cells is reported [35].

The essential oil of *Cymbopogon nardus* is found to inhibit HIV1 reverse transcriptase with a IC50 value of 1.2 mg/mL. *Thymus vulgaris*, *Cymbopogon citratus* and *Rosmarinus officinali* essential oils effectively inhibited HIV1 Tat/TAR RNA interaction with an IC50 ranging from 0.05 to 0.83 μg/mL [36].

Bicchi et al. (2009) reported the HIV 1 inhibitory properties of the essential oils of *Ridolfia segetum* and *Oenanthe crovata* [37]. Mediouni et al. (2020) found that the most abundant components of oregano oil, carvacrol and its isomer, thymol, are shown to block virus-target cell fusion while not perturbing other stages of the virus life cycle. The changes in virus particle density were observed since cholesterol depletion from the HIV-1 envelope membrane reduces virus entry. Furthermore, infection was rescued by adding exogenous cholesterol. Key motifs of carvacrol and thymol neutralized HIV and identified previously unknown active analogs. Oregano oil components carvacrol and thymol analogs could be considered to supplement current HIV therapeutics [38].

6.2.5 Dengue

Castro et al. (2015) reported that all DENV proteins are potential targets for docking by essential oil components which showed significant docking: (-)-α-copaene, (+)-α-copaene, β-bourbonene, (-)-germacrene D, (+)-germacrene D, (-)-spathulenol, (+)-spathulenol, β-caryophillene, caryophillene oxide and (+)-epibicyclosesquiphellandrene. β-caryophyllene showed promising antiviral activity at a concentration of 22.5 ± 5.6 μM [39]. Qadir et al. (2015) reviewed anti-dengue activities of plants from around the globe. Sixty-nine studies from 1997 to 2012 and 31 species from 24 families possessed anti-dengue activities. Ten phytochemicals have been isolated from 11 species, among which are compounds with the potential for development of dengue treatment [40].

6.2.6 Chikungunya

Kothandan and Swaminathan R (2014) [41] studied the *in vitro* cytotoxicity of the extracts of *Hyptis suaveolens* in vero cells for the determination of maximum non-toxic concentration and *in vitro* anti chikungunya activity. *Hyptis suaveolens* aqueous and aqueous ethanolic extracts partially inhibit the virulence of the Asian strain of chikungunya virus. Of all the three plants tested for antiviral activity against two lineages of chikungunya virus, *Hyptis suaveolens* was found to be effective against the Asian strain of chikungunya virus [41].

[6]-gingerol has antiviral activity against CHIKV infection in *in vitro* human hepatocyte HepG2 cells. The antiviral activity mechanism was investigated using direct virucidal and four indirect (pre-, post-, full-, and prevention) treatment assays. [6]-Gingerol showed weak virucidal activity but significant indirect antiviral activity against CHIKV through post- and full treatment with of 0.038 mM and 0.031 mM, respectively, without showing cell cytotoxicity (Hayati et al. 2021) [42].

6.2.7 HPV

Li et al. (2018) [43] investigated the effects of dehydrocostus lactone on the cell growth and apoptosis of recombinant human papilloma virus (HPV)-18 [43]. The HPV-18 genome was transfected into HaCaT cells, which were used for analysis. Dehydrocostus lactone reduced the cell proliferation and induced apoptosis of HPV-18 HaCaT cells. On treatment with dehydrocostus lactone, caspase-3/9 activities and the protein expression of Bcl-2-associated X and p53 in HPV-18 HaCaT cells were significantly increased and cyclin D1 protein expression was suppressed. Dehydrocostus lactone upregulated the expression of phosphatase and tensin homolog and inhibited the phosphatidylinositol 3-kinase (PI3K)/Akt signaling pathway in HPV-18 HaCaT cells. Dehydrocostus lactone may suppress cell growth and induce apoptosis in recombinant HPV-18 HaCaT cells via the PI3K/Akt signaling pathway [43].

A study by Moutinho et al. (2018) [44] addressed the effects of two nutraceuticals on the expression of COX2 and tumor-associated inflammation in human papillomavirus type 16 (HPV16)-transgenic mice. Six-week-old FVB/n mice were administrated with rutin or curcumin. Both the compounds reduced leukocytic infiltration, but neither prevented epidermal dysplasia. COX2 expression in HPV16-induced lesions may be modulated by nutraceuticals, reducing tumor-associated inflammation [44].

6.2.8 Zika Virus

The medicinal plant *Ayapana triplinervis* essential oil and its component thymohydroquinone dimethyl inhibited propagation of the Zika virus [45]. Assays were performed on human epithelial A549 cells infected with either GFP reporter Zika virus (ZIKV) or viral strain. Acute toxicity of THQ was studied in Zebra Fish. ZIKV infection was inhibited by the essential oil and THQ in human cells with IC_{50} values of 38 and 45 μg/mL, respectively. At the non-cytotoxic concentrations, essential oil and THQ reduced virus propagation by 3-log. This study revealed that THQ could act as a viral entry inhibitor. At the antiviral concentration, THQ injection in zebrafish did not cause any signs of stress and effect on fish survival, demonstrating the absence of acute toxicity for THQ [45].

Lippia alba is a species widely used in folk medicine for various health treatments. Essential oil from *L. alba* and its major compound β-caryophyllene were examined for its antiviral activity against ZIKV *in vitro* and *in silico*. In addition, the larvicidal activity on the *Aedes aegypti* and *Aedes albopictus* larvae of the third instar was investigated. Anti-zika activity of these products was determined at different stages of ZIKV infection and replication cycle [46].

6.2.9 Hepatitis Virus

Parvez et al. (2016) reviewed the efficacy of molecules of plant origins like flavonoids (e.g., Vogonin), terpenes (e.g., Artemisinin), alkaloids (e.g., Oxymatrine), polyphenolics (e.g., geraniin), saponins (e.g., Astragaloside IV) and lignans (e.g., Helioxanthin) against HBV *in vitro* as well as *in vivo*. These compounds possessed different and overlapping mechanisms of action like inhibition of viral antigens or suppression of DNA replication [47].

Chaieb et al. (2007) studied the biological activity of clove, *Eugenia caryophyllata* on Hepatitis C virus. In addition to antiviral activity, clove essential oil possesses anti-inflammatory, cytotoxic, insect repellent and anesthetic properties [48]. Giraud-Robert (2005) conducted a study on patients who were chronic carriers of Hepatitis C. Essential oils such as ravintsara, Labrador tea, carrot seed, thyme, thujanol, laurel, niaouli and helichrysum were used orally either in monotherapy or as a complement to allopathic treatment. In patients with Hepatitis C virus (HCV), treated with essential oils, tolerance and response to treatment was improved. For patients with HCV treated with monotherapy (essential oils), an improvement in hepatitis was noted in 64% of the cases [49]. Nanoemulsion formulations (NE) of *Jasminum humile* and *Jasminum grandiflorum* EO were prepared and examined for their cytotoxic and antiviral activities. NE characterization and stability examination tests were performed to ensure formula stability. The antiviral activity was determined against hepatitis A (HAV) and herpes simplex type-1 (HSV-1) viruses using MTT assay [50].

Quercetin is a natural flavonoid, which has been shown to have anti-HCV properties [51]. The mode of action of quercetin on HCV life cycle are not clearly understood. These workers have studied the effect of quercetin on various steps in the HCV life cycle in Huh-7.5 cells and primary human hepatocytes (PHH) infected with HCV. Quercetin treatment caused considerable decrease in the viral genome replication, production of infectious HCV particles and specific infectivity of the newly produced viral particles by 85% and 92%, Huh7.5 and PHH respectively. When treated directly on HCV particles, quercetin reduced their infectivity by 65%, suggesting that it affects the integrity of the virion. Quercetin inhibited HCV-induced up-regulation of diacylglycerol acyltransferase (DGAT) and the typical localization of the HCV core protein to the surface of lipid droplets, which is known to be mediated by DGAT [51].

6.2.10 Japanese Encephalitis Virus

Roy et al. (2015) [52] extracted Ajwain oil and *in vitro* cytotoxicity assay was performed in vero cell line [52]. *In vitro* antiviral activity of Ajwain oil was quantified by the plaque reduction neutralization test (PRNT). Cytotoxic concentration of the oil was 1 mg/ml. The titer of the virus pool was 50×10^7 PFU/ml. These researchers observed 80% and 40% virus inhibition in 0.5mg/ml of Ajwain oil [52]. The antiviral activity of citronella essential oil against Japanese encephalitis virus is reviewed by Sharma et al. (2019) [53].

6.3 CONCLUSIONS AND FUTURE PROSPECTS

Limited literature is available on the antiviral properties of essential oils and little information on their components. More than 123 essential oils have been shown to possess antiviral properties. Even though the essential oils are reported to possess antiviral properties, because of the variations in the composition, reproducibility of the results is questionable. As such, the use of components of essential oils is advisable. Around 45 essential oil components (EOC) are reported to exhibit antiviral properties. The molecular mode of action is rarely investigated. Most of the data is available on HSV viruses. Sixteen EOCs are reported to inhibit HSV 1 viruses followed by six molecules inhibiting HSV 2 virus and avian influenza virus (AIV subtype H9N2). Ten EOCs are known to inhibit various types of influenza viruses. Four EOCs are reported to inhibit Japanese encephalitis virus. Very few studies address the potential of EOCs as inhibitors of SARS-CoV-2. A small percentage of the phytochemicals from plants are systematically investigated for their antiviral properties. The COVID-19 pandemic has positively influenced the growth of the market of essential oils globally. Essential oils possess many bioactive properties including antiviral, antibacterial, antifungal, anticancer, immune boosting, antioxidant, anti-depressant and alleviating sleeping disorders properties. With the pandemic, enhanced awareness around the globe and health awareness have become important. Exploration of essential oils as sources of antivirals seems to be a good alternative. A coordinated effort from natural product chemists, organic chemists, phramacologists, clinicians and basic scientists may be necessary to take full advantage of the potential of essential oils. There is necessity for detailed scientific studies on the quality, pharmacological activities, safety and therapeutic benefits of essential oil components.

REFERENCES

[1] https://reportocean.com/industry-verticals/sample-request?report_id=BWCC889 (downloaded 26 May 2022).

[2] Hensel A, Bauer R, Heinrich M, Spiegler V, Kayser O, Hempel G, Kraft K. Challenges at the Time of COVID-19: Opportunities and Innovations in Antivirals from Nature. Planta med. 2020 Jul;86(10):659–64.

[3] Yang J, Zhao L, Li R, Yan Y, Yin J, Dai Q, Guo X, Li W, Li Y, Liu M, Ren X. In vitro and in vivo antiviral activity of Maqian (*Zanthoxylum myriacanthum var. pubescens*) essential oil and its major constituents against strains of influenza virus. Ind Crops Prod. 2022 Mar 1;177:114524.

[4] Barkat MA, Kaushik P, Barkat HA, Khan MI, Ab Hadi H. Phytoconstituents in the management of Covid-19: Demystifying the fact. Drug Res. 2022 Jan 3.

[5] Ma L, Yao L. Antiviral effects of plant-derived essential oils and their components: An updated review. *Molecules*. 2020 Jan; 25(11):2627.

[6] Wińska K, Mączka W, Łyczko J, et al. Essential oils as antimicrobial agents—Myth or real alternative?. *Molecules*. 2019 Jan;24(11):2130.

[7] Raut JS, Karuppayil SM. A status review on the medicinal properties of essential oils. Ind. Crops Prod. 2014 Dec 1;62:250–64.

[8] Swamy MK, Akhtar MS, Sinniah UR. Antimicrobial properties of plant essential oils against human pathogens and their mode of action: an updated review. Evid.-Based Complementary Altern. Med. 2016.

[9] da Silva JK, Figueiredo PL, Byler KG, Setzer WN. Essential oils as antiviral agents, potential of essential oils to treat SARS-CoV-2 infection: An *in-silico* investigation. Int. J. Mol. Sci. 2020 Jan;21(10):3426.

[10] Sharifi-Rad J, Quispe C, Kumar M, Akram M, Amin M, Iqbal M, Koirala N, Sytar O, Kregiel D, Nicola S, Ertani A. Hyssopus essential oil: An update of its phytochemistry, biological activities, and safety profile. Oxid Med Cellul Long. 2022 Jan 13. Article ID 8442734.

[11] Corrêa AN, Ferreira CD. Essential oil for the control of fungi, bacteria, yeasts and viruses in food: an overview. Crit Rev Food Sci Nutr. 2022 Apr 5:1–5.

[12] Jadhav AK, Karuppayil SM. *Andrographis paniculata* (Burm. F) Wall ex Nees: Antiviral properties. Phytother Res. 2021 Oct;35(10):5365–5373.

[13] Tomas M, Capanoglu E, Bahrami A, Hosseini H, Akbari Alavijeh S, Shaddel R, Rehman A, Rezaei A, Rashidinejad A, Garavand F, Goudarzi M. The direct and indirect effects of bioactive compounds against coronavirus. Food Front. 2022 Mar;3(1):96–123.

[14] Kumar A, Choudhir G, Shukla SK, Sharma M, Tyagi P, Bhushan A, Rathore M. Identification of phytochemical inhibitors against main protease of COVID-19 using molecular modeling approaches. J. Biomol. Struct. Dyn. 2020 May 201–21.

[15] Thuy B, My T, Hai N et al. Investigation into SARS-CoV-2 resistance of compounds in garlic essential oil. *ACS omega*. 2020 Mar 31;*5*(14):8312–8320.

[16] Domingues JM, Teixeira MO, Teixeira MA, Freitas D, Silva SF, Tohidi SD, Fernandes RD, Padrão J, Zille A, Silva C, Antunes JC. Inhibition of Escherichia virus MS2, surrogate of SARS-CoV-2, via essential oils-loaded electrospun fibrous mats: Increasing the multifunctionality of antivirus protection masks. Pharmaceu. 2022 Jan 27;14(2):303.

[17] Loizzo M, Saab A, Tundis R, Statti G et al. Phytochemical analysis and in vitro antiviral activities of the essential oils of seven Lebanon species. *Chem Biodivers*. 2008 Mar;5(3):461–470.

[18] Bouchentouf S, Missoum N. Identification of Compounds from *Nigella sativa* as New Potential Inhibitors of 2019 Novel Coronasvirus (Covid-19): Mol docking Study. Chemvrxiv.org.2020.

[19] Sharma AD, Kaur I. Eucalyptol (1, 8 cineole) from eucalyptus essential Oil a potential inhibitor of COVID 19 corona virus infection by molecular docking studies. Preprints. 2020 Mar 31.

[20] Minami M, Kita M, Nakaya T, Yamamoto T, Kuriyama H, Imanishi J. The inhibitory effect of essential oils on herpes simplex virus type-1 replication *in vitro*. Microbiol. Immunol. 2003 Sep;47(9):681–4.

[21] Astani A, Reichling J, Schnitzler P. Comparative study on the antiviral activity of selected monoterpenes derived from essential oils. Phytother Res. 2010 May;24(5):673–679.

[22] Schuhmacher A, Reichling J, Schnitzler P. Virucidal effect of peppermint oil on the enveloped viruses herpes simplex virus type 1 and type 2 *in vitro*. Phytomed. 2003 Jan 1;10 (6-7):504–510.

[23] Mieres-Castro D, Ahmar S, Shabbir R, Mora-Poblete F. Antiviral activities of eucalyptus essential oils: Their effectiveness as therapeutic targets against human viruses. Pharmaceuticals. 2021 Dec;14(12):1210.

[24] Reichling J, Schnitzler P, Suschke U, Saller R. Essential oils of aromatic plants with antibacterial, antifungal, antiviral, and cytotoxic properties—an overview. Complement. Med. Res. 2009;16(2):79–90.

[25] Brun P, Bernabè G, Filippini R, Piovan A. In vitro antimicrobial activities of commercially available tea tree (*Melaleuca alternifolia*) essential oils. Curr. Microbiol. 2019 Jan 15;76(1):108–116.

[26] Vimalanathan S, Hudson J. Anti-influenza virus activity of essential oils and vapors. J. Essent. Oil Res. 2014;2(1):47–53.

[27] Liao Q, Qian Z, Liu R, An L, Chen X. Germacrone inhibits early stages of influenza virus infection. Antivir. Res. 2013 Dec 1;100(3):578–588.

[28] Paulpandi M, Kannan S, Thangam R, Kaveri K, Gunasekaran P, Rejeeth C. *In vitro* anti-viral effect of β-santalol against influenza viral replication. Phytomed. 2012, 19, 231–235.

[29] Yang J, Zhao L, Li R, Yan Y, Yin J, Dai Q, Guo X, Li W, Li Y, Liu M, Ren X. In vitro and in vivo antiviral activity of Maqian (Zanthoxylum myriacanthum var. pubescens) essential oil and its major constituents against strains of influenza virus. Ind Crop Prod. 2022 Mar 1;177:114524.

[30] Fadilah NQ, Jittmittraphap A, Leaungwutiwong P, Pripdeevech P, Dhanushka D, Mahidol C, Ruchirawat S, Kittakoop P. Virucidal activity of essential oils from citrus x aurantium L. against influenza A virus H 1 N 1: Limonene as a potential household disinfectant against virus. Nat. Prod. Comm. 2022 Jan;17(1).
[31] Brochot A, Guilbot A, Haddioui L, Roques C. Antibacterial, antifungal, and antiviral effects of three essential oil blends. Microbiol. 2017 Aug;6(4):e00459.
[32] Sood R, Swarup D, Bhatia S, Kulkarni DD, Dey S, Saini M, Dubey SC. Antiviral activity of crude extracts of *Eugenia jambolana* Lam. against highly pathogenic avian influenza (H5N1) virus. Indian J. Exp. Biol. 2016.March 50.179–186.
[33] Choi HJ. Chemical constituents of essential oils possessing anti-influenza A/WS/33 virus activity. Osong Public Health Res Perspect. 2018 Dec;9(6):348.
[34] Salehi B, Kumar NV, Şener B, Sharifi-Rad M, Kılıç M, Mahady GB, Vlaisavljevic S, Iriti M, Kobarfard F, Setzer WN, Ayatollahi SA. Medicinal plants used in the treatment of human immunodeficiency virus. Int. J. Mol. Sci.2018 May;19(5):1459.
[35] Zubair MS, Khairunisa SQ, Widodo A, Pitopang R. Antiviral screening on *Alpinia eremochlamys, Etlingera flexuosa*, and *Etlingera acanthoides* extracts against HIV-infected MT-4 cells. Heliyon. 2021 Apr 1;7(4):e06710.
[36] Feriotto G, Marchetti N, Costa V et al. Chemical composition of essential oils from *Thymus vulgaris, Cymbopogon citratus, and Rosmarinus officinalis,* and their effects on the HIV 1 Tat protein function. Chem. Biodivers. 2018 Feb;15(2):e1700436.
[37] Bicchi C, Rubiolo P, Ballero M, Sanna C, Matteodo M, Esposito F, Zinzula L, Tramontano E. HIV-1-inhibiting activity of the essential oil of Ridolfia segetum and Oenanthe crocata. Planta medica. 2009 Oct;75(12):1331–1335.
[38] Mediouni S, Jablonski JA, Tsuda S et al. Oregano oil and its principal component carvacrol inhibit HIV-1 fusion into target cells. J. Virol. 2020 May 27.
[39] Castro NP, Flechas MC, Ocazionez R, Stashenko E, Olivero-Verbel J. Potential interaction of components from essential oils with dengue virus proteins. B Latinoam Caribe Pl. 2015;14(3):141–155.
[40] Qadir MI, Abbas K, Tahir M et al. Dengue fever: Natural management. Pakistan J. Pharm. Sci. 2015 Mar 1;28(2).
[41] Kothandan S, Swaminathan R. Evaluation of *in vitro* antiviral activity of Vitex Negundo L., Hyptis suaveolens (L) poit., *Decalepis hamiltonii* Wight & Arn., to Chikungunya virus. Asian Pac. J. Trop. Dis. 2014 Jan 1;4:S111–1115.
[42] Hayati RF, Better CD, Denis D, Komarudin AG, Bowolaksono A, Yohan B, Sasmono RT. [6]-Gingerol Inhibits Chikungunya Virus Infection by Suppressing Viral Replication. BioMed Res Int. 2021 Mar 27;2021.
[43] Li W, Ma YB, Mao YQ et al. Dehydrocostus lactone suppresses cell growth and induces apoptosis in recombinant human papilloma virus-18 HaCaT cells via the PI3K/Akt signaling pathway. Mol. Med. Rep. 2018 Jun 1;17(6):7925–7930.
[44] Moutinho M, Aragao S, Carmo D et al. Curcumin and Rutin down-regulate cyclooxygenase-2 and reduce tumor-associated inflammation in HPV16-Transgenic Mice. Anticancer Res. 2018 Mar 1;38(3):1461–1466.
[45] Haddad JG, Picard M, Bénard S, Desvignes C et al. *Ayapana triplinervis* essential oil and its main component thymohydroquinone dimethyl ether inhibit Zika virus at doses devoid of toxicity in Zebrafish. Molecules. 2019 Jan;24(19):3447.
[46] Sobrinho AC, de Morais SM, Marinho MM, de Souza NV, Lima DM. Antiviral activity on the Zika virus and larvicidal activity on the Aedes spp. of Lippia alba essential oil and β-caryophyllene. Ind. Crop. Prod. 2021 Apr 1;162:113281.
[47] Parvez M, H Arbab A, S Al-Dosari M, J Al-Rehaily A. Antiviral natural products against chronic hepatitis B: Recent developments. Curr. Pharm. Des. 2016 Jan 1;22(3):286–293.
[48] Chaieb K, Hajlaoui H, Zmantar T et al. The chemical composition and biological activity of clove essential oil, *Eugenia caryophyllata* (*Syzigium aromaticum* L. Myrtaceae): a short review. Phytother Res. 2007 June; 21(6):501–506.
[49] Giraud-Robert A. The role of aromatherapy in the treatment of viral hepatitis. Int. J. Aromather. 2005 Jan 1;15(4):183–192.

[50] Mansour KA, El-Neketi M, Lahloub MF, Elbermawi A. Nanoemulsions of Jasminum humile L. and *Jasminum grandiflorum* L. essential oils: An approach to enhance their cytotoxic and antiviral effects. Molecules. 2022 June 6;27(11):3639.

[51] Rojas Á, Del Campo JA, Clement S et al. Effect of quercetin on hepatitis C virus life cycle: From viral to host targets. Sci. Rep. 2016 Aug 22;6(1):1–9.

[52] Roy S, Chaurvedi P, Chowdhary A. Evaluation of antiviral activity of essential oil of *Trachyspermum Ammi* against Japanese encephalitis virus. *Pharmac Res.* 2015 July;7(3):263.

[53] Sharma R, Rao R, Kumar S, Mahant S, Khatkar S. Therapeutic potential of citronella essential oil: A review. Curr. Drug Disc. Technol. 2019 Dec 1;16(4):330–339.

[54] Astani A, Schnitzler P, Reichling J. Screening for antiviral activities of isolated compounds from essential oils: Evidence-Based Complement. Altern. Med. 2011,2011:253643.

[55] Sivropoulou A, Nikolaou C, Papanikolaou E, Kokkini S, Lanaras T, Arsenakis M. Antimicrobial, cytotoxic, and antiviral activities of *Salvia fructicosa* essential oil. J. Agric. Food Chem. 1997,45:3197–3201.

[56] Pilau MR, Alves SH, Weiblen R, Arenhart S, Cueto AP, Lovato LT. Antiviral activity of the *Lippia graveolens* (Mexican oregano) essential oil and its main compound carvacrol against human and animal viruses. Brazilian J. Microbiol. 2011,42:1616–1624.

[57] Garozzo A, Timpanaro R, Bisignano B, Furneri PM, Bisignano G, Castro A. In vitro antiviral activity of *Melaleuca alternifolia* essential oil. Lett. Appl. Microbiol. 2009,49:806–808.

[58] Benencia F, Courrèges MC. In vitro and in vivo activity of eugenol on human herpesvirus. Phyther. Res. 2000,14:495–500.

[59] Armaka M, Papanikolaou E, Sivropoulou A, Arsenakis M. Antiviral properties of isoborneol, a potent inhibitor of herpes simplex virus type 1. Antiviral Res. 1999,43:79–92.

[60] Ryabchenko B, Tulupova E, Schmidt E, Wlcek K, Buchbauer G, Jirovetz L. Investigation of anticancer and antiviral properties of selected aroma samples. Nat. Prod. Commun. 2008,3:1085–1088.

[61] Civitelli L, Panella S, Marcocci ME, De Petris A, Garzoli S, Pepi F, Vavala E, Ragno R, Nencioni L, Palamara AT et al. In vitro inhibition of herpes simplex virus type 1 replication by *Mentha suaveolens* essential oil and its main component piperitenone oxide. Phytomed. 2014,21: 857–865.

[62] Hayashi K, Imanishi N, Kashiwayama Y, Kawano A, Terasawa K, Shimada Y, Ochiai H. Inhibitor effect of cinnamaldehyde, derived from Cinnamomi cortex, on the growth of influenza A/PR/8 virus *in vitro and in vivo.* Antiviral Res. 2007,74:1–8.

[63] Gómez LA, Stashenko E, Ocazionez RE. Comparative study on in vitro activities of citral, limonene and essential oils from *Lippia citriodora* and L. alba on yellow fever virus. Nat. Prod. Commun. 2013,8:249–252.

[64] Alburn HE, Chester W, Greenspan G. Thymol as an anti-influenza agent. U.S. Patent 3,632,782, 4 January 1972.

[65] Liao Q, Qian Z, Liu R, An L, Chen X. Germacrone inhibits early stages of influenza virus infection. Antiviral Res. 2013,100:578–588.

[66] Kiyohara H, Ichino C, Kawamura Y, Nagai T, Sato N, Yamada H. Patchouli alcohol: In vitro directanti-influenza virus sesquiterpene in *Pogostemon cablin* Benth. J. Nat. Med. 2012,66:55–61.

[67] Wu H, Li B, Wang X, Jin M, Wang G. Inhibitory effect and possible mechanism of action of patchouli alcohol against influenza a (H2N2) virus. Molecules 2011,16:6489–6501.

7 Antimicrobials for Sepsis Management

Where Do We Stand?

Shikha Joon,[*1] *Enty Tjoa,*[2] *Mohit Mishra,*[1] *Akash Tripathi Satsangi,*[3] *Ketan Parmar,*[4] *and Rakesh Bhatnagar*[1]
[1]Institutes for Systems Genetics, Frontiers Science Center for Disease-related Molecular Network, West China Hospital, Sichuan University, Chengdu, Sichuan, China.
[2]Department of Microbiology, School of Medicine and Sciences, Atma Jaya Catholic University of Indonesia, Jakarta, Indonesia.
[3]Department of Biotechnology, School of Engineering and Technology, Sharda University, Knowledge Park III, Greater Noida, Uttar Pradesh, India
[1]Molecular Biology and Genetic Engineering Laboratory, School of Biotechnology, Jawaharlal Nehru University, New Delhi, India.
[4]Department of chemistry, Sir P. T. Sarvajanik College of Science, Veer Narmad South Gujarat University, Surat, India.
[*]**Corresponding author:** (Shikhasriv86@gmail.com)

CONTENTS

DOI: 10.1201/9781003268932-7

7.1 INTRODUCTION

Sepsis is a life-threatening emergency and a major public health issue that influences millions of people across the world and it is one of the top causes of mortality [1]. It is a potentially life-threatening illness characterized by an organ malfunction because of a dysregulated host response to infection [2]. Furthermore, the prevalence rate of sepsis is escalating, which is likely due to population aging, increased comorbidity, and better detection of the syndrome [3]. Despite tremendous progress in our knowledge of sepsis pathogenesis, multiple clinical studies have failed to find novel treatments capable of altering disease progression and outcome [4, 5]. It is vital to recognize sepsis as a medical issue since, in the absence of a specific treatment. Moreover, effective treatments need early infection management and organ support [6]. It is recommended that intravenous broad-spectrum antibiotics should be started immediately as feasible, ideally within an hour post sepsis diagnosis [7]. Various investigations on sepsis and septic shock have revealed that delaying antibiotic administration has adverse implications [8–11].

Broad-spectrum antibiotics, despite their apparent benefits, can pose serious health risks, notably antibiotic-associated potential complications and life-threatening issues linked to antimicrobial resistance [12–14]. Multidrug-resistant (MDR) pathogen infections have expanded substantially around the world, restricting our therapeutic approaches. As a result, improving antimicrobial efficacy while limiting the evolution of resistant strains during therapy is a real concern in sepsis management. In this scenario, antimicrobial stewardship (AS) is a realistic sepsis management approach. It relies on multidisciplinary teamwork, and the adequacy, appropriateness, and optimization of antimicrobial therapy [15]. AS teams, comprise pharmacists, infectious disease specialists, and microbiologists and are now the standard of care for patients diagnosed with sepsis. Pharmacodynamics (PD) and antimicrobial pharmacokinetics (PK) are indeed key factors to consider for better treatment outcomes. The individualized dose should be used instead of prescribed dosing in critically ill individuals. AS also entails the change to or suspension of a drug class, culminating in selective drugs and treatment shortening. This chapter highlights various critical factors that must be taken into consideration before the administration of antimicrobial therapy.

7.2 WHAT DETERMINES THE SUCCESS OF AN ANTIMICROBIAL THERAPY IN SEPSIS MANAGEMENT?

Early diagnosis of sepsis severity and timely administration of anti-sepsis antimicrobials are important elements in the success of antimicrobial therapy in sepsis management. The ability to ascribe organ dysfunction to infection is tricky for clinicians and involves a combination of extensive clinical information and sound clinical judgment. The Sequential Organ Failure Assessment (SOFA) score is a valuable tool for assessing organ dysfunction [2]. In addition, the quick SOFA (q-SOFA) score is useful to screen patients with infection and a substantial risk of death [16]. A retrospective examination of adult sepsis patients to see whether the one-hour target from Emergency Department (ED) triage to the administration of antibiotic treatment could be met [17]. Reportedly, non-specific symptoms at onset, the shift from triage to non-acute regions, and poorer SOFA scores were all individually linked to delayed antibiotic administration [17]. Further, there have been investigations focusing on the factors related to the patient and institutional policies that restrict early antibiotic administration [18–20]. Furthermore, an earlier administration of antibiotics was done in patients, who received out-of-hospital care. These patients were then admitted to ED. In another study, the q-SOFA score at triage proved ineffective in detecting early sepsis [21]. Moreover, no improvement in the timing of the antibiotics regarding patient outcomes (diagnosed with bacterial infections) was observed [21]. Notably, antibiotic administration (even early administration) was done often for patients having a q-SOFA score of equal to or more than 2 in comparison to those having a q-SOFA score of less than 2. Soon, emerging technology will help patients with suspected sepsis get a faster diagnosis and choose the best treatment [21].

Multi-professional collaboration and implementing programs to improve sepsis management are two viable methods for eliminating antibiotic delays. The appropriateness of antimicrobial therapy in sepsis has progressed due to educational and training programs targeted at improving sepsis management [22, 23]. Improved communication among medical workers and the deployment of automated electronic sepsis alert systems are also advantageous [24–26]. Irrespective of the patient's location, hospital administrative structures and alignments must be adjusted to allow the delivery of critical treatment procedures, particularly when patients are scheduled to be relocated to the Intensive Care Unit (ICU) [27]. In this context, the adoption of a "sepsis code" protocol has yielded great results [28]. A diligent pharmacist should include appropriate antibiotic distribution and immediate availability of the most commonly used antibiotics [29]. The accessibility of premixed antimicrobial solutions was investigated to minimize the period between the first antimicrobial drugs being administered in the ED [30].

7.3 IS EMPIRIC ANTIMICROBIAL THERAPY A FEASIBLE OPTION FOR SEPSIS MANAGEMENT?

For improved patient outcomes, it is critical to prescribe an empiric antibiotic treatment that addresses the most likely microorganisms responsible. There is a substantial amount of research on the deleterious repercussions of inadequate empiric therapy in sepsis [10, 15, 27, 31–34]. Thus, it is recommended that a single or a combination of broad-spectrum antibiotics against all potential infections should be used as the first line of defense [7]. Quite dismally, insufficient empiric therapy is still being prescribed [22, 33–36]. Lower survival rates have been linked to the confluence of postponement and paucity of therapy. Reportedly, individuals with greater severity-of-illness (SOI) scores gain the most from effective antibiotic treatment. Further, in nearly 5,000 patients with septic shock, it was found that inadequate empiric antimicrobial therapy was linked to a five-fold decline in their survival (from 55% to roughly 11%) [11]. In a prospective observational study, the effect of delayed administration of antibiotics and their inefficiency on the survival of 342 patients was assessed (at the stage of septic shock) [32]. Conclusively, the patient survival was negatively affected by delayed administration of antibiotics as well as their inefficiency. This was regardless of patient characteristics or sepsis severity [32]. Another single-center retrospective cohort study was undertaken to determine adequate antimicrobial therapy as an outcome predictor in patients (2,594 patients) with severe sepsis and septic shock [34]. As reported, adequate antimicrobial therapy was determined to be an important outcome predictor in patients with severe sepsis and septic shock. In conclusion, the authors suggested that enhanced empiric antimicrobial targeting for MDR bacteria, *candida* species, MRSA (methicillin-resistant *Staphylococcus aureus*), and *Pseudomonas aeruginosa* might strongly influence the diminished mortality in these patients (that arise from inappropriate antimicrobial treatment) [34].

When choosing an appropriate empiric antimicrobial therapy, make sure to include all possible microorganisms that are causing the infection. When selecting an initial antimicrobial therapy, consider the factors associated with both the patient and the pathogen [15]. Age, weight, allergies, and other comorbidities are all issues to consider when treating a patient. Immunosuppressive therapy, chronic organ failure (such as hepatic or renal dysfunction), and antimicrobial therapy administered recently must all be taken into consideration. In addition, it is important to evaluate the risk associated with the MDR strains. These include exposure to the antibiotics, duration of stay in the hospital, hospitalization of the patient in recent times, use of invasive therapy; whether the patient was exposed to MDR strains previously, resistance trends, and so on [15].

Various methods to improve empiric antibiotic therapy in sepsis patients have been proposed. These include but are not limited to infectious diseases consultation (IDC), automated antibiotic alerts (AAA), local antibiotic prescription guidelines (LAPG), and computerized clinical decision support systems (CCDSSs) [25, 26, 37]. Antibiotic empirical regimens that are prescribed must be established. Further, these antibiotic empirical regimens need to be upgraded for catering to individual institutional healthcare protocols [26].

7.4 HOW IMPORTANT IS A COMBINATION ANTIMICROBIAL THERAPY FOR SEPSIS MANAGEMENT?

For a sustainable antibiotic stewardship strategy, customized empiric therapy is crucial. Minimizing antimicrobial resistance and drug-associated detrimental effects are the main objectives of the de-escalation programs. In addition, it is recommended that when the causative organisms and their susceptibility to an antibiotic are established, or when a positive clinical trend is evident, the spectrum of antibiotics should be shortened [7].

De-escalation therapy has been demonstrated to be a safe approach with reduced mortality rates in sepsis patients [38, 39, 40]. Reportedly, there was no evidence that the de-escalation had any prognostic value in an observational study [41]. This study suggested that in cancer patients with severe sepsis (neutropenic), the de-escalation of empirical antibiotic therapy is often used in the ICU. Both the effectiveness of empiric antibiotic treatment and adherence to clinical recommendations were found to be independently related to the probability of de-escalation in this study [41]. Another study (randomized controlled trial) compared the effects of de-escalation over a continued empirical antimicrobial treatment in patients diagnosed with severe sepsis [42]. An approach based on antibiotics led to a prolonged duration of stay in the ICU over the group and a continued empirical antimicrobial treatment. The mortality rate, however, was unaffected [42]. Even though antibiotic de-escalation is a safe strategy, its execution has been inconsistent across settings [43]. In children diagnosed with sepsis (severe sepsis and septic shock), a multifaceted educational intervention reduced the time it took to administer antibiotics and even improved empirical antibiotic treatment in adult sepsis patients as reported in independent ABISS Edusepsis studies [22, 44]. These studies strengthen the notion that educational interventions can escalate the sepsis treatment approaches.

7.5 WHAT ARE THE STANDARD PRACTICES AND GUIDELINES FOR SEPSIS SAMPLE HANDLING AND DIAGNOSIS?

The proper collection of cultures from blood and the infection site is critical. The gold standard for detecting bacteremia is still blood cultures. Further, before commencing antimicrobial therapy, adequate cultures should be collected [7]. Optimization of antibiotic treatment is aided by the identification of the pathogen. This is followed by culture sterilization [7]. Procuring blood cultures can often cause antibiotic therapy to be delayed, thereby compromising outcomes for patients. The sensitivity of blood cultures collected soon after the commencement of antimicrobial therapy was investigated in a prospective study [45]. Therefore, procuring blood cultures before initial antibiotic therapy is recommended by this study [45]. In cases where this practice is not feasible, the individualized decision for prompt antimicrobial administration is advisable [45].

The time from the withdrawal of samples to culture findings may be exhaustive. Furthermore, getting sufficient samples might be challenging, and a large number of patients have negative results. By minimizing the time to directed therapy and swiftly detecting resistant species, technological improvements may boost the effectiveness of antimicrobial therapy. These include mass spectrometry, polymerase chain reaction (PCR), the matrix-assisted laser desorption ionization-time of flight (MALDI-TOF), and next-generation sequencing (NGS). Unfortunately, these techniques are not commonly available, and clinicians must make an effort to translate findings into quick and appropriate interventions. These fast diagnostic tests should be used with prudence and in conjunction with educational and AS initiatives [46].

7.6 FOR HOW LONG TO ADMINISTER THE ANTIMICROBIAL THERAPY TO PATIENTS WITH SEPSIS?

Antimicrobial therapy over a longer period is linked to increasing resistance, an elevated risk of drug-related side effects, and increased costs. There is a shred of accumulating clinical evidence on the safety of short treatments over the longer ones while the treatment outcomes remain unaffected [47–50]. Reportedly, short-duration treatments enhanced the number of days without antibiotics while MDR-related ventilator-associated pneumonia (VAP) infections declined. In addition, there were no adverse effects on the mortality or the treatment outcomes (failure or success rates) [51]. Further, a shorter anti-bacteremia regimen (against Enterobacteriaceae) exhibited treatment outcomes comparable to prolonged therapy. However, a shorter anti-bacteremia regimen conferred protection against the MDR strains of Gram-negative bacteria [52]. Similarly, a shorter anti- meningococcal regimen was a success. In addition, it did not affect the risk of future recurrences of the infection [53]. In another clinical study, a pre-determined duration of antimicrobial therapy for intra-abdominal infections was evaluated against the standard treatment regimen. When a four-day course was compared to an eight-day course, there was no difference in the patient outcomes [48]. In yet another clinical trial (prospective, randomized), an 8-day course was compared to a 15-day course. It was observed that the patients (with recurring lung infections) who received a short 8-day course were less likely to have resistant bacteria over a longer 15-day course. Moreover, the duration of stay in the ICU or mortality rates remained unaffected [47].

However, the length of therapy cannot be precisely determined in diverse situations because antimicrobial prescription modifications depend on an effective source control as found in an observational study [55]. Although starting antibiotic medication can be a challenging issue, discontinuing antibiotic therapy might be even more challenging in certain circumstances. Infected necrotizing pancreatitis or intra-abdominal infections have a difficult clinical course. In some circumstances, effective infection control is not possible promptly, promoting the development of antibiotic resistance. Antibiotic therapy duration decisions must be made on an individual basis. It, therefore, becomes pertinent to take into account factors that are associated with the individual patient (for example, type of clinical condition and its severity and response to the treatment regimen), infection type, and the diagnostic efficiency and type [56].

7.7 WHAT ARE THE POTENTIAL BIOMARKERS FOR SEPSIS MANAGEMENT USING AN ANTIBIOTIC THERAPY?

At present, there are various diagnostic, therapeutic, and prognostic biomarkers employed in sepsis management. Amongst these, the most common studies are the C-reactive protein (CRP) and procalcitonin (PCT). For bacterial infections, PCT is more sensitive than CRP, and its use has been suggested as a way to reduce the overuse of antibiotics [57]. PCT has a half-life of about 24 hours, and successful antibiotic therapy is linked with a steady decline in its plasma levels. Various studies have investigated the role of algorithms based on PCT plasma levels in reducing antibiotic treatment time without compromising clinical results [58]. In critically ill patients, the advantages of PCT-based algorithms have been established. Yet, the PCT-guided approach may not be effective in patients with lower respiratory tract infections [59]. When the levels of PCT are less than 0.5 ng/ml or there is a decline of 80% in the concentration levels of the highest PCT peak, it is advisable to discontinue antibiotic therapy in patients with sepsis or septic shock. This was also observed to be a key determiner of positive patient outcomes [60].

Inflammatory diseases that mimic infection can influence PCT levels, which makes it a non-specific sepsis biomarker. This, in turn, affects treatment decisions for surgical patients. In another clinical trial (on patients diagnosed with intra-abdominal infection and septic shock), a PCT-based algorithm was not clinically useful [61]. In another clinical pediatric study (on-pump heart surgical patients), PCT (cut-off level = 2 ng/mL) was reported to be an efficient diagnostic biomarker against bacteremia [62]. CRP is also an integral component of the AS program. A clinical trial on patients with sepsis demonstrated the significance of CRP at par with PCT in minimizing antibiotic use [63]. According to a meta-analysis, the PCT-guided treatment algorithm had a good effect on reducing and minimizing the antibiotic treatment duration and stay in the ICU stay without affecting mortality rates [64].

At present, there is no substantial evidence strengthening the reliability of PCT in critically ill patients for supporting the antibiotic de-escalation decision post-culture availability. As a result, using PCT levels to minimize the antibiotic spectrum is not a viable option. Infection biomarkers are progressively being used in clinical settings to help sepsis patients in de-escalating antibiotics. Although biomarkers present excellent diagnostic, therapeutic, and prognostic tools for sepsis, clinical decisions must not solely rely upon them.

7.8 HOW TO OPTIMIZE THE ANTIBIOTIC DOSAGE ADMINISTRATION?

The recommendation of an optimal dose and the existence of acceptable tissue concentrations at the target site dictate antibiotic effectiveness. Antibiotic dosage methods that maintain therapeutic levels of antibiotics improve clinical outcomes reduce antibiotic toxicity and resistance, and lower mortality in sepsis patients [65]. Furthermore, an elevation in renal clearance has been linked to a rise in the elimination of the drug by the kidneys [66–68].

Because of these changes, the loading dose of commonly used anti-sepsis antibiotics must be increased, which include colistin, glycopeptides, and aminoglycosides. Post-initial monitoring of plasma levels of antimicrobials, 50% of patients (diagnosed with sepsis and critically ill) in a prospective study needed a rise in the regular dose of beta-lactam antibiotics [69]. Further, it was revealed that the standard initial dose for several beta-lactams was unlikely to achieve the targeted PK/PD on the first day [70]. Generally, a starting dose of 1.5 times the regular dose is recommended [71].

Multi-organ dysfunction caused by sepsis usually follows a typical sequence, with reduced renal and hepatic function. As a result, the antibiotic clearance pathway influences maintenance dosages, and PD must be considered owing to the growing number of MDR infections. Another study revealed that one-fifth of the patients (amongst 248, diagnosed with severe sepsis and given a beta-lactam antibiotic treatment) failed to achieve the targeted PK/PD. In addition, there occurred a difference of 500 folds in antibiotic concentration in blood samples. Far more antibiotic exposure was linked to better clinical results [72].

Prolonged antibiotic infusions, according to several studies, are related to diminished mortality risk over infusions intended for short-term or intermittent-bolus dosing. With longer infusions, beta-lactam antibiotics have shown to be more effective at targeting MDR pathogens [73–75]. Extended infusions are also beneficial for achieving effective PK/PD levels and reducing linezolid and vancomycin toxicity [76]. Yet, when it comes to lowering fatality rates, the evidence is inconclusive [77].

7.9 CONCLUSION

Sepsis is a life-threatening medical condition with alarmingly high mortality and detrimental consequences. In this situation, one of the foundations of therapy to help save lives and life-threatening consequences is reliable and sustainable antibiotic therapy. To minimize the risk of mortality in suspected or confirmed instances of sepsis, doctors must administer a broad-spectrum antibiotic within an hour of diagnosis. In addition, antibiotic use must also be optimized to achieve positive outcomes and to decrease antibiotic-related morbidity and the emergence of antibiotic resistance. In this scenario, AS must be considered a critical component of sepsis management. It entails tailoring antimicrobial therapy to the person. To facilitate empirical antibiotic prescribing, hospitals should establish and update particular empirical regimens for specific locations of infection regularly, combining regional and national microbiology, and patterns of resistance based on practice standards. Antimicrobial sepsis care necessitates culture collection. AS also entails de-escalation and a reduction in treatment time. Individual dosing should substitute the notion of conventional dosing in critically ill patients. Most patients will require elevated loading doses along with an extended or continued beta-lactam perfusion. Patients with sepsis should get the routine treatment that includes ongoing monitoring of illness progression and effective antimicrobial therapy. Recommendations of the clinical guidelines recommendations, regarding ideal treatment duration, algorithms based on PCT, de-escalation, novel diagnostic stewardship approaches, and AS frameworks will all aid in improving patient care quality.

REFERENCES

1. Prescott HC, Angus DC. Enhancing recovery from sepsis: A review. JAMA 2018;319(1):62–75.
2. Singer M, Deutschman CS, Seymour CW et al. The Third International Consensus Definitions for Sepsis and Septic Shock (Sepsis-3). JAMA. 2016;315(8):801–10.
3. Fleischmann C, Scherag A, Adhikari NK et al. International Forum of Acute Care Trialists. Assessment of Global Incidence and Mortality of Hospital-treated Sepsis. Current Estimates and Limitations. Am J Respir Crit Care Med 2016;193(3):259–72.
4. Sprung CL, Annane D, Keh D et al. CORTICUS Study Group. Hydrocortisone therapy for patients with septic shock. N Engl J Med 2008;358(2):111–24.
5. Ranieri VM, Thompson BT, Barie PS et al. PROWESS-SHOCK Study Group. Drotrecogin alfa (activated) in adults with septic shock. N Engl J Med 2012;366(22):2055–64.
6. Plata-Menchaca EP, Ferrer R. Life-support tools for improving performance of the Surviving Sepsis Campaign Hour-1 bundle. Med Intensiva (Engl Ed) 2018;42(9):547–550. English, Spanish.
7. Rhodes A, Evans LE, Alhazzani W et al. Surviving Sepsis Campaign: International Guidelines for Management of Sepsis and Septic Shock: 2016. Intensive Care Med 2017;43(3):304–377.
8. Ferrer R, Martin-Loeches I, Phillips G et al. Empiric antibiotic treatment reduces mortality in severe sepsis and septic shock from the first hour: results from a guideline-based performance improvement program. Crit Care Med 2014;42(8):1749–55.
9. Liu VX, Fielding-Singh V, Greene JD et al. The timing of early antibiotics and hospital mortality in sepsis. Am J Respir Crit Care Med 2017;196(7):856–63.
10. Kumar A, Roberts D, Wood KE et al. Duration of hypotension before initiation of effective antimicrobial therapy is the critical determinant of survival in human septic shock. Crit Care Med 2006;34(6):1589–96.
11. Kumar A, Ellis P, Arabi Y et al. Cooperative Antimicrobial Therapy of Septic Shock Database Research Group. Initiation of inappropriate antimicrobial therapy results in a fivefold reduction of survival in human septic shock. Chest 2009;136(5):1237–48.
12. Bernhard M, Lichtenstern C, Eckmann C et al. The early antibiotic therapy in septic patients—milestone or sticking point? Crit Care 2014;18(6):671.
13. Mi MY, Klompas M, Evans L. Early administration of antibiotics for suspected sepsis. N Engl J Med 2019;380(6):593–6.
14. Beumier M, Casu GS, Hites M et al. Elevated β-lactam concentrations associated with neurological deterioration in ICU septic patients. Minerva Anestesiol. 2015;81(5):497–506.
15. Ulldemolins M, Nuvials X, Palomar M et al. Appropriateness is critical. Crit Care Clin 2011;27(1):35–51.
16. Seymour CW, Liu VX, Iwashyna TJ et al. Assessment of clinical criteria for sepsis: For the Third International Consensus Definitions for Sepsis and Septic Shock (Sepsis-3). JAMA 2016;315(8):762–74.
17. Filbin MR, Thorsen JE, Zachary TM et al. Antibiotic delays and feasibility of a 1-hour-from-triage antibiotic requirement: Analysis of an emergency department sepsis quality improvement database. Ann Emerg Med 2020;75(1):93–99.
18. Amaral AC, Fowler RA, Pinto R et al. Cooperative Antimicrobial Therapy of Septic Shock Database Research Group. Patient and organizational factors associated with delays in antimicrobial therapy for septic shock. Crit Care Med 2016;44(12):2145–53.
19. Peltan ID, Mitchell KH, Rudd KE et al. Prehospital care and emergency department door-to-antibiotic time in sepsis. Ann Am Thorac Soc 2018;15(12):1443–50.

20. Vattanavanit V, Buppodom T, Khwannimit B. Timing of antibiotic administration and lactate measurement in septic shock patients: A comparison between hospital wards and the emergency department. Infect Drug Resist 2018;11:125–32.
21. Petit J, Passerieux J, Maître O et al. qSOFAST study group. Impact of a qSOFA-based triage procedure on antibiotic timing in ED patients with sepsis: A prospective interventional study. Am J Emerg Med 2020;38(3):477–84.
22. Ferrer R, Martínez ML, Gomà G et al. ABISS-Edusepsis Study group. Improved empirical antibiotic treatment of sepsis after an educational intervention: the ABISS-Edusepsis study. Crit Care 2018;22(1):167.
23. Alam N, Oskam E, Stassen PM et al. PHANTASi Trial Investigators and the ORCA (Onderzoeks Consortium Acute Geneeskunde) Research Consortium the Netherlands. Prehospital antibiotics in the ambulance for sepsis: a multicentre, open label, randomised trial. Lancet Respir Med 2018;6(1):40–50.
24. Kim M, Song KH, Kim CJ et al. Electronic alerts with automated consultations promote appropriate antimicrobial prescriptions. PLoS One 2016;11(8):e0160551.
25. Micek ST, Heard KM, Gowan M et al. Identifying critically ill patients at risk for inappropriate antibiotic therapy: a pilot study of a point-of-care decision support alert. Crit Care Med 2014;42(8):1832–8.
26. Barlam TF, Cosgrove SE, Abbo LM et al. Executive Summary: Implementing an antibiotic stewardship program: Guidelines by the Infectious Diseases Society of America and the Society for Healthcare Epidemiology of America. Clin Infect Dis 2016;62(10):1197–202.
27. Garnacho-Montero J, Gutiérrez-Pizarraya A, Escoresca-Ortega A et al. Adequate antibiotic therapy prior to ICU admission in patients with severe sepsis and septic shock reduces hospital mortality. Crit Care 2015;19(1):302.
28. Walsh D, Gekle R, Bramante R et al. Emergency department sepsis huddles: Achieving excellence for sepsis benchmarks in New York State. Am J Emerg Med 2020;38(2):222–4.
29. Hitti EA, Lewin JJ 3rd, Lopez J et al. Improving door-to-antibiotic time in severely septic emergency department patients. J Emerg Med 2012;42(4):462–9.
30. Kufel WD, Seabury RW, Meola GM et al. Impact of premix antimicrobial preparation and time to administration in septic patients. CJEM 2018;20(4):565–71.
31. Vallés J, Rello J, Ochagavía A et al. Community-acquired bloodstream infection in critically ill adult patients: Impact of shock and inappropriate antibiotic therapy on survival. Chest 2003;123(5):1615–24.
32. Suberviola Cañas B, Jáuregui R, Ballesteros MÁ et al. Effects of antibiotic administration delay and inadequacy upon the survival of septic shock patients. Med Intensiva 2015;39(8):459–66. English, Spanish.
33. Zilberberg MD, Shorr AF, Micek ST et al. Multi-drug resistance, inappropriate initial antibiotic therapy and mortality in Gram-negative severe sepsis and septic shock: A retrospective cohort study. Crit Care 2014;18(6):596.
34. Vazquez-Guillamet C, Scolari M, Zilberberg MD et al. Using the number needed to treat to assess appropriate antimicrobial therapy as a determinant of outcome in severe sepsis and septic shock. Crit Care Med 2014;42(11):2342–9.
35. Garnacho-Montero J, Garcia-Garmendia JL, Barrero-Almodovar A et al. Impact of adequate empirical antibiotic therapy on the outcome of patients admitted to the intensive care unit with sepsis. Crit Care Med 2003;31(12):2742–51.
36. Battleman DS, Callahan M, Thaler HT. Rapid antibiotic delivery and appropriate antibiotic selection reduce length of hospital stay of patients with community-acquired pneumonia: link between quality of care and resource utilization. Arch Intern Med 2002;162(6):682–8.
37. Pogue JM, Mynatt RP, Marchaim D et al. Automated alerts coupled with antimicrobial stewardship intervention lead to decreases in length of stay in patients with gram-negative bacteremia. Infect Control Hosp Epidemiol 2014;35(2):132–8.
38. Turza KC, Politano AD, Rosenberger LH et al. De-escalation of antibiotics does not increase mortality in critically ill surgical patients. Surg Infect (Larchmt) 2016;17(1):48–52.
39. Garnacho-Montero J, Gutiérrez-Pizarraya A, Escoresca-Ortega A et al. De-escalation of empirical therapy is associated with lower mortality in patients with severe sepsis and septic shock. Intensive Care Med 2014;40(1):32–40.
40. Gutiérrez-Pizarraya A, Leone M, Garnacho-Montero J et al. Collaborative approach of individual participant data of prospective studies of de-escalation in non-immunosuppressed critically ill patients with sepsis. Expert Rev Clin Pharmacol 2017;10(4):457–65.

41. Mokart D, Slehofer G, Lambert J et al. De-escalation of antimicrobial treatment in neutropenic patients with severe sepsis: results from an observational study. Intensive Care Med 2014;40(1):41–9.
42. Leone M, Bechis C, Baumstarck K et al. AZUREA Network Investigators. De-escalation versus continuation of empirical antimicrobial treatment in severe sepsis: a multicenter non-blinded randomized noninferiority trial. Intensive Care Med 2014;40(10):1399–408.
43. van den Bosch CM, Hulscher ME, Natsch S et al. Dutch Sepsis QI expert panel. Development of quality indicators for antimicrobial treatment in adults with sepsis. BMC Infect Dis 2014;14:345.
44. Esteban E, Belda S, García-Soler P et al. A multifaceted educational intervention shortened time to antibiotic administration in children with severe sepsis and septic shock: ABISS Edusepsis pediatric study. Intensive Care Med 2017;43(12):1916–8.
45. Cheng MP, Stenstrom R, Paquette K et al. FABLED Investigators. Blood culture results before and after antimicrobial administration in patients with severe manifestations of sepsis: A diagnostic study. Ann Intern Med 2019;171(8):547–54.
46. Richter DC, Heininger A, Brenner T et al. Bacterial sepsis: Diagnostics and calculated antibiotic therapy. Anaesthesist 2019;68(Suppl 1):40–62. English.
47. Chastre J, Wolff M, Fagon JY et al. PneumA trial group. Comparison of 8 vs 15 days of antibiotic therapy for ventilator-associated pneumonia in adults: A randomized trial. JAMA 2003;290(19):2588–98.
48. Sawyer RG, Claridge JA, Nathens AB et al. STOP-IT trial investigators. Trial of short-course antimicrobial therapy for intraabdominal infection. N Engl J Med 2015;372(21):1996–2005.
49. Hepburn MJ, Dooley DP, Skidmore PJ et al. Comparison of short-course (5 days) and standard (10 days) treatment for uncomplicated cellulitis. Arch Intern Med 2004;164(15):1669–74.
50. el Moussaoui R, de Borgie CA, van den Broek P et al. Effectiveness of discontinuing antibiotic treatment after three days versus eight days in mild to moderate-severe community acquired pneumonia: randomised, double blind study. BMJ 2006;332(7554):1355.
51. Pugh R, Grant C, Cooke RP et al. Short-course versus prolonged-course antibiotic therapy for hospital-acquired pneumonia in critically ill adults. Cochrane Database Syst Rev 2015;2015(8):CD007577.
52. Chotiprasitsakul D, Han JH, Cosgrove SE et al; Antibacterial resistance leadership group. comparing the outcomes of adults with enterobacteriaceae bacteremia receiving short-course versus prolonged-course antibiotic therapy in a multicenter, propensity score-matched cohort. Clin Infect Dis 2018;66(2):172–7.
53. Cabellos C, Pelegrín I, Benavent E et al. Invasive meningococcal disease: Impact of short course therapy. A DOOR/RADAR study. J Infect 2017;75(5):420–5.
54. Sawyer RG, Claridge JA, Nathens AB et al. STOP-IT trial investigators. Trial of short-course antimicrobial therapy for intraabdominal infection. N Engl J Med 2015;372(21):1996–2005.
55. Martínez ML, Ferrer R, Torrents E et al. Edusepsis Study Group. Impact of source control in patients with severe sepsis and septic shock. Crit Care Med 2017;45(1):11–19.
56. Vincent JL, Bassetti M, François B et al. Advances in antibiotic therapy in the critically ill. Crit Care 2016;20(1):133.
57. Saeed K, González Del Castillo J, Backous C et al. Hot topics on procalcitonin use in clinical practice, can it help antibiotic stewardship? Int J Antimicrob Agents 2019;54(6):686–96.
58. de Jong E, van Oers JA, Beishuizen A et al. Efficacy and safety of procalcitonin guidance in reducing the duration of antibiotic treatment in critically ill patients: a randomised, controlled, open-label trial. Lancet Infect Dis 2016;16(7):819–27.
59. Huang DT, Yealy DM, Filbin MR et al. ProACT investigators. Procalcitonin-guided use of antibiotics for lower respiratory tract infection. N Engl J Med 2018;379(3):236–49.
60. Bouadma L, Luyt CE, Tubach F et al; PRORATA trial group. Use of procalcitonin to reduce patients' exposure to antibiotics in intensive care units (PRORATA trial): A multicentre randomised controlled trial. Lancet 2010;375(9713):463–74.
61. Jung B, Molinari N, Nasri M et al. Procalcitonin biomarker kinetics fails to predict treatment response in perioperative abdominal infection with septic shock. Crit Care 2013;17(5):R255.
62. Garcia IJ, Gargallo MB, Torné EE et al. Procalcitonin: a useful biomarker to discriminate infection after cardiopulmonary bypass in children. Pediatr Crit Care Med 2012;13(4):441–5.
63. Oliveira CF, Botoni FA, Oliveira CR et al. Procalcitonin versus C-reactive protein for guiding antibiotic therapy in sepsis: a randomized trial. Crit Care Med 2013;41(10):2336–43.

64. Iankova I, Thompson-Leduc P, Kirson NY et al. Efficacy and safety of procalcitonin guidance in patients with suspected or confirmed sepsis: A systematic review and meta-analysis. Crit Care Med 2018;46(5):691–8.
65. Roberts JA, Roger C, De Waele JJ. Personalized antibiotic dosing for the critically ill. Intensive Care Med 2019;45(5):715–8.
66. Udy AA, Lipman J, Jarrett P et al. Are standard doses of piperacillin sufficient for critically ill patients with augmented creatinine clearance? Crit Care 2015;19(1):28.
67. Huttner A, Von Dach E, Renzoni A et al. Augmented renal clearance, low β-lactam concentrations and clinical outcomes in the critically ill: An observational prospective cohort study. Int J Antimicrob Agents 2015;45(4):385–92.
68. Carlier M, Carrette S, Roberts JA et al. Meropenem and piperacillin/tazobactam prescribing in critically ill patients: Does augmented renal clearance affect pharmacokinetic/pharmacodynamic target attainment when extended infusions are used? Crit Care 2013;17(3):R84.
69. Roberts JA, Ulldemolins M, Roberts MS et al. Therapeutic drug monitoring of beta-lactams in critically ill patients: Proof of concept. Int J Antimicrob Agents 2010;36(4):332–9.
70. Taccone FS, Laterre PF, Dugernier T et al. Insufficient β-lactam concentrations in the early phase of severe sepsis and septic shock. Crit Care 2010;14(4):R126.
71. De Backer D, Cecconi M, Lipman J et al. Challenges in the management of septic shock: A narrative review. Intensive Care Med 2019;45(4):420–33.
72. Roberts JA, Paul SK, Akova M et al. DALI Study. DALI: Defining antibiotic levels in intensive care unit patients: are current β-lactam antibiotic doses sufficient for critically ill patients? Clin Infect Dis 2014;58(8):1072–83.
73. Abdul-Aziz MH, Lipman J, Akova M. et al. DALI Study Group. Is prolonged infusion of piperacillin/tazobactam and meropenem in critically ill patients associated with improved pharmacokinetic/pharmacodynamic and patient outcomes? An observation from the Defining Antibiotic Levels in Intensive care unit patients (DALI) cohort. J Antimicrob Chemother 2016;71(1):196–207.
74. Roberts JA, Abdul-Aziz MH, Davis JS et al. Continuous versus intermittent β-lactam infusion in severe sepsis. A meta-analysis of individual patient data from randomized trials. Am J Respir Crit Care Med 2016;194(6):681–91.
75. Vardakas KZ, Voulgaris GL, Maliaros A et al. Prolonged versus short-term intravenous infusion of antipseudomonal β-lactams for patients with sepsis: a systematic review and meta-analysis of randomised trials. Lancet Infect Dis 2018;18(1):108–20.
76. Taubert M, Zander J, Frechen S et al. Optimization of linezolid therapy in the critically ill: The effect of adjusted infusion regimens. J Antimicrob Chemother 2017;72(8):2304–10.
77. Dulhunty JM, Roberts JA, Davis JS et al. BLING II Investigators for the ANZICS Clinical Trials Group*. A multicenter randomized trial of continuous versus intermittent β-lactam infusion in severe sepsis. Am J Respir Crit Care Med 2015;192(11):1298–305.

8 Bacteriocins as Biotechnological Tools in Food and Pharmaceuticals

Applications and Future Prospects

Nagella Praveen,[1] *Arpita Jayan,*[1] *Suresh Nithya,*[1] *Udisha Singh,*[1] *BaraiyaDivyeksha Harishchandra,*[2] *Veerappa Lakshmaiah Vasantha,*[1] *and Jameel Mohammed Al-Khayri* [3*]
[1]Department of Life Sciences, CHRIST (Deemed to be University), Hosur Road, Bangalore, India
[2]Institute for Stem Cell Science and Regenerative Medicine (inStem), GKVK Campus, Bangalore, India
[3]Department of Agricultural Biotechnology, College of Agriculture and Food Sciences, King Faisal University, Al-Ahsa, Saudi Arabia
*Corresponding author: jkhayri@kfu.edu.sa; jmkhayri@yahoo.com

CONTENTS

DOI: 10.1201/9781003268932-8

8.1 INTRODUCTION

The World Health Organization (WHO) and FAO have defined probiotics as non-pathogenic living organisms that greatly benefit host cells and have several positive outcomes at the level of gut. The intake of probiotics at an adequate amount confers good health and many times is used for several treatments (Hill et al. 2014; Gibson et al. 2017). Not only the microorganisms as a whole, but the proteins or peptides secreted by these species have tremendous applications in food spoilage, pharmaceuticals, antibiotic development, and much more. Thus, antimicrobial peptides from bacteria have drawn more attention for their wide range of applications.

Antimicrobial proteins (AMPs), also termed as antimicrobial peptides (which are comparatively smaller in size), are produced by major groups of microorganisms. Two types of AMPs are produced by bacteria: a) Synthesized by ribosomes (termed bacteriocins) and b) AMPs which do not have structural gene coding for protein and, thus, are not synthesized by ribosomes (Chikindas et al. 2018). Bacteriocins are multifunctional, ribosomally generated proteinaceous antibacterial agents that are products of members of the phylum Firmicutes and Actinobacteria (Negash and Tsehai 2020; Lajis 2020a). Bacteriocins were discovered to exhibit antibacterial action against a wide range of pathogenic organisms. Bacteriocins which inhibit bacteria belong to the same species are termed narrow-spectrum bacteriocins and if they are found to inhibit bacteria belonging to a range of species they are considered broad-spectrum bacteriocins. Mostly cell membrane-associated molecules become an easy and accessible target for bacteriocin action (Daba et al. 2018; Daba et al. 2017). Interestingly, host cells producing bacteriocins are resistant to their own antimicrobial peptides due to specific immune proteins produced by themselves (Juturu and Wu 2018). Furthermore, structurally it can be linear or globular, and the properties of bacteriocins are determined by sequential arrangement of amino acids (Lajis 2020). There are many beneficial properties of bacteriocins produced by lactic acid bacteria (LAB), it is known for its tolerance to high thermal stress, and stable over a wide range of pH. These antimicrobial peptides are colorless, odorless, and tasteless and resulted in its potential application (Negash and Tsehai 2020; Perez et al. Zendo and Sonomoto 2014a). Structurally, bacteriocins are cationic, amphipathic molecules which possess more amounts of lysine and arginyl residues, and can readily get suspended into aqueous solutions. They can change conformation to helical when exposed to solvent-like trifluoroethanol or in combination with anion in phospholipid membranes (Vivek Kumar 2020). Bacteriocins being proteinaceous are prone to be inactivated by proteolytic enzymes like trypsin and chymotrypsin in the human gut. This minimizes the interaction of bacteriocins with gut microbiota or target strains (Perez, Zendo, and Sonomoto 2014a; Gálvez et al. 2014; Egan et al. 2016).

The time when first bacteriocins were reported in the 1920s from *E. coli*, as colicin V (Negash and Tsehai 2020; Daba and Elkhateeb 2020), has greatly attracted the worldwide scientific community to explore its therapeutic application. The actual nomenclature of these antimicrobial peptides was referred to as "bacteriocins" by Jacob and colleagues (Jacob et al. 1953). Gram-positive and Gram-negative bacteria both produce bacteriocins, but majorly the Gram-positive bacteria LAB is known for its extensive application in many fields. Bacteriocins from LAB are considered GRAS by the American FDA (Zacharof and Lovitt 2012). Bacteriocins are the best choice as commercial food preservation. Nisin A is one of the most studied bacteriocins, and 60 countries across the world have legally approved bacteriocin to commercially use as a food preservative (de Arauz et al.

2009a). Studies confirm food fermenting LAB as a significant producer of antimicrobial molecules like organic acetoin, acids, hydrogen peroxide, diacetyl, antifungal peptides, and bacteriocins. Therefore, LABs are used on an industrial scale for a wider range of bacteriocin production.

With regard to its genomic analysis, functions are determined by genes encoded by plasmid. The genes that code for bacteriocin participates in immunity are usually grouped as operon and can be found on elements like chromosomes in conjunction with transposons, or plasmids (Kumariya et al. 2019; Zacharof and Lovitt 2012). This also facilitates genetic engineering and strain improvement which can yield a higher amount of bacteriocins. Plasmid encoding bacteriocin genes can be transferred by natural means for higher production, but the disadvantage can be loss of the plasmid during transferring process (Gálvez et al. 2014). According to genetics, there are two main modes of biosynthesis, that is, circular and leaderless, which have been described in this chapter. For genes to be expressed there are several modes of regulation undergoing to produce functional and active bacteriocins. Sometimes, organisms have their own way of regulating the expression depending upon how the genes are encoded. The produced bacteriocins pose a differential mode of action against Gram-positive or Gram-negative bacteria, wherein they either act upon cell-envelop or inhibits the gene expression and protein production (Cotter et al. 2013).

The functions of bacteriocins are known worldwide and are applicable in many fields. The antimicrobial peptides from certain bacteria such as *Bacillus* and *Lactobacillus* are highly used for food preservation and in pharmaceuticals for several therapies. Food-spoilage bacteria are a threat to the quality of food and degrade them into toxic substances which can be fatal sometimes. Pathogens such as *Listeria monocytogenes*, *Clostridium botulinum*, *Streptococcus*, *Escherichia coli*, and so on, can be highly neurotoxic because of their secretive toxins which can cause inflammatory disease, intestinal disorders, several respiratory problems, and can be linked to cancer (Kumariya et al. 2019). In these conditions, antimicrobial peptides, that is, bacteriocins, have proven to be ideal for its application in food spoilage. Cheese spreads which contain the germination *C. botulinum* spores were found to be inhibited by nisin. Pediocin PA-1, also known as Alta™ 2341, is generated by *Pediococcus acidilactici* and has the capacity to increase the shelf life of ready-to-eat foods while arresting the growth of *L. monocytogenes* (Kumariya et al. 2019). Nisin finds its best application as a potential inhibitor of growth and colonization of Helicobacter which causes ulcers. Many pathogens such as *Streptococcus agalactiae*, *Micrococcus luteus*, and *L. monocytogenes* can be targeted by subtilosin A bacteriocin. This way, not only in pharmaceuticals or food preservation, it is widely applicable in livestock, aquaculture, and agriculture (Prasad et al. 2020). Thus, this chapter focuses on bacteriocins, its classification, biosynthesis, and the mode of action to its application majorly in the food industry and pharmaceuticals.

8.2 CLASSIFICATION OF BACTERIOCINS

Scientists from around the world have proposed different schemes for the classification of antimicrobial agents called bacteriocins. However, classification has proven to be difficult due to the sheer number of bacteriocins reported to date. This is because many properties of bacteriocins may coincide, like structure, function, genetic characters, among others (Simons et al. 2020a). Although bacteriocins can be grouped based on their dimensions, structure, chemical composition, thermal stability or modification (Simons et al. 2020; Cui et al. 2021a), they can be primarily differentiated based on the organism that produces them (Simons et al. 2020a). A general summary of the several classes of bacteriocins and their distinguishing features are detailed in Table 8.1.

Hence, bacteriocins are characterized as:

Bacteriocin obtained from Gram-positive bacteria (BGPB)

Bacteriocin obtained from Gram-negative bacteria (BGNB)

TABLE 8.1
Overview of Bacteriocin Classification

Group	Characteristics	Examples	Reference
Bacteriocins acquired from Gram-positive bacteria			
Class I	<5 kDa size, resistance to heat, pH and proteolysis, subdivided into 3 subclasses.	Nisin	(Wencewicz and Miller 2018; Ongey et al. 2017; Kumariya et al. 2019; Simons, Alhanout, and Duval 2020a; Cui et al. 2021a; GradisteanuPircalabioru et al. 2021)
Subclass I a	Long peptides, positively charged, possess unique amino acid composition (lanthionine bridges), pleated sheet N-terminal, α-helical C-terminal, type of lantibiotics.	Nisin, epidermin and gallidermin	(Balciunas et al. 2013; Kumariya et al. 2019; Simons, Alhanout, and Duval 2020a; Cui et al. 2021a)
Subclass I b	Labyrinthopeptins with post-translationally modified labionin and labyrinthine, globular peptides that are negatively charged, inflexible, with maze-like structure.	Lacticin 481, cytolysin and salivaricin	(Simons et al. 2020, Cui et al. 2021)
Subclass I c	It is comprised of sulphur-α-carbon cross-links, resemble sactibiotics.	Thuricin CD and subtilosin A	(Cui et al. 2021a)
Class II	<10 kDa size, heat stable, positively charged, subdivided into 4 subclasses.	Pediocin AcH	(Yang et al. 2014a; Wencewicz and Miller 2018; Wiebach et al. 2018; Simons, Alhanout, and Duval 2020a)
Subclass II a	Dominant class, anti-listerial bacteriocins, linear bisulfide bridge present, N-terminal conservative sequence of YGNGV present, pediocin-like bacteriocins, highly homologous, amphipathic, cationic.	Leucocin A, and pediocin PA-1	(Simons, Alhanout, and Duval 2020a; Zacharof and Lovitt 2012; Kumariya et al. 2019; Cui et al. 2021a; Collins et al. 2017)
Subclass II b	Two-peptide (heterodimeric) bacteriocins necessary for the antibiotic activity (α/β), cationic, encoded by genetically closely-related genes, possess GxxxG-motifs structure, show high conformational lability.	Lactococcin G, lactococcin Q and plantaricin NC8	(Simons, Alhanout, and Duval 2020a; Balciunas et al. 2013; Ekblad, Nissen-Meyer, and Kristensen 2017; Mokoena 2017; Cui et al. 2021a; Yang et al. 2014a)
Subclass II c	Small circular bacteriocins, leaderless, contain 1 (cystibiotics) or 2 (thiolbiotics) cysteine residues, covalent bonding between N-terminal and C-terminal.	Lactococcin A, divergicin A and acidocin B	(Yang et al. 2014a; Simons, Alhanout, and Duval 2020a; Cui et al. 2021a)
Subclass II d	Single, linear, absence of post-translational modification, lacks fixed conservative sequence and structural features, non-pediocin-like, leaderless peptides.	Epidermicin N I 01, lactococcin A and bactofencin A	(Cui et al. 2021a; Yang et al. 2014a; Kumariya et al. 2019)
Class III	>30 kDa size, heat susceptible, subdivided into 2 subclasses.	Zoocin A	(Bennallack and Griffitts 2017; Garvey and Rowan 2019; Simons, Alhanout, and Duval 2020a)
Subclass III a	Lytic peptides, bacteriolysins, possess hydrolase-like activity.	Enterolisin A and lysostaphin	(Yang et al. 2014a; Ceotto-Vigoder et al. 2016)
Subclass III b	Non-lytic peptides, hydrolase-like activity absent.	Helveticin J, caseicin 80 and Enterolysin A	(Yang et al. 2014a; Cui et al. 2021a)
Class IV	It is a large complex; properties are determined by lipid and carbohydrate parts.	Pediocin N5p and lactocin 27	(Maky et al. 2015; Simons, Alhanout, and Duval 2020a; Cui et al. 2021a)

TABLE 8.1 (Continued)
Overview of Bacteriocin Classification

Group	Characteristics	Examples	Reference
Bacteriocins acquired from Gram-negative bacteria			
Colicins	30-80 kDa size, heat and protease labile, subdivided into 3 subclasses.	Colicin K	(Simons, Alhanout, and Duval 2020a; Zimina et al. 2020)
(Based on mode of action)			
Subclass I	Forms pore/channels in the bacterial cell wall, disturb electrochemical gradient, loss of ions leading to cell death.	Colicins A, B and E1	(Simons, Alhanout, and Duval 2020a)
Subclass II	Degrades nucleic acid structures.	Colicins E2 to E9	(Simons, Alhanout, and Duval 2020a)
Subclass III	Exhibit peptidoglycanase activity causing cell death.	Colicin M	(Zimina et al. 2020)
(Based on translocator system)			
Group A	Tol proteins are required for translocation over the outer bacterial membrane, generated on tiny plasmids and coupled with the lysis gene.	Colicins A, E1 and K	(Kleanthous 2010; Camargo et al. 2018; Yang et al. 2014a)
Group B	Utilize Ton proteins for translocation across outer bacterial membrane, formed on larger plasmids, absence of lysis gene.	Colicin 5, 10 and B	(Yang et al. 2014; Kleanthous 2010; Camargo et al. 2018)
Colicin-like	>10 kDa size, resemble colicins, categorized into 2 subclasses.	Klebicins	(Simons, Alhanout, and Duval 2020a; Zimina et al. 2020)
Subclass I	Induces the growth of pores in the target cell's wall.	Pyocin S5	(Brown et al. 2012; Simons, Alhanout, and Duval 2020a)
Subclass II	Causes degradation of nucleic acid structures.	Pyocin S1 and S4	((Brown et al. 2012; Simons, Alhanout, and Duval 2020a)
Microcins	<10 kDa size, highly protease, pH and temperature stable, 2 subclasses.	Microcin D93	(Simons, Alhanout, and Duval 2020a; Zimina et al. 2020)
Subclass I	Post-translational modification, molecular weight <5 kDa.	Microcins B17 and J25	(Simons, Alhanout, and Duval 2020a)
Subclass II	Unmodified/minimal modification, 5-10 kDa, linear, chromosomally encoded, a siderophore with a C-terminal positioning is present, further subdivided.	Microcins E492, V and L	(Simons, Alhanout, and Duval 2020a; Zimina et al. 2020)
(Subdivision of Subclass II)			
Group II a	Three genes are involved in the synthesis and assembly of functional peptides, peptides encoded by plasmids, post-translational modification absent, disulphide bonds might be present.	Microcins L, V and N	(Camargo et al. 2018; Yang et al. 2014a)
Group II b	Linear, C-terminal end displays post-translational modifcations (siderophores), chromosomes encode peptides.	Microcins E492, M and H47	(Camargo et al. 2018; Yang et al. 2014)
Phage tail-like	20-100 kDa size, subdivided into 2 subclasses.	R pyocins and F pyocins	(Zimina et al. 2020)
R type	Evolutionarily resemblance with phage tails in the *Myoviridae* family, long shell-encircled tube present, basal plate possesses receptor-binding proteins (RBP).	R pyocins	(Simons, Alhanout, and Duval 2020a; Zimina et al. 2020)
F type	Resemble phage tails in *Siphoviridae* family, shell absent.	F pyocins	(Zimina et al. 2020)

TABLE 8.2
Comparison between Aantibiotics and Bacteriocins

Characters	Antibiotics	Bacteriocins	Reference
Etymology	Originated in the mid-19th century from two Greek words *anti*meaning "against" and *bios* meaning "life."	It was derived in the 1950s from the French word *bactériocine*, formed by the Greek word *bactērion* for "small cane" andcolicin.	(Yang et al. 2014)
Discovery	In 1928, Alexander Fleming discovered antibiotics that ultimately led to the development of penicillin.	Discovered in 1925 by André Gratia.	(Yang et al. 2014; Castellano et al. 2017b; Gaynes 2017)
Size	The molecular weight of <1 kDa is usually observed.	Usually possess low molecular weight. Sometimes can be over 10 kDa.	(Simons, Alhanout, and Duval 2020b; Bhatti, Veeramachaneni, and Shelef 2004)
Stability to temperature	Shows low thermal stability.	Shows high thermal stability.	(Perez, Zendo, and Sonomoto 2014b)
Functional pH range	Can work in limited pH range.	Can work over greater pH range.	(Perez, Zendo, and Sonomoto 2014b)
Other properties	May exhibit characteristic color, odor, and taste.	Do not exhibit any characteristic color, odor or taste.	(Perez, Zendo, and Sonomoto 2014b)
Origin	The origin of antibiotics is non-ribosomal.	Ribosomal in origin.	(Cleveland et al. 2001)
Production	The soil bacteria and fungi produce it as a secondary metabolite.	The primary growth phase in bacteria has ribosomes acting as the production centers of bacteriocins by translation.	(Beasley and Saris 2004; Zacharof and Lovitt 2012)
Nature	These drugs are antibacterial.	These substances are proteinaceous toxins.	(Cleveland et al. 2001)
Activity	Antibiotics monitor bacterial populations by utilizing the structural difference between bacteria and host cell walls and inhibiting their development or killing them.	Action by inhibiting the growth of bacteria.	(Cotter, Ross, and Hill 2013)
Classification	It is categorized based on its effectiveness against a range of pathogens as narrow spectrum (affecting fewer species) and a broad spectrum (affecting a more extensive range of species).	Among the various schemes existing for classification, bacteriocins are primarily grouped based on origin as- bacteriocins acquired from Gram-positive bacteria (BGPB) and bacteriocins obtained from Gram-negative bacteria (BGNB).	(Simons, Alhanout, and Duval 2020b; Acar 1997)
Modification	Post-translational changes are not essential.	Post-translationally modified peptides.	(Cotter, Ross, and Hill 2013)
Cost	Industrially expensive drug.	Not as expensive as antibiotics.	(Hahn-Löbmann et al. 2019)
Target	Antibiotics affect a broad range of bacteria equally without preferentially attacking related strains.	Bacteriocins show antibacterial activity either against weaker strains of their producing species or closely-related species.	(Beasley and Saris 2004) (Hahn-Löbmann et al. 2019)
Attachment	Cause antibacterial action by binding to specific receptors provided on target organisms' cell walls.	Bind to the target via attachment to the cell wall to bring about antibacterial action.	(Alavi and Ghoroghi 2018b; Cleveland et al. 2001)

TABLE 8.2 (Continued)
Comparison between Aantibiotics and Bacteriocins

Characters	Antibiotics	Bacteriocins	Reference
Mode of action	Antibiotics inhibit vital processes like the synthesis of cell walls. By making use of different mechanisms, like DNA degradation, they efficiently carry out bactericidal and bacteriostatic action.	The strategy of action of bacteriocins necessitates the formation of openings in the bacterial cell wall. These holes form due to an ionic imbalance, resulting in cell leakage and the death of the target cell.	(Alavi and Ghoroghi 2018b; GradisteanuPircalabioru et al. 2021a)
Dosage	Requires higher concentration to carry out the antibacterial activity (50 mg/kg).	Shows optimum activity even at low concentrations (0.5–5 mg/kg).	(Perez, Zendo, and Sonomoto 2014b)
Degradability	It may or may not be easily degraded by proteolytic enzymes.	Proteolytic enzymes quickly degrade them.	(Perez, Zendo, and Sonomoto 2014b)
Effect on beneficial microbes	Harm the beneficial organisms present in humans owing to their broad-spectrum action.	Do not cause any damage to the beneficial microbes in humans.	(Ghoson M. Daba and Elkhateeb 2020)
Toxicity	May cause damage to eukaryotic cells.	Usually do not bring harm to eukaryotic cells.	(Perez, Zendo, and Sonomoto 2014b)
Susceptibility to bactericidal agent	Producers show high susceptibility to bactericidal agents.	Producers are not sensitive to the action of these agents.	(Alavi and Ghoroghi 2018a)
Development of resistance	Antimicrobial resistance develops quickly in response to antibiotics by inactivation of active compounds.	Resistance to bacteriocins may take more time to develop. It may occur by changing cell membrane composition.	(Perez, Zendo, and Sonomoto 2014c)
Example	Penicillin.	Nisin.	(Cui et al. 2021b)
Applications	They are usually used in the medical field.	They have utilized in the food as well as the medical field.	(Perez, Zendo, and Sonomoto 2014b)

8.2.1 Bacteriocins Acquired from Gram-positive Bacteria (BGPB)

Bacteria that are Gram-positive synthesize low molecular weight bacteriocins. Typically lactic acid bacteria, or LAB, is bacteria producing a wide array of bacteriocins that differ in size, structure, properties, and inhibitory action (Yang et al. 2014a). Between 2–21 genes may be involved in their production (Zimina et al. 2020). Due to this substantial diversity of bacteriocins, different scientists have classified them differently (Yang et al. 2014a; Cotter et al. 2005). According to Klaenhammer classification, there are four classes of bacteriocins (Klaenhammer 1993) in contrast to the two classes described by the Cotter classification (Cotter et al. 2005). Here, an alternate classification that incorporates the Klaenhammer and Cotter schemes (Balciunas et al. 2013), and classifies the bacteriocins into Class Ⅰ to Ⅳ, according to dimensions, biochemistry, action strategy, structure, and other attributes of the genes (Ahmad et al. 2017; Antonio et al. 2021).

Class Ⅰ —these are small-sized, ribosomally synthesized bacteriocins (19–50 amino acids) and possess membrane-active peptides and show post-translational modifications (dehydration and cyclization), forming non-standard amino acids present as polycyclic ethers or unsaturated forms (Zacharof and Lovitt 2012; Simons, Alhanout, and Duval 2020a). These unusual amino acids (Kumariya et al. 2019) form multiple ring structures, thus providing rigidity and functionality

(Zacharof and Lovitt 2012; Simons et al. 2020a; Zimina et al. 2020). Nisin is best known amongst several members of this division (Cui et al. 2021a). Bacteriocins organized under Class I are separated to form smaller divisions according to their respective structural and functional attributes (Meade et al. 2020; Antonio et al. 2021).

Class II—this group contains small, ribosomally synthesized bacteriocins (30-60 amino acids) made of amphiphilic helical structure and lacking unusual amino acids (Simons et al. 2020a; Cui et al. 2021a). However, some members of this class, like pediocin PA-1 and pediocin AcH, may be minimally modified. These membrane-active, hydrophobic, non-lanthionine possessing bacteriocins can be further subdivided based on structural features (Cui et al. 2021a; Meade et al. 2020; Mokoena 2017).

Class III—the large, unmodified bacteriocins belonging to this group have different domains and are subdivided based on whether these are lytic or non-lytic (Yang et al. 2014a; Antonio et al. 2021; Cui et al. 2021a). Some scientists have placed the Gram-positive bacteria derived tailocins in this class as well (Garvey and Rowan 2019).

Class IV—here, the peptides are found along with lipid or carbohydrate moieties (O'Bryan et al. 2018; Simons et al. 2020a). However, some authors do not consider Class IV as a category of BGPB and are referred to as bacteriolysins (Kumariya et al. 2019; Antonio et al. 2021).

8.2.2 Bacteriocins Acquired from Gram-negative Bacteria (BGNB)

The narrow-spectrum antimicrobial activity of BGNB serves as a vital drawback behind its restricted utilization compared to BGPB (Balciunas et al. 2013). Although most BGNB was isolated from *E. coli*, many were also obtained from genera like *Pseudomonas* and *Klebsiella*. Here, we outline the four major classes of BGNB, ranging from colicins and colicin-like to bacteriocins classified as microcins and phage tail-like (Simons et al. 2020a; Zimina et al. 2020).

Colicins—these *E. coli* synthesized bacteriocins serve as models to understand the structure of bacteriocins, as well as function and evolution (Acedo et al. 2018). They are produced by colicinogenic plasmids with few present on chromosomes during stress conditions. This depends on a cluster of three related and closely-placed genes that encode toxin, immunity, and lysis protein (BRP—bacteriocin release protein) (Chikindas et al. 2018; Simons et al. 2020a; Zimina et al. 2020). Based on the action mechanism and translocator system, diverse types of these bacteriocins exist (Zimina et al. 2020).

Colicins-like—these bacteriocins are similar in structure, size, and function to colicins; however, they are produced by bacteria other than *E. coli*. Similar to the colicins, this class is also subdivided according to the mechanism of action. Unlike colicins, these bacteriocins, like pyocin S3, may lack the lysis gene (Simons et al. 2020a).

Microcins—these peptides are hydrophobic and ribosomally synthesized mainly by *Enterobacteriaceae* during stress conditions where they take part in competitive interactions (Camargo et al. 2018; Zimina et al. 2020). Usually, the plasmids encode these microcins. Genomic DNA seldom contributes to the same. Although variable in number, the genes present are conserved (Zimina et al. 2020; Camargo et al. 2018). They lack the gene causing lysis and are transported outside the bacterial cell via a transporter system like Type I ABC (ATP binding cassette) transporter (Yang et al. 2014a). This class of bacteriocins can be subdivided based on properties like molecular mass, disulphide bonds, organization of gene cluster, leader peptide sequence, and post-translational modifications (Yang et al. 2014a; Zimina et al. 2020).

Phage tail-like—also called tailocins, consists of a cylindrical structure bearing 8–14 different polypeptide units that resemble a phage tail structure and can be further subdivided. This class of bacteriocins is encoded by a larger gene cluster (>40 kbp) that encodes other vital proteins that help in releasing bacteriocins into the environment (Zimina et al. 2020). The overview of bacteriocin classification is presented in Table 8.1.

Bacteriocins may demonstrate comprehensive antibacterial action depending on the target. This broad-spectrum nature helps them to act on various pathogens. This is deemed helpful in treating infections when the cause is unspecified. On the other hand, narrow-spectrum bacteriocins attack specific pathogens, thus protecting the beneficial organisms (Cotter et al. 2005).

8.3 ANTIBIOTIC VS BACTERIOCIN

Alexander Fleming discovered the penicillin drug in 1928, now being extensively used to control bacterial growth, and known as antibiotics. These antimicrobials act by either killing, as in the case of bactericidal antibiotics, or inhibiting bacterial growth like in bacteriostatic antibiotics. The discovery of penicillin kickstarted the development of many drugs. These are being utilized today to treat a sizable range of bacteria. However, they are ineffective against other microbes like fungi, viruses, and others (Cleveland et al. 2001).

Despite the similarities in their production (both are acquired from bacteria) and activity (potent antimicrobials being developed as drugs against bacteria), both antibiotics and bacteriocins are in stark contrast from one another. Here, we have highlighted some significant differences between antibiotics and bacteriocins.

The common antimicrobial agents being used today comprise antibiotics, bacteriocins, and bacteriophages. Even though all have been studied equally, their simplicity in production and ability to attack a considerable range of bacteria, irrespective of their Gram status, has led to the overconsumption of antibiotics. The detrimental effect of this industrially drug is the development of resistance in bacteria that renders them useless and threatens our health. Another way it harms our health is by eradicating the population of beneficial bacteria owing to their broad-spectrum action of these antibiotics. These limitations have paved the path for bacteriocins to emerge as an appealing alternative, thereby reducing the dependence on antibiotics (Gholami et al. 2018; Alanis 2005).

8.4 MODE OF ACTION OF BACTERIOCINS

Bacteriocins stop the target organism from growing. The cationic bacteriocins target the anionic phosphatidylethanolamine (PE), phosphatidylglycerol (PG), lipopolysaccharide (LPS), lipoteichoic acid (LTA), and cardiolipin (CL) present on the bacterial cell surface (Rashid et al. 2016). Bacteriocins exhibit electrostatic attraction to the negatively charged cell surface target molecules since they pose a net positive charge. On other hand, hydrophobic surfaces align with the membrane and pass through the lipid bilayer. It causes the cell membrane of the microorganism to permeabilize, causing proton motive forces to dissipate, intracellular ATP to deplete, cellular content leakage, and cell death (Kumariya et al. 2019; Yang et al. 2014).

The "barrel-stave" concept, in which antimicrobial peptides form transmembrane holes by inserting into the target's lipid core, was the first proposed mechanism. The "toroidal" method, on the other hand, entails bundling peptides and injecting them into the membrane. In the "carpet" model, bacteriocins will cover the bacterial cell surface, changing its shape; this contact is driven by electrical attraction, causing in cell disruption (Wiebach et al. 2018). Bacteriocins can result in bacteriolysis in both Gram-positive and Gram-negative bacteria. Gram-positive bacteria are attacked by disrupting the cell wall, while Gram-negative bacteria are targeted by affecting gene expression (Negash and Tsehai 2020).

8.4.1 Gram-positive Bacteria

The bacterial cell wall is an important component that keeps the cell's integrity and shape. It prevents cytoplasmic osmotic pressure from causing cell lysis and allows membrane components to be attached. The cell walls of Gram-positive bacteria have thicker peptidoglycan outer cell coats.

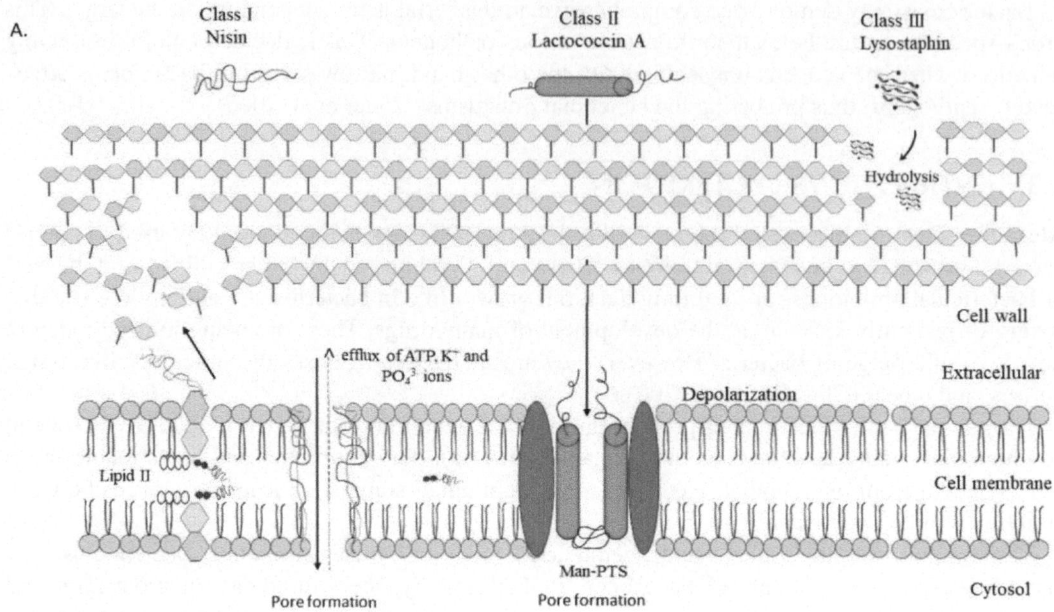

FIGURE 8.1 In Gram-positive bacteria, nisin attaches to lipid II resulting in pores on the cell membrane, loss of amino acids, potassium ions, and ATP and disruption of teichoic acid. Lactococcin A bind to cell envelope associated mannose phosphotransferase system leading to pores and destabilizing the membrane. Lysostaphin help in bacterial cell wall disruption due to enzymatic activity.

A Gram-negative bacterial cell wall is made up of thin peptidoglycan layers. Lipid II is a potent peptide attractant and a critical precursor of peptidoglycan synthesis. Bacteriocins of class I kill bacteria by preventing the development of cell walls while enabling the efflux of ATP, K^+, and PO_4^{3-} ions (Wiebach et al. 2018).

Nisin, a common food-grade class I bacteriocin, attaches to lipid II first, generating membrane holes and the loss of amino acids, potassium ions, and ATP. Teichoic acid and lipoteichoic acid are rendered ineffective by nisins binding to lipid III and IV. It also causes cell wall disintegration, lysis, and death by increasing the action of the autolysin N-acetylmuramoyl-L-alanine amidase. Mersacidin, a type I antibiotic, binds to sugar residues on the lipid II binding site, while nisin binds to phosphate, ensuring peptidoglycan inhibition (Wiebach et al. 2018).

Lactococcin A127 and microcin E492 are examples of class II bacteriocins which target the cell membrane-bound mannose phosphotransferase system (Man-PTS), causing the membrane to destabilize and leak (Wiebach et al. 2018; Simons et al. 2020a). Class III bacteriocins have antibacterial activity due to enzymatic activity (such as endopeptidase), which aids in bacterial cell wall disintegration. Bacteriocins of the category IIIa lyses the cell walls of susceptible strains, killing them. Class IV bacteriocins help by breaking the bacteria cell membrane with their lipid and carbohydrate moieties. Because of their structural specificity, these bacteriocins are sensitive to a wide range of glycolytic and lipolytic enzymes (Simons et al. 2020b). The mechanism of action for Gram-positive is shown in Figure 8.1.

8.4.2 Gram-Negative Bacteria

Gram-negative bacteriocins destroy prey by affecting the metabolism of protein, DNA, and RNA (Cotter et al. 2013). Buforin II peptides can bypass the cell membrane and assemble in the cytoplasm

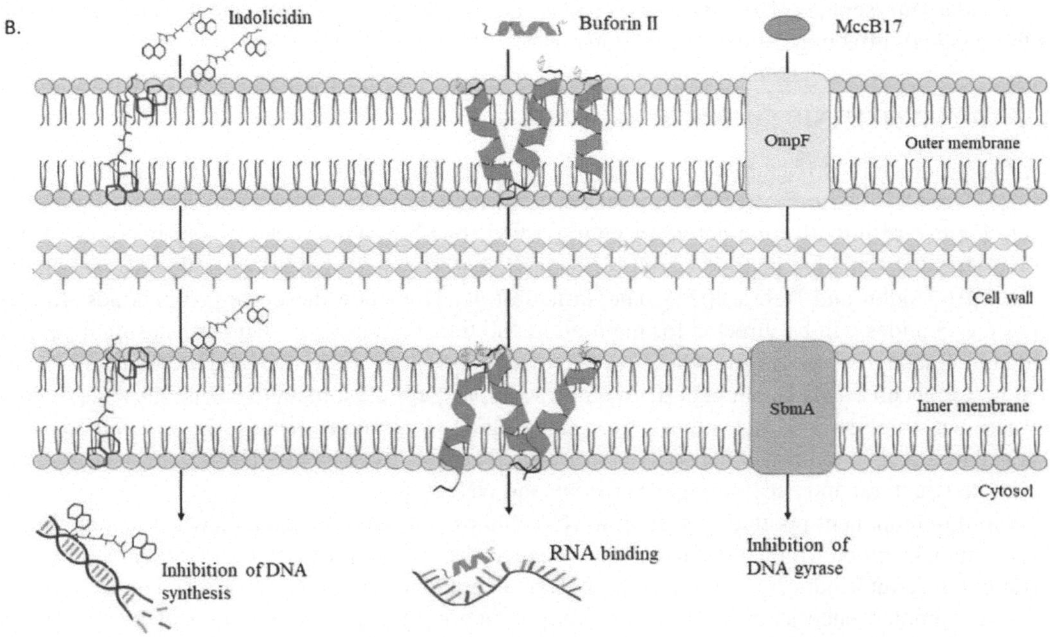

FIGURE 8.2 In Gram-negative bacteria, Idolicidin leads to memebrane depolarization and DNA synthesis inhibition. Buforin II attaches to RNA and DNA preventing mutilication of microbes. MccB17 passes through the outer membrane of the cell with help of porin OmpFand passes across via SbmA inhibiting DNA gyrase thereby interfereing with DNA replication.

without being permeabilized. This has a strong preference for nucleic acids. The proline hinge permits peptide to enter the cell and attach to DNA and RNA in the cytoplasm, preventing microbes from multiplying. Indolicidin, the tiniest natural cationic peptide, causes membrane depolarization and inhibits DNA synthesis. Puroindoline, a peptide with a high tryptophan residue count, functions by interfering with DNA synthesis (Wiebach et al. 2018).

Colicins peptides have three domains: amino-end translocation (T) domain, central receptor-binding (R) domain, and an antibacterial carboxy-end cytotoxic (C) domain (Yang et al. 2014). This antibacterial activity can be detected in sensitive species strains as well as strains that are closely related to them (Yang et al. 2014c). The strain that produces the colicin-like bacteriocins is dangerous because when they are released, they cause membrane lysis and cell death. They, like colicins, act against the weaker strains of the generating species (Cascales et al. 2007).

Bacteriocins that look like phage tails work by either perforating the membrane or inducing cell death (Scholl 2017) or membrane potential disturbance resulting in membrane pore creation (Michel-Briand and Baysse 2002). MccJ25, a bacteriocin, suppresses transcription by inhibiting RNA polymerase channels. MccB17 peptide prevents mRNA synthesis by preventing DNA supercoiling caused by DNA gyrase. MccC7-C51 prevents DNA replication by inhibiting aspartyl-tRNA synthetase (Negash and Tsehai 2020; Cotter et al. 2013). The mechanism of action for Gram-negative bacteria is shown in Figure 8.2.

8.4.3 Induction of Cell Death

The peptides' distinguishing feature is that they generate pores in the cell wall, allowing the cell to leak. The outflow of K^+ initiates apoptosis, with calcium ions and reactive oxygen species functioning as secondary messengers. The caspase cascade is activated by the release of cytochrome E as a result of these signals affecting mitochondrial homeostasis (Wiebach et al. 2018).

8.5 BIOSYNTHESIS OF BACTERIOCINS

For the production of bacteriocins, four types of genes are required such as structural genes, an immune gene, ABC transporter which is associated with membrane, and several accessory proteins. Bacteriocins are initially non-active in nature when they are synthesized as pre-peptides, it has amino-terminal signaling peptide which is bonded to carboxy-terminal pro-peptide (Alvarez-Sieiro et al. 2016; Sidhu and Nehra 2021). The structural genes encode these pre-pro-peptides; further these pre-peptides will be directed for maturation and transportation will happen via signal peptide as that is a maker for recognition which keeps the peptide in an inactive state and protected inside the host bacterial cells. This signal peptide interacts with pro-peptide domain to ensure the state of conformational structure. The biosynthesis functions well due to immune genes asthe host cell creates a defence mechanism to protect themselves from the action of bacteriocins. These genes are encoded nearby to structural and accessory genes (Sidhu and Nehra 2021).

A multi-component regulatory system for biosynthesis of bacteriocins includes a) a signal peptide; b) signal peptide which binds to cell surface receptor; and c) a regulator for response. However, Nisin exhibits auto-inducer expression through two-component system activation. There are a few notable examples, such as class II bacteriocin production of LsbB, which controls its expression by increasing RNA stability thrice, leading to a 30-fold increase in expression (Uzelac et al. 2015). Additionally, the signal peptide's attachment to the histidine-kinase receptor results in phosphorylation and a signalling influx, which directs the phosphate residue to the response regulator, which in turn activates gene expression for increased bacteriocin production by attaching to the promoter region (Sidhu and Nehra 2021). Bacteriocins are generated as precursors, which must be processed and modified post-transcriptionally (Morton et al. 2015). The modified bacteriocins are then transported using transporters such as ABC, and sec-dependent exporters (Kumariya et al. 2019). The ABC transporter has two functions: it separates the substrate from the signal peptide and transports it across the cytoplasmic membrane. This helps to prevent the active and fully produced bacteriocins to withstand inside the cytoplasm. Biosynthesis of bacteriocins can be divided depending upon types of organism, that is, Bacteriocin Produced by Gram-Positive Bacteria (BGPB) and Bacteriocin Produced by Gram-Negative Bacteria (BGNB). The mode of action of antimicrobial peptides from Gram-positive bacteria in comparison to Gram-negative has a broad range of target. Thus, many researchers have focused on LAB for active biosynthesis of bacteriocins (Perez et al. 2014).

This was a general mechanism for biosynthesis, while its detailed mechanism is divided into two important groups named Circular and Leaderless Bacteriocin biosynthesis which will be discussed in this chapter.

8.5.1 Biosynthesis of Circular Bacteriocins

Structurally, these bacteriocins exhibit a signature structure with a backbone of conserved circular peptides due to covalent bonding of N and C-terminal residues. Circular bacteriocins are a part of family which is ribosomally synthesized and post-translationally modified peptides (RiPPs). The greatest advantage of circular bacteriocins over linear bacteriocins is their better stability. But probably the most promising aspect of this class of bacteriocins is a better knowledge of their biosynthetic mechanism, which remains the most difficult process (Perez et al. 2018).

The genetic cluster either on chromosome or plasmids is designed in such a way that it codes for biosynthesis of amylocyclicin (Scholz et al. 2014), acidocin B (Acedo et al. 2018), enterocin NKR-5-3B (Masuda et al. 2016), pumilarin (Auke J. van Heel et al. 2017), and plantaricyclin A (Borrero et al. 2017), and many more which is briefly described by Perez et al. (2018). Even though with greater understanding of these gene clusters, the mechanism and its regulation are still a challenge. Leader peptides from the same group will share a similar motif that is essential in the substrate-enzyme interaction process during the formation of circular bacteriocins (Oman and van der Donk 2010). Type I lantibiotic bacteriocin leader peptides feature a conserved FNLD box and helical structures that are important for interaction with their biosynthetic enzymes (Rink et al. 2007). Because of the length and sequence of leader peptide in circular bacteriocins differ so significantly, it's difficult to anticipate its involvement during biosynthesis.

The three major and primary processes in production of circular bacteriocins are cleavage of the leader peptide, circularization, and its export to the extracellular space as shown in Figure 8.3. Previously, it was assumed that these processes were linked, but it was shown that the cleavage of the leader peptide was independent of the remaining of the biosynthetic activities. Genes for biosynthesis which are responsible for producing precursor peptide and linear forms, but a single mutation can cause loss of function for circularization reaction. Similarly for other bacteriocin leucocyclin Q, circularization and secretion process are separate processes depicting different genes for its regulation (Gabrielsen et al. 2014; Mu et al. 2014). This data is also not enough to prove the independent nature of cleavage and circularization, thus according to the scientists the alterations might be caused by the separation of cleaved leader peptides and events of circularization. Considering the regulation of circular bacteriocin production, it is still unknown in many aspects. Regulator proteins and histidine kinase were found near biosynthetic gene cluster in the regulation of a few circular bacteriocins such as circularin, enterocin NKR-5-3B, butyrivibriocin AR10, and uberolysin. Also, one such regulation which is also found as quorum sensing using three component system (Bennallack and Griffitts 2017).

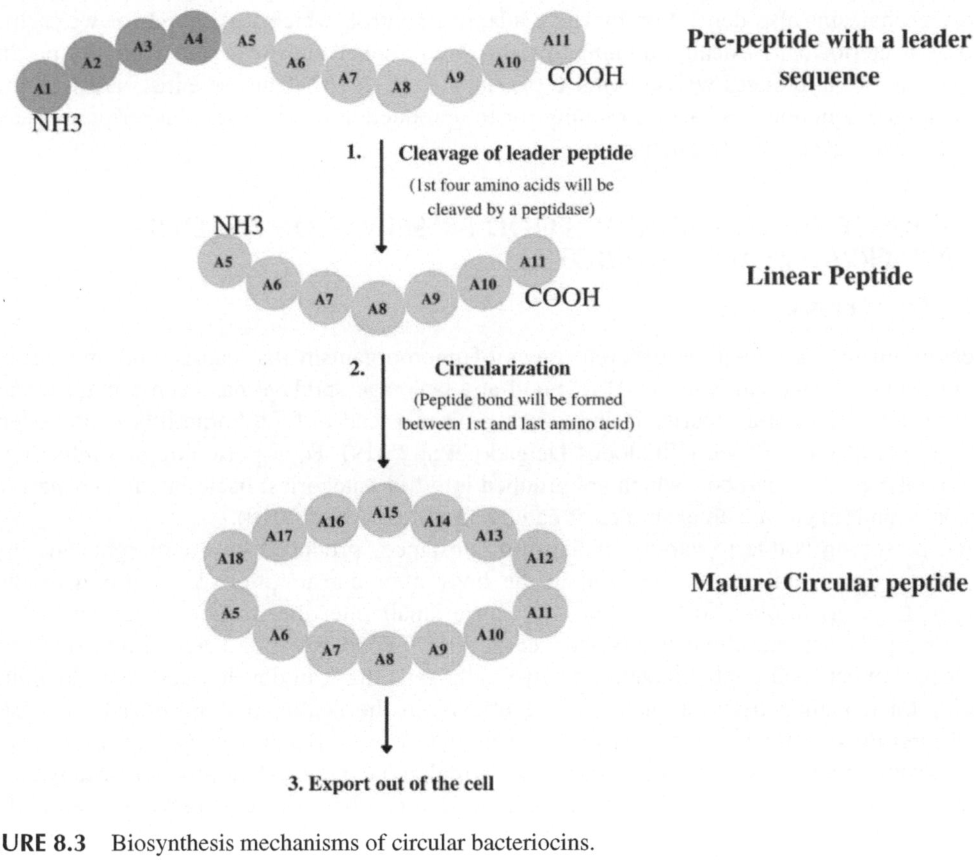

FIGURE 8.3 Biosynthesis mechanisms of circular bacteriocins.

8.5.2 Biosynthesis of Leaderless Bacteriocins

Leaderless bacteriocins are unique from other kinds of bacteriocins. According to term "Leaderless", they are synthesized without the leader sequence while all other inactive bacteriocins have leader sequence at their N-terminal. Due to this characteristic, they don't come under family of RiPPs because they don't follow post-translational modifications. But it is still considered a part of RiPPs because of its biosynthesis mode which is ribosomal in nature and its characteristic. Also, the simplicity of leaderless bacteriocins being the strongest resource, because it enables them to be expressed for large-scale production (Perez et al. 2018). The functions of the leader peptides are: a) acts as a biosynthetic enzyme recognition site; b) protects the producer strain from the bacteriocin's inhibitory action; and c) ensuring the stable and suitable conformation for the enzyme-substrate interaction, as it lacks post-translational modifications (PTM).

The genes for leaderless bacteriocins, which give immunity like lacticins Q and Z, are similar to circular bacteriocins (Iwatani et al. 2012 2013), aureocin A53 (Nascimento et al. 2012), and aureocin A70 (Coelho et al. 2014; Netz et al. 2002). However the common mechanism for its biosynthesis is still unknown. For enterocin L50 (A and B), its structural genes are encoded in 50 kb plasmid of pCIZ1, while for enterocin Q which is produced along enterocin L50 is encoded on 7.4-kb plasmid of pCIZ2 (Perez et al. 2018). While for lacticins Z and Q (Iwatani et al. 2013 2012), aureocin A53 (Nascimento et al. 2012), and aureocin A70 (Coelho et al. 2014; Netz et al. 2002).

Considering its mode of regulation, there is no general mechanism for leaderless bacteriocin production. Recently it was found that LnqR, a member of the TetR family of transcriptional regulators, is a positive regulator that can improve transcription of lacticin Q's biosynthetic genes lnqBCDEF for its synthesis (Iwatani et al. 2016). For aurecin A70, the whole production is a complex process which is regulated by alternate transcription factors and phage regulators (Coelho et al. 2016). Several mechanisms also depend on nutritive-adaptive control, which was found for weissellicins Y and M bacteriocins. Basically, thiamine presence will enhance growth and normally produces weissellicins M but reduces weissellicins Y (Masuda et al. 2016). Thus, scientists are working to understand the common mechanism running for the production of leaderless bacteriocins because of genetic heterogeneity in production.

8.6 ROLE OF BACTERIOCINS IN FOOD PRESERVATION AS FOOD PRESERVATIVE AND PROBIOTIC

8.6.1 Food Preservative

The environment has numerous different types of microorganism that cause food and beverage contamination (Haruta and Kanno 2015). Food and beverage spoilage has been a major concern to ensure food safety and security in industries, as it can lead to food-borne illness and degrade the flavor and quality of food (Villalobos-Delgado et al. 2019). Food poisoning and infection are caused by dangerous microbes, which are grouped into five categories: bacteria, viruses, parasites, protozoans, and fungus are all examples of pathogens (Bhalla et al. 2019).

Food poisoning is due to variety of harmful substances produced by microorganisms in the food, which causes an adverse reaction in the body after consumption. Enterotoxin produced by *Staphylococcus aureus*, for example, can cause small intestine irritation. Humans infected with *Salmonella* can cause diarrhea (Castro et al. 2018). Food infection leads to intestinal infection when you eat food contaminated with live pathogens that can develop and reproduce in the digestive tract. Pathogens have such adverse effects on the health, it is necessary to preserve food. Therefore, chemical preservatives have been widely used. Because of its harmful effects on human health, there is a significant demand for natural and chemical-free products. Many studies have been conducted to look into bio preservatives that can be utilized to prevent diseases from infecting food.

Bio-preservation is a method of preserving nutrient, quality and increasing the shelf life of food by utilizing non-pathogenic microorganisms or metabolites (Iordache et al. 2017). Bacteriocins are used in bio-preservation to address the demand for natural preservatives. When it comes to the preservation, bacteriocins are used in dairy, meat, seafood, beverages, and vegetables. Some of the commonly used bacteriocins are as follows.

8.6.1.1 Nisin

Nisin bacteriocin has been confirmed safe by the World Health Organization and the US Food and Drug Administration and suitable for use in the food sector (Ng et al. 2020). Nisin was discovered for the first time from fermented milk cultures in 1928 and sold in 1953 due to its ability of microbial static effect; it is used as a bio-preservative. Its molecular size is 3354kDa and made up of 34 amino acids (da Costa et al. 2019). Lactococcus and Streptococcus, both Gram-positive bacteria, are employed to manufacture it. Nisin is safe for humans to use at concentrations less than 83.25 mg/kg (Kitagawa et al. 2019). Nisin inhibits *B. cereus*, *C. botulinum*, *L. monocytogenes*, and *S. aureus*, among other Gram-positive bacteria present in food (Zhao et al. 2016). Gram-negative microorganisms can be killed or inhibited by a combination of nisin and antibiotics (Ng et al. 2020).

NisaplinTM, a commercially available version of nisin, inhibits pathogen growth in both liquid and solid form (Silva et al. 2018). It kills *L. monocytogenes* and *S. aureus*, and improves the shelf life of cheese. Nisin, when added to the food matrix, inhibits the growth of bacteria which are heat-resistant spore-forming organisms of the genera *B. cereus*, *C. botulinum*, and *C. perfringens*, and thereby preventing spoiling of milk and milk products (Negash and Tsehai 2020). In dairy, culinary, bakery, and beverage products, the pathogens *L. monocytogenes*, *B. cereus*, and *C. botulinum* are detected, whereas *C. botulinum* and *L. monocytogenes* cause spoilage in meat and meat products. Nisin is used as an excellent antimicrobial agent against them for extending the expiration date (Negash and Tsehai 2020; Gharsallaoui et al. 2016). The basic action of nisin as preservation is binding to anionic phospholipids in the cell membrane (particularly lipid II) and then insertion into the cell membrane. These mechanisms can halt all metabolic processes in Gram-positive bacteria, resulting in their death (Younes et al. 2017a).

8.6.1.2 Pediocin

Pediocin is a class II bacteriocin with molecular weight of 2.7–17 kDa (Ng et al. 2020). It is marketed in the name of Alta 2341TM or MicrogardTM (Ng et al. 2020; Garsa et al. 2014). Gram-negative bacteria like *Pseudomonas* and *E. coli* are more sensitive to this peptide than Gram-positive bacteria like *L. monocytogenes and S. aureus* (Ng et al. 2020). Derived from *Pediococcus* strains, it has a high heat, pH, and protease enzyme stability. By adsorbing the amino acid at the lipid bilayer of the target cells, they inhibit microorganisms that cause food spoilage, such as *C. perfringens and L. monocytogenes*. The bacteriocin is administered in the food matrix by being inoculated with *Pediococcus*, *Enterococcus*, or *Lactobacillus* strains to make pediocin, or directly employing pediocin in the food matrix to prevent pathogen growth (Silva et al. 2018).

Pediocin PA-1, derived from the *P. acidilactici* MCH14 bacteria, possesses antibacterial capabilities against *L. monocytogenes* and *C. perfringens*, enhancing the shelf life of sausages and fermented pork products. Pediocin PA-1 is a salad dressing ingredient that also has an antibacterial effect against *Leuconostoc mesenteroides*, making it effective for preserving highly perishable food such as cottage cheese, fresh meat, and vacuum-packed beef and *L monocytogenes* growth is inhibited in fish fillets (Ng et al. 2020). *Bacillus subtilis* and *B. licheniformis* can be hindered by pediocin PA-1 in sous vide products, pediocin PA-1 from *E. faecium* NCIM 5423 in fermented soy milk, pediocin 34 from *Pediococcus pentosaceous* 34 in meat and milk products can all extend shelf life (Egan et al. 2016). By creating pores in the cytoplasmic membrane, pediocin inhibits a variety of sensitive bacterial cells by targeting the cell, collapse intracellular pH of target, and inhibit proton motive force for energy production (Niamah 2018).

8.6.1.3 Enterocin

Enterocin is a bacteriocin that is produced by *Enterococcus* spp. Two species, namely *Enterococcus faecalis* and *Enterococcus faecium*, are potential isolates for enterocin production obtained from human intestine (Gálvez et al. 2014). There are four different classes of enterocin, among them class II and III enterocins like enterocin AS-48, *E. faecalis*, which have great importance as they inhibit pathogenic strains of *Bacillus* and *Clostridium* (Dubin and Pamer 2017). When people consume ready-to-eat veggies, they are susceptible to a range of illnesses; to combat this, enterocin AS-48 is effective against *B. cereus*, *Bacillus macroides*, *Paenibacillus spp.*, and *S.aureus* (Luna-Guevara et al. 2019). It is used to store soybeans, sprouts, canned fruits, and vegetables, inhibiting *L. monocytogenes* (Burgos et al. 2014; Gálvez et al. 2014). Because of the high nutritious composition, the dairy sector is readily polluted. It was found to have bactericidal properties against *Staphylococci* in skimmed milk, as well as *L. monocytogenes* in skim, half fat, full fat, and no-fat hard cheese (da Costa et al. 2019). To stop *L. monocytogenes* from multiplying, enterocin A from *E. faecium* CTC492 and enterocin 416K1 *from E. casseli-flavus* IM 416K1 were employed in cottage cheese (Vimont et al. 2017). The enterocins L50A and B from *E. faecium* F58, as well as enterocin CRL 35 from *E. faecium* CRL35, were used to prevent the growth of *L. monocytogenes* in goat milk for cheese preservation (Ng et al. 2020). Furthermore, enterocin CCM 4231 from *E. faecium* has been claimed to possess antibacterial properties in the preservation of skimmed milk, yogurt, and Saint-Paulin cheese against *S. aureus and L. monocytogenes.*

Lactobacillus sakei CTC746 was suppressed using enterocins A and B, as well as sakacin K from *E. faecium* CTC492 and *L. sakei* CTC494 was employed to keep up the quality of cooked pork. Pathogens detected in cooked ham and dried fermented sausages were shown to be resistant to enterocins A and B (Castellano et al. 2017a). In addition, enterocin AS-48 produced by *E. faecalis* A-48-32 is used extensively in the preservation of fruit juices, apple cider and infant rice-based food (Burgos et al. 2014). Enterocin AS-48 has a high pH and heat stability, making it useful in the food industry. It can be added to meals in two ways: by introducing enteroicn-producing strains or by adding pure or semi-purified enterocin (Ng et al. 2020).

8.6.1.4 Leucocin

Leucocin is a bacteriocin isolated from *Leuconostoc* spp. It belongs to the class IIa group. Leucocin A, a bactericidal bacteriocin produced by *Leuconostoc gelidum* UAL187, deprives the growth of *L. monocytogenes* in milk, meat, and sausages (Ng et al. 2020). The molecular weight of leucocin A is 3.93kDa (Balay et al. 2017). It prevents the growth of *C. divergens* UAL9 bacteria in meat products and *L. monocytogenes.* It was inhibited by leucocin K7, while *L. innocua* CIP 80.11 is detected in dry sausages and pork products that have been fermented restricted by leucocin B-KM432Bz (Ng et al. 2020).

8.6.1.5 Lacticins

Lacticins are produced by strains of *L. lactis*. Lacticin 3147 is useful in making buttermilk. This bacteriocin has antibacterial activity against a wide range of pathogens in addition to LAB which prevents deterioration of food. Studies had reported the application of a powder form of lacticin 3147 to inhibit *Listeria* and *Bacillus* in milk and milk products (Castro et al. 2018). Another example is lacticin 481, a mono peptide that exhibits average inhibition spectrum. It is efficiently employed against LAB, *Clostridium tyrobutyricum*, and *L. monocytogenes*. Administration of semi-purified lacticin 481 to fresh cheeses kept in the fridge reduced the amount of *L. monocytogenes* in the cheese (Ribeiro et al. 2017).

8.6.2 Probiotics

The term "probiotic" comes from the Greek word pro bio, which means "for life" or "life support" (Yang et al. 2014d). Probiotics help to keep the gut flora in check and provide additional health advantages. Probiotics are defined by the WHO as "live microorganisms that, when administered in suitable proportions, impart a health benefit on the host" (Dobson et al. 2012). Probiotics are a group of strains that are beneficial to the host animal and can survive in the environment and aid in metabolic activity, they are non-pathogenic and nontoxic, and remain stable and viable in the face of extreme conditions for long periods of time in the environment (Yang et al. 2014b). Probiotics regulate intestinal infections by the synthesis of antimicrobial substances, competitive pathogen exclusion, immune system modulation, and nutritional competition. Probiotics create bacteriocins, which suppress gut microorganisms. Purified bacteriocins and probiotics that produce bacteriocins have been shown to reduce the number of infections in mice, chickens, and pigs, as well as alters the composition of their intestinal microbiota (Bernbom et al. 2006).

Studies prove that *L. lactis* strain CHCH2862 that can produce nisin, as well as those that don't produce nisin, can have an influence on the microbiota composition in the intestine of rats. In rats fed with *L. lactis* nisin-producing and non-nisin-producing *L. lactis*, the number of bifidobacterium in their feces increased, but the number of *Enterococci* and *Streptococci* in the duodenum, ileum, cecum, and colon decreased. This clearly indicates its impact on the intestinal microbiota through competition for nutrients or adhesion sites. *Enterobacter*, *Escherichia*, *Klebsiella*, *Morganella*, *Salmonella*, *Shigella*, and *Yersinia* have all been demonstrated to be inhibited by colicin Ib, E1, and microcin C7 from *E. coli* strain H22 (Cursino et al. 2006).

E. coli strain H22 is one of the promising probiotics for both cattle and humans, *Lactobacillus salivarius* UCC118 produced bacteriocin Abp118 for inhibition of food-borne infection by *L. monocytogenes* in mice, mutation of which was unable to protect the mice from this pathogen. *L. lactis* MM19 and *P. acidilactici* MM33 isolated from the human feces showed the capability of reducing the number of vancomycin-resistant *Enterococci* (VRE) in mice (Cutler et al. 2007). Moreover, nisin Z and pediocin PA-1 bacteriocins, from *L. lactis* MM19 and *P. acidilactici* MM33 respectively, displayed strong antimicrobial activity against clinical VRE isolate. *L. lactis* and mutants of *P. acidilactici* MM33A that produce non-pediocin PA-1/AcH increased total LAB and anaerobe populations in healthy mouse feces, but *P. acidilactici* MM33 decreased enterobacteriaceae numbers (Yang et al. 2014b). Furthermore, VRE populations decreased after three days of feeding MM19 or MM33. The pediocin PA-1 generating probiotic *P. acidilactici* UL5 showed in vitro suppression of *L. monocytogenes* (Dabour et al. 2009).

8.7 ROLE OF BACTERIOCINS IN FOOD PROCESSING

Two notable applications of bacteriocins in dairy industry have to a greater extent prevented food-borne diseases. The simple practice of inoculating food with lactic acid bacteria that produce bacteriocins and adding pure or semi-purified bacteriocins directly to food improved the product life and also reduced food-borne diseases. Antibacterial starter cultures and protection cultures made from bacteriocin-producing LAB strains may have an advantage over semi-purified/purified bacteriocins. When bacteriocins are typically incorporated into food substrates they get quickly degraded and tend to lose their antibacterial activity and, as a result, including bacteriocins into food packaging films/coatings, which boosts their activity and stability in complex food systems, is a feasible alternative (Salgado et al. 2015). The possibility of employing immobilized bacteriocins in the production of antimicrobial packaging films to reduce food-borne pathogenic bacteria like *L. monocytogenes* (Sánchez-Hidalgo et al. 2011; Ibarguren et al. 2015; Cursino et al. 2006). The use of dairy products, particularly soft cheeses, has been connected to several listeriosis outbreaks, posing a challenge for the dairy sector and public health authorities (Melo et al. 2015). Despite the fact that most dairy products, particularly cheeses, are manufactured from pasteurized milk, *Listeria* infection still persists. Cheeses are ready-to-eat foods that are frequently stored at low temperatures that encourage psychrotrophic bacteria, such as *L. monocytogenes*, to survive and proliferate. This ultimately can lead to contamination in subsequent stages of dairy product processing (Melo et al. 2015; Carpentier and Cerf 2011). As a result, *Listeria* active bacteriocins appear to be an excellent option for inhibiting the pathogen's development following boiling or packing (Cotter et al. 2005). Bacteriocins are rendered useful to regulate non-starter flora in cheese and wine, such as non-starter lactic acid bacteria (NSLAB), and so improve the end product's quality (Oumer et al. 2001; O'Sullivan et al. 2003). They may also utilize promoting food fermentation by speeding up the ripening of cheese and even improving its flavor (Oumer et al. 2001).

8.7.1 Nisin

Nisin is a class-1 bacteriocin, and it is well-studied and economically valuable (Ross et al. 1999). The Joint Food and Agriculture Organization/World Health Organization (FAO/WHO) Expert Committee on Food Additives has approved nisin as a food preservative (E234) and has determined that it is safe (Favaro et al. 2015). Nisins A, Z, F, and Q, generated by *L. lactis*, and nisins U, U2, P, and H, produced by various *Streptococcus* strains, have been found and described to date (O'Connor et al. 2015). Nisaplin™ is one of the recommended food preservatives comprising nisin as the functional component (2.5 percent), as well as additional additives including sodium chloride in the case of non-fat dairy milk (Chen and Hoover 2003).

Nisin is an effective bacteriocin proven to inhibit a broad spectrum of Gram-positive bacteria, including the major food pathogens like *Listeria* and *Staphylococcus*, LAB, and spore-forming bacteria like *Bacillus* and *Clostridium* (Chen and Hoover 2003). Nisin was first used to prevent gas-producing Clostridium spp. from generating late blowing in cheese, which can result in a slit, crack, or eye in cheese with an abnormal flavour (Gálvez et al. 2014). It has also been extensively used to substitute nitrate in cheese and pasteurized cheese spreads to minimize the spread of Clostridia spores (Chen and Hoover 2003). In dairy products, nisin is demonstrated to be efficient in the prevention of *L. monocytogenes* and *S. aureus* (Sobrino-López and Martín-Belloso 2008). Nisin demonstrated antimicrobial property against *L. monocytogenes* and *S. aureus* in milk held at refrigeration temperatures (19°C to -4°C) (Arqués et al. 2011). Numerous studies have been conducted by adding nisin to various cheese varieties (cottage cheese, cheddar, and ricotta-type cheeses) and it was found that it can effectively reduce *L. monocytogenes* growth, however only by 1–3 log cycles (Chen and Hoover 2003). Its anti-listerial action is boosted by lowering the pH and adding NaCl (Chen and Hoover 2003; Kuo et al. 2013; Khan and Oh 2016).

After addition of nisin to Minas Frescal cheese, there was decrease in the growth of *S. aureus* by around 1.5 log cycles. Nisin inhibited *Bacillus cereus* and *Bacillus subtilis* in processed cheese (Sobrino-López and Martín-Belloso 2008). The antimicrobial treatment of nisin increased the shelf life of a Greek soft acid-curd cheese by more than 21 days (Kykkidou et al. 2007). Nisin at its concentration of 2.5 mg/l has been proven to extend the shelf life of Ricotta-type cheeses by preventing the growth of *L. monocytogenes* for more than 8 weeks (Davies et al. 1997). Furthermore, only 10–32 percent of nisin was lost throughout the 10-week storage duration, indicating a good degree of nisin retention. Ferreira and Lund (1996) investigated the impact of nisin on the survival of the most resistant strains of *L. monocytogenes* in cottage cheese, finding a decrease in log cycles in three days.

Nisin has been examined in various pasteurized dairy products, including clotted cream, chilled desserts, flavored milk, and canned evaporated milks, and has been proven to reduce post-process contaminating microorganisms like *L. monocytogenes* (Gálvez et al. 2014). Pimentel-Filho et al. (2014) investigated the efficiency of nisin and bovicin HC5 combination against *L. monocytogenes* and *S. aureus* in fresh cheese. *L. monocytogenes* was reduced to undetectable levels, after 9 days of storage at 4°C, however the combination of bacteriocins did not inhibit *S. aureus* development.

The antibacterial action of nisin is influenced by a variety of parameters, including temperature, pH, content, and the natural microbiome of food, according to various studies (Zhou et al. 2014; Cleveland et al. 2001). Proteolysis during the cheese-making process may potentially impact nisin's activity and antibacterial potency. However, after 10 weeks of storage, several writers reported a little reduction of activity of nisin (10–32%) in Ricotta cheese (Cleveland et al. 2001). Furthermore, in research on Emmental cheese, proteases had no effect on nisin activity (Favaro et al. 2015). Nisin use in dairy products is restricted to pH values lesser than 7, as nisin loses its activity at high pH levels (de Arauz et al. 2009b). Studies have found that using nisin in dairy products has various drawbacks owing to its interaction with food matrix components which includes fat. However, the significance of fat in nisin action is unclear, as experiments on heat-treated cream reveal that low amounts of nisin prevent *B. cereus* development (Nissen et al. 2001). In high-fat milk-based puddings, nisin was also proven to inhibit spoiling microorganisms and improve shelf life (Oshima et al. 2014). Meanwhile, homogenization of milk was shown to diminish the anti-listerial effects of nisin, indicating that food treatment has a key role in the effectiveness of bacteriocins (Bhatti et al. 2004).

8.7.2 Pediocins

Pediocins are a type IIa bacteriocin generated by *Pediococcus* spp. and are sold under the brand names Alta 2341™ and Microgard™ (Garsa et al. 2014). Food-borne pathogens, such as *L. monocytogenes* and *S. aureus* (Cintas et al. 1998), have been proven to be more effective than nisin. The stability of pediocins in aqueous solutions, their wide pH range, as well as their high tolerance to heating and freezing further boost their prospective use to dairy produce (Sobrino-López and Martín-Belloso 2008). Pediocins have been added to milk or dairy items in a few experiments, despite their high potential. In cottage cheese, cream, and cheese sauce, pediocin (PA-1) was observed to reduce *L. monocytogenes* concentrations (Pucci et al. 1988). The anti-listerial impact was observed throughout a wide temperature and pH range, and was especially functional at low levels of initial contamination of *L. monocytogenes* (102 cfu ml^{-1}).

8.7.3 Lacticins

Lacticins, such as lacticin 3147 and lacticin 481, are produced by some strains of *L. lactis*. The two-component lantibiotic, lacticin 3147, requires both structural proteins to have complete biological activity. It was isolated from an Irish kefir grain used to make buttermilk (Ross et al. 1999). In addition to other LAB, lacticin has antibacterial action against a variety of food pathogenic and food-spoilage bacteria (Sobrino-López and Martín-Belloso 2008; Carmen Martínez-Cuesta et al. 2010). In the control of *Listeria* and *Bacillus* in newborn milk formulation, cottage cheese, and yogurt, lacticin 3147 powder preparation was demonstrated to be beneficial (Morgan et al. 2001). Studies show that concentrations of lacticin powder utilized were high, which accounted for 10% of the product weight, were impractical, and were uneconomic (Morgan et al. 2001). Lacticin 481 is a single-peptide lantibiotic with a broad inhibitory range, primarily effective against other LAB, *C. tyrobutyricum* (O'Sullivan et al. 2003), and *L. monocytogenes* (Ribeiro et al. 2017). Arqués et al. (2011) found that using non-purified lacticin 481 had a minor bacteriostatic effect in milk held at refrigeration temperatures. However, applying semi-purified lacticin 481 to new cheeses kept in the refrigerator for 3–7 days decreased *L. monocytogenes* by 3 log cycles (Ribeiro et al. 2017). Further, using lacticins in food complexes is unlikely to result in the total eradication of pathogens like *L. monocytogenes*.

8.7.4 Enterocins

Enterocins are bacteriocins generated by *Enterococcus* species, and they are a varied category in terms of taxonomy and inhibitory spectrum (Egan et al. 2016). Despite the fact that most LAB are GRAS-certified and may be safely in food, bacteriocinogenic *Enterococci* create certain risks. Enterococci are a type of bacteria that has been connected to a number of human illnesses (Ross et al.

1999). Additionally, certain *Enterococci* may carry virulence factors along with antibiotic resistance genes (Younes et al. 2017b). Food *Enterococci* are recognized for their ability to share genetic information, while having fewer virulence factors than clinical strains. Pure enterocins may be deemed more suited for consumption, given the safety issues associated with utilizing bacteriocinogenic *Enterococci*.

Enterocin AS-48 is a class IIc cyclic bacteriocin generated by *Enterococcus faecalis* that is effective against a numerous *Bacillus* and *Clostridium spp* (Egan et al. 2016). Enetrocin is a most researched bacteriocin, with excellent pH and heat stability, making it a good option for food use. Muñoz et al. (2007) investigated the ability of enterocin AS-48 to control the growth of *Staphylococci* in skimmed milk. They discovered a bactericidal property proportionate to the bacteriocin concentration (10–50 g/ml), however *Staphylococci* were not completely eliminated at any of the doses examined. Yildirim et al. (2016) evaluated the antibacterial properties of Enterocin KP in skim, half fat, and full fat milks against *L. monocytogenes*. Enterocin KP showed strong anti-listerial activity, but when the fat content of milk and the number of *L. monocytogenes* inoculated rose, so did the bactericidal impact. In addition, Gradisteanu Pircalabioru et al. (2021a) found that pure enterocin CCM 4231 (3200 AU/ml) inhibited *S. aureus* and *L. monocytogenes* growth in skimmed milk and yogurt. Even yet, after 24 hours, enterocin's antagonistic impact on *S. aureus* in yogurt was less significant. Enterocins are usually effective against food-borne pathogens such as *Listeria spp*. and *Clostridium spp*., but their use in food systems is unlikely to prevent pathogen regrowth during storage (Zacharof and Lovitt 2012; Cursino et al. 2006). However, Arqués et al. (2011) found that the combined impact of two bacteriocins, reuterin and enterocin AS-48, reduced *L. monocytogenes* levels in milk below the detection limit after 4 and 24 hours. Ribeiro et al. (2017) recently demonstrated that a semi-purified enterocin generated by an *E. faecalis* strain was effective in decreasing *L. monocytogenes* contamination (in a dose-dependent manner) in fresh cheese. Furthermore, the maximum dose administered to cheese (about 2000 AU/g of cheese) led in a pathogen decrease below levels of detectability that lasted throughout the storage period (72 h).

8.7.5 Other Bacteriocins

Many bacteriocins with a high specific activity range can be used as an alternative (Abee et al. 1995). As a result, there has been a lot of interest in finding novel bacteriocins with a stability in various dietary settings, broad range of antibacterial activity, heat tolerance, and proteolytic enzyme resistance.

8.8 ROLE OF BACTERIOCINS IN THE PHARMACEUTICAL INDUSTRY

8.8.1 Treatment of Pathogen-associated Diseases

The ability of antibiotics to limit the growth of clinically relevant pathogens such as multidrug-resistant *Staphylococci*, *Streptococci*, *Enterococci*, and *Clostridia*, has enable it to be a promising antimicrobial medicine under development.

Furthermore, several antibiotics are specifically active against a few Gram-negative bacteria species, including *Neisseria* and *Helicobacter* strains. These lantibiotics have the ability to operate as antibacterial agents while also having a low tendency to build resistance, making them ideal for medicinal use (Auke Johan van Heel, Montalban-Lopez, and Kuipers 2011; Brown et al. 2012). Several lantibiotics are now being studied in vitro and in vivo for antibacterial studies, and a few others are in preclinical and clinical development. For example, a phase I clinical study for the treatment of *Clostridium difficile* infections (Kuo et al. 2013) is presently underway with the semi-synthetic carboxy-amide derivative of the globular class II lantibiotic deoxyactagardine B (NVB302, Novacta Biosystems Limited). Antibiotics were originally licenced by the USFDA in 1951 and were used in animal feed, resulting in a considerable decrease in the number of deaths caused by bacterial infections. However, due to concerns about antibiotic overuse, the problem of multiple drug resistance bacteria has grown increasingly significant (Joerger 2003). Bacteriocins have been shown to suppress the action of pathogens such as Shiga toxin-producing *E. coli* (STEC), enterotoxigenic *E. coli* (ETEC), methicillin-resistant *Staphylococcus aureus* (MRSA), VRE, *Agrobacterium, and Brenneria spp.* in animals and plants (Grinter et al. 2012; Cotter et al. 2013). The bactericidal action of bacteriocins is mostly observed in the receptor-binding region of bacterium surfaces, followed by cytotoxicity across the membrane. Bacteriocins are also low-toxic peptides or proteins that are highly susceptible to proteases like trypsin and pepsin (Cleveland et al. 2001). Jordi et al., (2001) discovered that 20 different *E. coli* strains could produce colicin, which inhibited five different Shiga toxin-producing *E. coli* strains (O26, O111, O128, O145, and O157: H7). This E. coli can cause diarrhoea and hemolytic uremic syndrome in humans. *E. coli* producing colicin E1, E4, E8-J, K, and S4 can considerably inhibit STEC growth in a simulated cow rumen environment. In-vitro efficacy of pure colicin E1 and colicin N against enterotoxigenic *E. coli* pathogens F4 (K88) and F18, which caused post-weaning diarrhoea in pigs. In addition, pure colicin E1 proteins were put into the feed of young pigs. F-18 positive *E. coli* reduced the frequency of post-weaning diarrhea (Cutler et al. 2007). As a result, the piglets' growth was improved. Józefiak et al. (2013) fed broiler chickens a nisin-supplemented bird meal and discovered a lower number of bacteroides and enterobacteriacae in the ileal digesta of the chicken supplemented with nisin. Nisin has a similar mechanism of action as salinomycin. The mean body weight gain of nisin-supplemented (2700 IU nisin/g) chickens after 35 days is 1918 g/bird, compared to 1729 g for non-nisin-supplemented chickens and 1763 g for salinomycin-supplemented chickens. Class II low molecular mass bacteriocin OR-7 was isolated from the *Lactobacillus salivarius* strain NRRL B-30514 (Stern et al. 2006). *Campylobacter jejuni*, a human gastroenteritis pathogen, was shown to be resistant to this bacteriocin. OR-7 remained stable when heated to 90°C with lysozyme and lipase, or at pH ranges of 3.0 to 9.1. For chicken feed, pure OR-7 was encapsulated in polyvinylpyrrolidone. In the fecal substances of OR-7-treated hens, *C. jejuni* population were decreased by at least one million times compared to non-OR-7 supplemented animals. These findings indicate that nisin, OR-7, and other bacteriocins have promise as antibiotic substitutes in chicken and other animal diets (Yang et al. 2014).

8.8.2 Cancer Therapy

Cancer has evolved into a severe disease and a hazard to human health during the last half-century. According to new data from the WHO, 8.2 million people died from cancer in 2012, with 14.1 million new cancer cases, with Africa, Asia, and Central and South America accounting for 60% of the world's total new cancer cases. Lung cancer (2.7 percent, including trachea and bronchus cancer) was the 7th greatest cause of death in 2011, with 1.5 million (2.7 percent) fatalities, up from 1.2 million (2.2 percent, 9th) deaths in 2000. In 2013, there were 1,660,290 new cancer cases in the United States, with 580,350 cancer deaths expected, making cancer the second biggest cause of death, only after cardiac disease (Siegel et al. 2013).

Bacteriocins have been shown to have effect against tumor cells in various studies in cancer treatment. They may be ideal as a possible anti-tumor medication option because they are naturally and lawfully added to foods. Chumchalova and Smarda (2003) found that pore-forming colicin A and E1 hindered the proliferation of one standard human fibroblast line, MRC5, and 11 human tumor-cell lines. Colicin D, E2, E3, and pore-forming colicin A significantly inhibited the viability of P388, whereas pore-forming colicin E1 and colicin E3 repressed v-myb-transformed chicken monoblasts. Bures et al. (1986) isolated *E. coli* from the feces of 77 colorectal cancer patients, finding bacteriocin-producing *E. coli* in 32 patients (41.6%). The presence of bacteriocin-producing *E. coli* in the feces of 160 healthy adults was found in 102 (63.8 percent), indicating that colicins from bacteria in the colon may be one of the reasons in lowering human colorectal cancer. Colicins have the potential to be a moderately effective anti-cancer medication. Supplementing with bacteriocin-producing probiotics might also help to prevent cancer. Joo et al. (2012) discovered that nisin has the ability to suppress cancer cell proliferation in recent research. Nisin was used to treat three types of head and neck squamous cell carcinoma (HNSCC): 17B, HSC, and 14A, at doses of 40 and 80 g/ml. Nisin promoted DNA fragmentation or apoptosis, stopped the cell cycle, and inhibited HNSCC cell growth after 24 hours. The anti-HNSCC function of nisin was tested using a floor-of-mouth oral cancer xenograft mice model. When the mice were given a 150-mg/kg oral dosage of nisin daily for three weeks, the tumor mass was dramatically decreased in the nisin-treated animals compared to the control mice that just had water. These findings suggest that nisin is a risk-free and effective treatment for HNSCC.

8.9 CONCLUSION AND FUTURE PROSPECTS

The potency of bacteriocins in biotechnology has been known for years, as it is employed in several fields such as agriculture, aquaculture, food preservation, pharmaceutical industries, and health care. Due to the current spread of lethal diseases with a limited number of approved drugs, serious steps for discovering of novel compound-like bacteriocins can improve its role as antimicrobial agent. Bacteriocins are antimicrobial peptides, which can be produced on a large scale with greater application as therapeutic agent and as preservative agent. With the advent of recombinant DNA technology and bioprocess engineering, gene coding for bacteriocin is cloned and regulated for industrial production using prokaryotic host. In in vivo studies, the active bacteriocins have shown promising results. Evaluating the safety of these bacteriocins and authorized pharmaceutical with human trials should be done for its further applications. Further, in food preservation it has shown great outcomes to eliminate spores, and increasing the shelf life of food. Until now nisin has only been authorized to use as a food preservative. But there are also chances of few more bacteriocins whose efficiency needs to be verified.

Bacteriocin has a potent activity and might be used to replace antibiotics in the future because of its less toxicity and its proteinaceous nature which can help to solve the problem of antibiotic resistance. Besides the synergic effect of bacteriocins with a reduced dose of antibiotics can give greater killing activity on pathogens. While exploring the use of bacteriocins, care should be taken for its appropriate use as we are already dealing with antibiotic resistance. Bacteriocins have huge potential in field of pharmaceuticals and food preservation.

REFERENCES

Abee, Tjakko, Lothar Krockel, and Colin Hill. 1995. "Bacteriocins: Modes of Action and Potentials in Food Preservation and Control of Food Poisoning." *International Journal of Food Microbiology* 28 (2): 169–85. doi:10.1016/0168-1605(95)00055-0.

Acar, J. 1997. "Broad- and Narrow-Spectrum Antibiotics: An Unhelpful Categorization." *Clinical Microbiology and Infection* 3 (4). Blackwell Publishing Ltd: 395–96. doi:10.1111/j.1469-0691.1997.tb00274.x.

Acedo, Jeella Z., Sorina Chiorean, John C. Vederas, and Marco J. van Belkum. 2018. "The Expanding Structural Variety among Bacteriocins from Gram-Positive Bacteria." *FEMS Microbiology Reviews*, 42 (6): 805–28. doi:10.1093/femsre/fuy033/5063573.

Ahmad, Varish, Mohd Sajid Khan, Qazi Mohammad Sajid Jamal, Mohammad A. Alzohairy, Mohammad A. al Karaawi, and Mughees Uddin Siddiqui. 2017. "Antimicrobial Potential of Bacteriocins: In Therapy, Agriculture and Food Preservation." *International Journal of Antimicrobial Agents*, 49 (1): 1–11. doi:10.1016/j.ijantimicag.2016.08.016.

Alanis, Alfonso J. 2005. "Resistance to Antibiotics: Are We in the Post-Antibiotic Era?" *Archives of Medical Research,* 36 (6): 697–705. doi:10.1016/j.arcmed.2005.06.009.

Alvarez-Sieiro, Patricia, Manuel Montalbán-López, Dongdong Mu, and Oscar P. Kuipers. 2016. "Bacteriocins of Lactic Acid Bacteria: Extending the Family." *Applied Microbiology and Biotechnology*. 100 (7): 2939–51. doi:10.1007/s00253-016-7343-9.

Antonio, Linelle A, Princess A. Alfonso, Erika Joyce C. Tolentino, Chester P. Tabios, Prince Dean S. Sebastian, Gabriel V. Salamat, Danica C Gracila, Mariousz Ian N. Tabug, and Bernadino M. Hagosojos. 2021. "A Review of Lactic Acid Bacteria and Their Bacteriocins: Classification, Biosynthesis, and Mechanism against Oral Pathogens." *Asian Journal of Biological and Life Sciences* 10 (2): 291–99. doi:10.5530/ajbls.2021.10.41.

Arauz, Luciana Juncioni de, Angela Faustino Jozala, Priscila Gava Mazzola, and Thereza Christina Vessoni Penna. 2009a. "Nisin Biotechnological Production and Application: A Review." *Trends in Food Science and Technology* 20 (34): 146–54. doi:10.1016/j.tifs.2009.01.056.

Arqués, Juan L., Eva Rodríguez, Manuel Nuñez, and Margarita Medina. 2011. "Combined Effect of Reuterin and Lactic Acid Bacteria Bacteriocins on the Inactivation of Food-Borne Pathogens in Milk." *Food Control* 22 (3–4): 457–61. doi:10.1016/j.foodcont.2010.09.027.

Balay, Danielle R., Ramana V. Dangeti, Kamaljit Kaur, and Lynn M. McMullen. 2017. "Purification of Leucocin A for Use on Wieners to Inhibit *Listeria monocytogenes* in the Presence of Spoilage Organisms." *International Journal of Food Microbiology* 255: 25–31. doi:10.1016/j.ijfoodmicro.2017.05.016.

Balciunas, Eduardo Marcos, Fabio Andres Castillo Martinez, Svetoslav Dimitrov Todorov, Bernadette Dora Gombossy de Melo Franco, Attilio Converti, and Ricardo Pinheiro de Souza Oliveira. 2013. "Novel Biotechnological Applications of Bacteriocins: A Review." *Food Control* 32 (1): 134–42. doi:10.1016/j.foodcont.2012.11.025.

Beasley, Shea S., and Per E. J. Saris. 2004. "Nisin-Producing *Lactococcus lactis* Strains Isolated from Human Milk." *Applied and Environmental Microbiology* 70 (8): 5051–3. doi:10.1128/AEM.70.8.5051-5053.2004.

Bennallack, Philip R., and Joel S. Griffitts. 2017. "Elucidating and Engineering Thiopeptide Biosynthesis." *World Journal of Microbiology and Biotechnology* 33 (6): 119. doi:10.1007/s11274-017-2283-9.

Bernbom, Nete, Tine Rask Licht, Carl Henrik Brogren, Birthe Jelle, Anette H. Johansen, Iker Badiola, Finn K. Vogensen, and Birgit Nørrung. 2006. "Effects of *Lactococcus lactis* on Composition of Intestinal Microbiota: Role of Nisin." *Applied and Environmental Microbiology* 72 (1): 239–44. doi:10.1128/AEM.72.1.239-244.2006.

Bhalla, Tek Chand, Monika, Sheetal, and Savitri. 2019. "International Laws and Food-Borne Illness." *Food Safety and Human Health*, 319–71. doi:10.1016/B978-0-12-816333-7.00012-6.

Bhatti, Meena, Aparna Veeramachaneni, and Leora A. Shelef. 2004. "Factors Affecting the Antilisterial Effects of Nisin in Milk." *International Journal of Food Microbiology* 97 (2): 215–19. doi:10.1016/j.ijfoodmicro.2004.06.010.

Borrero, Juan, Eoin Kelly, Paula M. O'connor, Philip Kelleher, Colm Scully, Paul D. Cotter, Jennifer Mahony, and Douwe van Sinderen. 2017. "Plantaricyclin A, a Novel Circular Bacteriocin Produced by *Lactobacillus plantarum* NI326: Purification, Characterization, and Heterologous Production". *Appl Environ Microbiol.* 15: 84 (1): e01801–17. doi: 10.1128/AEM.01801-17.

Brown, Carla L., Karen Smith, Laura McCaughey, and Daniel Walker. 2012. "Colicin-like Bacteriocins as Novel Therapeutic Agents for the Treatment of Chronic Biofilm-Mediated Infection". *Biochemical Society Transactions*, 40: 1549–52. doi:10.1042/BST20120241.

Bures, J., V. Horák, B. Fixa, O. Komárková, K. Zaydlar, V. Lonský, and V. Masurka. 1986. "Colicinogeny in Colorectal Cancer." *Neoplasma* 33 (2): 233–7.

Burgos, María José Grande, Rubén Pérez Pulido, María del Carmen López Aguayo, Antonio Gálvez, and Rosario Lucas. 2014. "The Cyclic Antibacterial Peptide Enterocin AS-48: Isolation, Mode of Action, and Possible Food Applications." *International Journal of Molecular Sciences.* 15 (12): 22706–27. doi:10.3390/ijms151222706.

Camargo, Anderson C., Svetoslav D. Todorov, N. E. Chihib, D. Drider, and Luís A. Nero. 2018. "Lactic Acid Bacteria (LAB) and Their Bacteriocins as Alternative Biotechnological Tools to Control *Listeria monocytogenes* Biofilms in Food Processing Facilities." *Molecular Biotechnology.* 60 (9): 712–26. doi:10.1007/s12033-018-0108-1.

Carmen Martínez-Cuesta, M., José Bengoechea, Irene Bustos, Beatriz Rodríguez, Teresa Requena, and Carmen Peláez. 2010. "Control of Late Blowing in Cheese by Adding Lacticin 3147-Producing *Lactococcus lactis* IFPL 3593 to the Starter." *International Dairy Journal* 20 (1): 18–24. doi:10.1016/j.idairyj.2009.07.005.

Carpentier, Brigitte, and Olivier Cerf. 2011. "Review – Persistence of *Listeria monocytogenes* in Food Industry Equipment and Premises." *International Journal of Food Microbiology.* 145 (1): 1–8. doi:10.1016/j.ijfoodmicro.2011.01.005.

Cascales, Eric, Susan K. Buchanan, Denis Duché, Colin Kleanthous, Roland Lloubès, Kathleen Postle, Margaret Riley, Stephen Slatin, and Danièle Cavard. 2007. "Colicin Biology." *Microbiology and Molecular Biology Reviews* 71 (1): 158–229. doi:10.1128/mmbr.00036-06.

Castellano, Patricia, Mariana Pérez Ibarreche, Mariana Blanco Massani, Cecilia Fontana, and Graciela M. Vignolo. 2017a. "Strategies for Pathogen Biocontrol Using Lactic Acid Bacteria and Their Metabolites: A Focus on Meat Ecosystems and Industrial Environments." *Microorganisms* 5 (3): 38. doi:10.3390/microorganisms5030038.

Castro, Ana, Joana Silva, and Paula Teixeira. 2018. "*Staphylococcus aureus*, a Food Pathogen: Virulence Factors and Antibiotic Resistance." *Foodborne Diseases*, 15: 213–38. doi:10.1016/B978-0-12-811444-5.00008-7.

Ceotto-Vigoder, H., S. L. S. Marques, I. N. S. Santos, M. D. B. Alves, E. S. Barrias, A. Potter, D. S. Alviano, and M. C. F. Bastos. 2016. "Nisin and Lysostaphin Activity against Preformed Biofilm of *Staphylococcus aureus* Involved in Bovine Mastitis." *Journal of Applied Microbiology* 121 (1): 101–14. doi:10.1111/jam.13136.

Chen, H., and D. G. Hoover. (2003). "Bacteriocins and their Food Applications." *Comprehensive Reviews in Food Science and Food Safety,* 2(3): 82–100. doi:10.1111/j.1541-4337.2003.tb00016.x

Chikindas, Michael L., Richard Weeks, Djamel Drider, Vladimir A. Chistyakov, and Leon M. T. Dicks. 2018. "Functions and Emerging Applications of Bacteriocins." *Current Opinion in Biotechnology*. 49: 23–28. doi:10.1016/j.copbio.2017.07.011.

Chumchalova, J., and J. Smarda. 2003. "Human Tumor Cells Are Selectively Inhibited by Colicins." *Folia Microbiol.* 48(1): 111–5. doi: 10.1007/BF02931286.

Cintas, L M, P Casaus, M F Ferna ´ Ndez, and P E Herna. 1998. "Comparative Antimicrobial Activity of Enterocin L50, Pediocin PA-1, Nisin A and Lactocin S against Spoilage and Foodborne Pathogenic Bacteria." *Food Microbiology* 15 (3): 289–98. doi.org/10.1006/fmic.1997.0160.

Cleveland, Jennifer, Thomas J. Montville, Ingolf F. Nes, and Michael L. Chikindas. 2001. "Bacteriocins: Safe, Natural Antimicrobials for Food Preservation." *International Journal of Food Microbiology*. 71 (1): 1–20. doi: 10.1016/s0168-1605(01)00560-8.

Coelho, Marcus Lívio Varella, Bruna Gonçalves Coutinho, Olinda Cabral da Silva Santos, Ingolf F. Nes, and Maria do Carmo de Freire Bastos. 2014. "Immunity to the *Staphylococcus aureus* Leaderless Four-Peptide Bacteriocin Aureocin A70 Is Conferred by AurI, an Integral Membrane Protein." *Research in Microbiology* 165 (1): 50–9. doi:10.1016/j.resmic.2013.11.001.

Coelho, Marcus Lívio Varella, Luana Rocha Fleming, and Maria do Carmo de Freire Bastos. 2016. "Insights into Aureocin A70 Regulation: Participation of Regulator AurR, Alternative Transcription Factor ΣB and Phage Φ11 Regulator CI." *Research in Microbiology* 167 (2): 90–102. doi:10.1016/j.resmic.2015.10.004.

Collins, Fergus W. J., Paula M. O'Connor, Orla O'Sullivan, Beatriz Gómez-Sala, Mary C. Rea, Colin Hill, and R. Paul Ross. 2017. "Bacteriocin Gene-Trait Matching across the Complete Lactobacillus Pan-Genome." *Scientific Reports* 7 (1): 3481. doi:10.1038/s41598-017-03339-y.

Costa, Roger J. da, Flávia L. S. Voloski, Rafael G. Mondadori, Eduarda H. Duval, and Ângela M. Fiorentini. 2019. "Preservation of Meat Products with Bacteriocins Produced by Lactic Acid Bacteria Isolated from Meat." *Journal of Food Quality*. 1–12. doi:10.1155/2019/4726510.

Cotter, P., Hill, C., and Ross, R. 2005. Bacteriocins: Developing Innate Immunity for Food. *Nat Rev Microbiol* 3, 777–788.

Cotter, Paul D., R. Paul Ross, and Colin Hill. 2013. "Bacteriocins-a Viable Alternative to Antibiotics?" *Nature Reviews Microbiology*. 11 (2): 95–105. doi:10.1038/nrmicro2937.

Cui, Yanlong, Lingli Luo, Xin Wang, Yingying Lu, Yanglei Yi, Yuanyuan Shan, Bianfang Liu, Yuan Zhou, and L. Xin. 2021a. "Mining, Heterologous Expression, Purification, Antibactericidal Mechanism, and Application of Bacteriocins: A Review." *Comprehensive Reviews in Food Science and Food Safety* 20 (1): 863–899. doi:10.1111/1541-4337.12658.

Cursino, L., D. Šmajs, J. Šmarda, R. M. D. Nardi, J. R. Nicoli, E. Chartone-Souza, and A. M. A. Nascimento. 2006. "Exoproducts of the *Escherichia coli* Strain H22 Inhibiting Some Enteric Pathogens Both in Vitro and in Vivo." *Journal of Applied Microbiology* 100 (4): 821–9. doi:10.1111/j.1365-2672.2006.02834.x.

Cutler, S. A., S. M. Lonergan, N. Cornick, A. K. Johnson, and C. H. Stahl. 2007. "Dietary Inclusion of Colicin E1 Is Effective in Preventing Postweaning Diarrhea Caused by F18-Positive *Escherichia coli* in Pigs." *Antimicrobial Agents and Chemotherapy* 51 (11): 3830–5. doi:10.1128/AAC.00360-07.

Daba, G. M., N. Ishibashi, T. Zendo, and K. Sonomoto. 2017. "Functional Analysis of the Biosynthetic Gene Cluster Required for Immunity and Secretion of a Novel Lactococcus-Specific Bacteriocin, Lactococcin Z." *Journal of Applied Microbiology* 123 (5): 1124–32. doi:10.1111/jam.13564.

Daba, Ghoson M., and Waill A. Elkhateeb. 2020. "Bacteriocins of Lactic Acid Bacteria as Biotechnological Tools in Food and Pharmaceuticals: Current Applications and Future Prospects." *Biocatalysis and Agricultural Biotechnology*. 28: 101750. doi:10.1016/j.bcab.2020.101750.

Daba, Ghoson Mosbah, Naoki Ishibashi, Xiao Gong, Hiroya Taki, Keisuke Yamashiro, Yen Yi Lim, Takeshi Zendo, and Kenji Sonomoto. 2018. "Characterisation of the Action Mechanism of a Lactococcus-Specific Bacteriocin, Lactococcin Z." *Journal of Bioscience and Bioengineering* 126 (5): 603–10. doi:10.1016/j.jbiosc.2018.05.018.

Dabour, Nassra, Annina Zihler, Ehab Kheadr, Christophe Lacroix, and Ismail Fliss. 2009. "In Vivo Study on the Effectiveness of Pediocin PA-1 and *Pediococcusacidilactici* UL5 at Inhibiting Listeria Monocytogenes." *International Journal of Food Microbiology* 133 (3): 225–33. doi:10.1016/j.ijfoodmicro.2009.05.005.

Davies, E. A., H. E. Bevis, J. Delves-Broughton, V. I. Es, and H. E. B. Ev. 1997. "The Use of the Bacteriocin, Nisin, as a Preservative in Ricotta-Type Cheeses to Control the Food-Borne Pathogen *Listeria monocytogenes*." *Letters in Applied Microbiology* 24 (5): 343–6. doi: 10.1046/j.1472-765x.1997.00145.x.

Dobson, Alleson, Paul D. Cotter, R. Paul Ross, and Colin Hill. 2012. "Bacteriocin Production: A Probiotic Trait?" *Applied and Environmental Microbiology*. 78 (1): 1–6. doi:10.1128/AEM.05576-11.

Dubin, Krista, and Eric G. Pamer. 2017. "Enterococci and Their Interactions with the Intestinal Microbiome." *Microbiology Spectrum* 5 (6): 1–16. doi:10.1128/microbiolspec.bad-0014-2016.

Egan, Kevin, Des Field, Mary C. Rea, R. Paul Ross, Colin Hill, and Paul D. Cotter. 2016. "Bacteriocins: Novel Solutions to Age Old Spore-Related Problems?" *Frontiers in Microbiology* 7: 461. doi:10.3389/fmicb.2016.00461.

Ekblad, Bie, Jon Nissen-Meyer, and Tom Kristensen. 2017. "Whole-Genome Sequencing of Mutants with Increased Resistance against the Two-Peptide Bacteriocin Plantaricin JK Reveals a Putative Receptor and Potential Docking Site." *PLoS ONE* 12 (9): e0185279. doi:10.1371/journal.pone.0185279.

Favaro, Lorenzo, Ana Lúcia Barretto Penna, and Svetoslav Dimitrov Todorov. 2015. "Bacteriocinogenic LAB from Cheeses - Application in Biopreservation?" *Trends in Food Science and Technology*. 41 (1): 37–48. doi:10.1016/j.tifs.2014.09.001.

Ferreira', M. A. S. S., and B. M. Lund. 1996. "The Effect of Nisin on *Listeria monocytogenes* in Culture Medium and Long-Life Cottage Cheese." *Letters in Applied Microbiology*. 22 (6): 433–8. doi: 10.1111/j.1472-765x.1996.tb01197.x.

Gabrielsen, Christina, Dag A. Brede, Ingolf F. Nes, and Dzung B. Diep. 2014. "Circular Bacteriocins: Biosynthesis and Mode of Action." *Applied and Environmental Microbiology*. 80 (22): 6854–62. doi:10.1128/AEM.02284-14.

Gálvez, Antonio, Rosario Lucas López, Rubén Pérez Pulido, and María José Grande Burgos. 2014. "Application of Lactic Acid Bacteria and Their Bacteriocins for Food Biopreservation." *Springer Briefs in Food, Health, and Nutrition*, 15–22. doi:10.1007/978-1-4939-2029-7_3.

Garsa, Anita Kumari, Rashmi Kumariya, Anil Kumar, Puja Lather, Suman Kapila, and S. K. Sood. 2014. "Industrial Cheese Whey Utilization for Enhanced Production of Purified Pediocin PA-1." *LWT - Food Science and Technology* 59 (2P1): 656–65. doi:10.1016/j.lwt.2014.07.008.

Garvey, M., and N. J. Rowan. 2019. "Pulsed UV as a Potential Surface Sanitizer in Food Production Processes to Ensure Consumer Safety." *Current Opinion in Food Science* 26: 65–70. doi:10.1016/j.cofs.2019.03.003.

Gaynes, Robert. 2017. "The Discovery of Penicillin—New Insights after More than 75 Years of Clinical Use." *Emerging Infectious Diseases* 23 (5): 849–53. doi:10.3201/eid2305.161556.

Gharsallaoui, Adem, Nadia Oulahal, Catherine Joly, and Pascal Degraeve. 2016. "Nisin as a Food Preservative: Part 1: Physicochemical Properties, Antimicrobial Activity, and Main Uses." *Critical Reviews in Food Science and Nutrition* 56 (8): 1262–74. doi:10.1080/10408398.2013.763765.

Gholami, D., S. Aminzadeh, S. M. Alavi, N. Kazemipour, A. Ghoroghi, and Z. Emruzi. 2018. "Comparison of Antibiotics and Bacteriocins Antibacterial Activity on *Xanthomonas Citri* Subsp.Citri." *Iranian Journal of Fisheries Sciences*, 17 (1): 162–178. doi:10.22092/ijfs.2018.115592.

Gibson, Glenn R., Robert Hutkins, Mary Ellen Sanders, Susan L. Prescott, Raylene A. Reimer, Seppo J. Salminen, Karen Scott et al. 2017. "Expert Consensus Document: The International Scientific Association for Probiotics and Prebiotics (ISAPP) Consensus Statement on the Definition and Scope of Prebiotics." *Nature Reviews Gastroenterology and Hepatology*. 1–12. doi:10.1038/nrgastro.2017.75.

GradisteanuPircalabioru, Gratiela, Laura Ioana Popa, Luminita Marutescu, Irina Gheorghe, Marcela Popa, Ilda Czobor Barbu, Rodica Cristescu, and Mariana Carmen Chifiriuc. 2021a. "Bacteriocins in the Era of Antibiotic Resistance: Rising to the Challenge." *Pharmaceutics* 13 (2): 196. doi:10.3390/pharmaceutics13020196.

Grinter, Rhys, Joel Milner, and Daniel Walker. 2012. "Bacteriocins Active against Plant Pathogenic Bacteria." *Biochemical Society Transactions*, 40: 1498–1502. doi:10.1042/BST20120206.

Hahn-Löbmann, Simone, Anett Stephan, Steve Schulz, Tobias Schneider, Anton Shaverskyi, Daniel Tusé, Anatoli Giritch, and Yuri Gleba. 2019. "Colicins and Salmocins – New Classes of Plant-Made Non-Antibiotic Food Antibacterials." *Frontiers in Plant Science* 10: 437. doi:10.3389/fpls.2019.00437.

Haruta, Shin, and Nanako Kanno. 2015. "Survivability of Microbes in Natural Environments and Their Ecological Impacts." *Microbes and Environments* 30 (2): 123–5. doi:10.1264/jsme2.ME3002rh.

Heel, Auke J. van, Manuel Montalban-Lopez, Quentin Oliveau, and Oscar P. Kuipers. 2017. "Genome-Guided Identification of Novel Head-to-Tail Cyclized Antimicrobial Peptides, Exemplified by the Discovery of Pumilarin." *Microbial Genomics* 3 (10): e000134. doi:10.1099/mgen.0.000134.

Heel, Auke Johan van, Manuel Montalban-Lopez, and Oscar P. Kuipers. 2011. "Evaluating the Feasibility of Lantibiotics as an Alternative Therapy against Bacterial Infections in Humans." *Expert Opinion on Drug Metabolism and Toxicology*. 7 (6): 675–80. doi:10.1517/17425255.2011.573478.

Hill, Colin, Francisco Guarner, Gregor Reid, Glenn R. Gibson, Daniel J. Merenstein, Bruno Pot, Lorenzo Morelli, et al. 2014. "Expert Consensus Document: The International Scientific Association for Probiotics and Prebiotics Consensus Statement on the Scope and Appropriate Use of the Term Probiotic." *Nature Reviews Gastroenterology and Hepatology* 11 (8): 506–14. doi:10.1038/nrgastro.2014.66.

Ibarguren, Carolina, Gustavo Céliz, Antonela S. Díaz, M. Alejandra Bertuzzi, Mirta Daz, and M. Carina Audisio. 2015. "Gelatine Based Films Added with Bacteriocins and a Flavonoid Ester Active against Food-Borne Pathogens." *Innovative Food Science and Emerging Technologies* 28: 66–72. doi:10.1016/j.ifset.2015.01.007.

Iordache, Florin, Irina Gheorghe, Veronica Lazar, Carmen Curutiu, Lia Mara Ditu, Alexandru Mihai Grumezescu, and Alina Maria Holban. 2017. "Nanostructurated Materials for Prolonged and Safe Food Preservation." *Food Preservation*, 305–35. Elsevier. doi:10.1016/b978-0-12-804303-5.00009-2.

Iwatani, Shun, Fuminori Yoneyama, Shiho Miyashita, Takeshi Zendo, Jiro Nakayama, and Kenji Sonomoto. 2012. "Identification of the Genes Involved in the Secretion and Self-Immunity of Lacticin Q, an Unmodified Leaderless Bacteriocin from *Lactococcus lactis* QU 5." *Microbiology (United Kingdom)* 158 (12): 2927–35. doi:10.1099/mic.0.062943-0.

Iwatani, Shun, Naoki Ishibashi, Floirendo P Flores, Takeshi Zendo, Jiro Nakayama, and Kenji Sonomoto. 2016. "LnqR, a TetR-Family Transcriptional Regulator, Positively Regulates Lacticin Q Production in Lactococcus Lactis QU 5." FEMS Microbiol Lett. 363 (18): fnw200. doi: 10.1093/femsle/fnw200.

Iwatani, Shun, Yuko Horikiri, Takeshi Zendo, Jiro Nakayama, and Kenji Sonomoto. 2013. "Bifunctional Gene Cluster LnQBCDEF Mediates Bacteriocin Production and Immunity with Differential Genetic Requirements." *Applied and Environmental Microbiology* 79 (7): 2446–9. doi:10.1128/AEM.03783-12.

Jacob, F., A. Lwoff, A. Siminovitch, and E. Wollman. 1953. "Definition of Some Terms Relative to Lysogeny." *Annales de l'Institut Pasteur* 84 (1): 222–4.

Joerger, R. D. 2003. "Alternatives to Antibiotics: Bacteriocins, Antimicrobial Peptides and Bacteriophages." *Poultry Science* 82 (4): 640–7. doi: 10.1093/ps/82.4.640.

Joo, Nam E., Kathryn Ritchie, Pachiyappan Kamarajan, Di Miao, and Yvonne L. Kapila. 2012. "Nisin, an Apoptogenic Bacteriocin and Food Preservative, Attenuates HNSCC Tumorigenesis via CHAC1." *Cancer Medicine* 1 (3): 295–305. doi:10.1002/cam4.35.

Jordi, Bart J. A. M., Khalil Boutaga, Caroline M. E. van Heeswijk, Frans van Knapen, and Len J. A. Lipman. 2001. "Sensitivity of Shiga toxin-producing Escherichia coli (STEC) strains for colicins under different experimental conditions." *FEMS Microbiology Letters* 13; 204 (2): 329–34. doi: 10.1111/j.1574-6968.2001.tb10906.x.

Józefiak, Damian, Bartosz Kierończyk, Jerzy Juśkiewicz, Zenon Zduńczyk, Mateusz Rawski, Jakub Długosz, Anna Sip, and Ole Højberg. 2013. "Dietary Nisin Modulates the Gastrointestinal Microbial Ecology and Enhances Growth Performance of the Broiler Chickens." *PLoS ONE* 8 (12): e85347. doi:10.1371/journal.pone.0085347.

Juturu, Veeresh, and Jin Chuan Wu. 2018. "Microbial Production of Bacteriocins: Latest Research Development and Applications." *Biotechnology Advances*. 36: 2187–200. doi:10.1016/j.biotechadv.2018.10.007.

Khan, Imran, and Deog Hwan Oh. 2016. "Integration of Nisin into Nanoparticles for Application in Foods." *Innovative Food Science and Emerging Technologies*. 34: 376–84. doi:10.1016/j.ifset.2015.12.013.

Kitagawa, Norio, Takahito Otani, and Tetsuichiro Inai. 2019. "Nisin, a Food Preservative Produced by Lactococcus Lactis, Affects the Localization Pattern of Intermediate Filament Protein in HaCaT Cells." *Anatomical Science International* 94 (2): 163–71. doi:10.1007/s12565-018-0462-x.

Klaenhammer, Todd R. 1993. "Genetics of Bacteriocins Produced by Lactic Acid Bacteria." *FEMS Microbiology Reviews* 12 (1–3): 39–85. doi:10.1111/j.1574-6976.1993.tb00012.x.

Kleanthous, Colin. 2010. "Swimming against the Tide: Progress and Challenges in Our Understanding of Colicin Translocation." *Nature Reviews Microbiology* 8 (12): 843–48. doi:10.1038/nrmicro2454.

Kumariya, Rashmi, Anita Kumari Garsa, Y. S. Rajput, S. K. Sood, Nadeem Akhtar, and Seema Patel. 2019. "Bacteriocins: Classification, Synthesis, Mechanism of Action and Resistance Development in Food Spoilage Causing Bacteria." *Microbial Pathogenesis* 128: 171–7. doi:10.1016/j.micpath.2019.01.002.

Kuo, Yang Cheng, Cheng Feng Liu, Jhao Fen Lin, An Chieh Li, Ta Chun Lo, and Thy Hou Lin. 2013. "Characterization of Putative Class II Bacteriocins Identified from a Non-Bacteriocin-Producing Strain *Lactobacillus casei* ATCC 334." *Applied Microbiology and Biotechnology* 97 (1): 237–46. doi:10.1007/s00253-012-4149-2.

Kykkidou, Sonia, Nikolaos Pournis, Olga K. Kostoula, and Ioannis N. Savvaidis. 2007. "Effects of Treatment with Nisin on the Microbial Flora and Sensory Properties of a Greek Soft Acid-Curd Cheese Stored Aerobically at 4°C." *International Dairy Journal* 17 (10): 1254–8. doi:10.1016/j.idairyj.2007.02.006.

Lajis, Ahmad Firdaus B. 2020a. "Biomanufacturing Process for the Production of Bacteriocins from Bacillaceae Family." *Bioresources and Bioprocessing* 7 (8): 2–26. doi:10.1186/s40643-020-0295-z.

Luna-Guevara, J. J., M. M. P. Arenas-Hernandez, C. Martínez De La Peña, Juan L. Silva, and M. L. Luna-Guevara. 2019. "The Role of Pathogenic *E. Coli* in Fresh Vegetables: Behavior, Contamination Factors, and Preventive Measures." *International Journal of Microbiology* 1–10. doi:10.1155/2019/2894328.

Maky, Mohamed Abdelfattah, Naoki Ishibashi, Takeshi Zendo, Rodney Honrada Perez, Jehan Ragab Doud, Mohamed Karmi, and Kenji Sonomoto. 2015. "Enterocin F4-9, a Novel O-Linked Glycosylated Bacteriocin." *Applied and Environmental Microbiology* 81 (14): 4819–26. doi:10.1128/AEM.00940-15.

Masuda, Y., R. H. Perez, T. Zendo, and K. Sonomoto. 2016. "Nutrition-Adaptive Control of Multiple-Bacteriocin Production by *Weissellahellenica* QU 13." *Journal of Applied Microbiology* 120 (1): 70–79. doi:10.1111/jam.12997.

Meade, Elaine, Mark Anthony Slattery, and Mary Garvey. 2020. "Bacteriocins, Potent Antimicrobial Peptides and the Fight against Multi Drug Resistant Species: Resistance Is Futile?" *Antibiotics* 9(1): 32. doi:10.3390/antibiotics9010032.

Melo, J., P. W. Andrew, and M. L. Faleiro. 2015. "Listeria Monocytogenes in Cheese and the Dairy Environment Remains a Food Safety Challenge: The Role of Stress Responses." *Food Research International* 67: 75–90. doi:10.1016/j.foodres.2014.10.031.

Michel-Briand, Yvon, and Christine Baysse. 2002. "The Pyocins of *Pseudomonas aeruginosa*." 84 (5–6): 499–510. doi.org/10.1016/S0300-9084(02)01422-0.

Mokoena, Mduduzi Paul. 2017. "Lactic Acid Bacteria and Their Bacteriocins: Classification, Biosynthesis and Applications against Uropathogens: A Mini-Review." *Molecules*. 22 (8): 1255. doi:10.3390/molecules22081255.

Morgan, S. M., M. Galvin, R. P. Ross, and C. Hill. 2001. "Evaluation of a Spray-Dried Lacticin 3147 Powder for the Control of *Listeria monocytogenes* and *Bacillus cereus* in a Range of Food Systems." *Letters in Applied Microbiology* 33(5): 387–91. doi: 10.1046/j.1472-765x.2001.01016.x.

Morton, James T., Stefan D. Freed, Shaun W. Lee, and Iddo Friedberg. 2015. "A Large Scale Prediction of Bacteriocin Gene Blocks Suggests a Wide Functional Spectrum for Bacteriocins." *BMC Bioinformatics* 16 (1): 381. doi:10.1186/s12859-015-0792-9.

Mu, Fuqin, Yoshimitsu Masuda, Takeshi Zendo, Hiroshi Ono, Hiroshi Kitagawa, Haruo Ito, Jiro Nakayama, and Kenji Sonomoto. 2014. "Biological Function of a DUF95 Superfamily Protein Involved in the Biosynthesis of a Circular Bacteriocin, Leucocyclicin Q." *Journal of Bioscience and Bioengineering* 117 (2): 158–64. doi:10.1016/j.jbiosc.2013.06.023.

Muñoz, A., S. Ananou, A. Gálvez, M. Martínez-Bueno, A. Rodríguez, M. Maqueda, and E. Valdivia. 2007. "Inhibition of *Staphylococcus aureus* in Dairy Products by Enterocin AS-48 Produced in Situ and Ex Situ: Bactericidal Synergism with Heat." *International Dairy Journal* 17 (7): 760–9. doi:10.1016/j.idairyj.2006.09.006.

Nascimento, Janaína dos Santos, Marcus Lívio Varella Coelho, Hilana Ceotto, Amina Potter, Luana Rocha Fleming, Zhian Salehian, Ingolf F. Nes, and Maria do Carmo de Freire Bastos. 2012. "Genes Involved in Immunity to and Secretion of Aureocin A53, an Atypical Class II Bacteriocin Produced by *Staphylococcus aureus* A53." *Journal of Bacteriology* 194 (4): 875–83. doi:10.1128/JB.06203-11.

Negash, Abebe Worku, and Berhanu AndualemTsehai. 2020. "Current Applications of Bacteriocin." *International Journal of Microbiology*. 1–7. doi:10.1155/2020/4374891.

Netz, Daili Jacqueline Aguilar, Maria do Carmo de Freire Bastos, and Hans Georg Sahl. 2002. "Mode of Action of the Antimicrobial Peptide Aureocin A53 from *Staphylococcus aureus*." *Applied and Environmental Microbiology* 68 (11): 5274–80. doi:10.1128/AEM.68.11.5274-5280.2002.

Ng, Zhang Jin, Mazni Abu Zarin, Chee Keong Lee, and Joo Shun Tan. 2020. "Application of Bacteriocins in Food Preservation and Infectious Disease Treatment for Humans and Livestock: A Review." *RSC Advances* 10 (64): 38937–64. doi:10.1039/d0ra06161a.

Niamah, Alaa K. 2018. "Structure, Mode of Action and Application of Pediocin Natural Antimicrobial Food Preservative: A Review." *Basrah Journal of Agricultural Sciences* 31 (1): 59–69. doi:10.33762/bagrs.2018.160126.

Nissen, H., H. Holo, L. Axelsson, and H. Blom. 2001. "Characterization and Growth of Bacillus spp. in Heat-treated Cream with and without Nisin." *Journal of Applied Microbiology* 90 (4): 530–4. doi: 10.1046/j.1365-2672.2001.01279.x.

O'Bryan, Corliss A., Ok Kyung Koo, Michael L. Sostrin, Steven C. Ricke, Philip G. Crandall, and Michael G. Johnson. 2018. "Characteristics of Bacteriocins and Use as Food Antimicrobials in the United States." *Food and Feed Safety Systems and Analysis*, 273–86. doi:10.1016/B978-0-12-811835-1.00015-4.

O'Connor, Paula M., R. Paul Ross, Colin Hill, and Paul D. Cotter. 2015. "Antimicrobial Antagonists against Food Pathogens: A Bacteriocin Perspective." *Current Opinion in Food Science* 2: 51–7. doi:10.1016/j.cofs.2015.01.004.

O'Sullivan, Lisa, Maire P. Ryan, R. Paul Ross, and Colin Hill. 2003. "Generation of Food-Grade Lactococcal Starters Which Produce the LantibioticsLacticin 3147 and Lacticin 481." *Applied and Environmental Microbiology* 69 (6): 3681–85. doi:10.1128/AEM.69.6.3681-3685.2003.

Oman, Trent J., and Wilfred A. van der Donk. 2010. "Follow the Leader: The Use of Leader Peptides to Guide Natural Product Biosynthesis." *Nature Chemical Biology* 6 (1): 9–18. doi:10.1038/nchembio.286.

Ongey, Elvis Legala, Hüseyin Yassi, Stephan Pflugmacher, and Peter Neubauer. 2017. "Pharmacological and Pharmacokinetic Properties of Lanthipeptides Undergoing Clinical Studies." *Biotechnology Letters* 39 (4): 473–82. doi:10.1007/s10529-016-2279-9.

Oshima, S., A. Hirano, H. Kamikado, J. Nishimura, Y. Kawai, and T. Saito. 2014. "Nisin A Extends the Shelf Life of High-Fat Chilled Dairy Dessert, a Milk-Based Pudding." *Journal of Applied Microbiology* 116 (5): 1218–28. doi:10.1111/jam.12454.

Oumer, A., S. Garde, P. Gaya, M. Medina, M. Nuñez, and Nuñež Nuñez. 2001. "The Effects of Cultivating Lactic Starter Cultures with Bacteriocin-Producing Lactic Acid Bacteria." *Journal of Food Protection* 64 (1): 81–6. doi: 10.4315/0362-028x-64.1.81.

Perez Rodney H., Takeshi Zendo, and Kenji Sonomoto. 2018. "Circular and Leaderless Bacteriocins: Biosynthesis, Mode of Action, Applications, and Prospects." *Frontiers in Microbiology* 9: 2085. doi:10.3389/fmicb.2018.02085.

Perez, Rodney H., Takeshi Zendo, and Kenji Sonomoto. 2014. "Novel Bacteriocins from Lactic Acid Bacteria (LAB): Various Structures and Applications." *Microbial Cell Factories* 13: 53. doi:10.1186/1475-2859-13-S1-S3.

Pimentel-Filho, Natan de J., Hilário C. Mantovani, Antônio F. de Carvalho, Ricardo S. Dias, and Maria Cristina D. Vanetti. 2014. "Efficacy of Bovicin HC5 and Nisin Combination against *Listeria monocytogenes* and *Staphylococcus Aureus* in Fresh Cheese." *International Journal of Food Science & Technology* 49 (2): 416–22. doi:10.1111/ijfs.12316.

Prasad Ram, Vivek Kumar, Joginder Singh, Chandrama Prakash Upadhyaya 2020. *Environmental and Microbial Biotechnology Recent Developments in Microbial Technologies*. Springer Publishers.

Pucci, M. J., E. R. Vedamuthu, B. S. Kunka, and P. A. Vandenbergh. 1988. "Inhibition of *Listeria monocytogenes* by Using Bacteriocin PA-1 Produced by *Pediococcusacidilactici* PAC 1.0." *Applied and Environmental Microbiology* 54 (10): 2349–53. doi:10.1128/aem.54.10.2349-2353.1988.

Rashid, Rafi, Mark Veleba, and Kimberly A. Kline. 2016. "Focal Targeting of the Bacterial Envelope by Antimicrobial Peptides." *Frontiers in Cell and Developmental Biology* 4: 55. doi:10.3389/fcell.2016.00055.

Ribeiro, Susana C., R. Paul Ross, Catherine Stanton, and Célia C.G. Silva. 2017. "Characterization and Application of AntilisterialEnterocins on Model Fresh Cheese." *Journal of Food Protection* 80 (8): 1303–16. doi:10.4315/0362-028X.JFP-17-031.

Rink, Rick, Leon D. Kluskens, Anneke Kuipers, Arnold J. M. Driessen, Oscar P. Kuipers, and Gert N. Moll. 2007. "NisC, the Cyclase of the Lantibiotic Nisin, Can Catalyze Cyclization of Designed Nonlantibiotic Peptides." *Biochemistry* 46 (45): 13179–89. doi:10.1021/bi700106z.

Ross, R Paul, Mary Galvin, Olivia Mcauliffe, Sheila M Morgan, Maire P Ryan, Denis P Twomey, William J Meaney, and Colin Hill. 1999. "Developing Applications for Lactococcal Bacteriocins." *Antonie van Leeuwenhoek* 76(1–4): 337–46.

Salgado, Pablo R., Cristian M. Ortiz, Yanina S. Musso, Luciana di Giorgio, and Adriana N. Mauri. 2015. "Edible Films and Coatings Containing Bioactives." *Current Opinion in Food Science* 5: 86–92. doi:10.1016/j.cofs.2015.09.004.

Sánchez-Hidalgo, Marina, Manuel Montalbán-López, Rubén Cebrián, Eva Valdivia, Manuel Martínez-Bueno, and Mercedes Maqueda. 2011. "AS-48 Bacteriocin: Close to Perfection." *Cellular and Molecular Life Sciences* 68 (17): 2845–57. doi:10.1007/s00018-011-0724-4.

Scholl, Dean. 2017. "Phage Tail-Like Bacteriocins." *Annual Reviews in Virology*. 29; 4 (1): 453–67. doi: 10.1146/annurev-virology-101416-041632.

Scholz, Romy, Joachim Vater, Anto Budiharjo, Zhiyuan Wang, Yueqiu He, Kristin Dietel, Torsten Schwecke, Stefanie Herfort, Peter Lasch, and Rainer Borriss. 2014. "Amylocyclicin, a Novel Circular Bacteriocin Produced by *Bacillus amyloliquefaciens* FZB42." *Journal of Bacteriology* 196 (10): 1842–52. doi:10.1128/JB.01474-14.

Sidhu Parveen Kaur, and Kiran Nehra. 2021. "Bacteriocins of Lactic Acid Bacteria as Potent Antimicrobial Peptides against Food Pathogens." Biomimetics, Edited by Maki K. Habib and César Martín-Gómez, IntechOpen, doi: 10.5772/intechopen.95747.

Siegel, R., Naishadham, D., Jemal, A. Cancer statistics, 2013. CA Cancer J Clin. 2013 Jan;63(1):11-30. doi: 10.3322/caac.21166. Epub 2013 Jan 17. PMID: 23335087.

Silva, Célia C. G., Sofia P. M. Silva, and Susana C. Ribeiro. 2018. "Application of Bacteriocins and Protective Cultures in Dairy Food Preservation." *Frontiers in Microbiology* 9: 594. doi:10.3389/fmicb.2018.00594.

Simons, Alexis, Kamel Alhanout, and Raphaël E. Duval. 2020. "Bacteriocins, Antimicrobial Peptides from Bacterial Origin: Overview of Their Biology and Their Impact against Multidrug-Resistant Bacteria." *Microorganisms* 8 (5): 639. doi:10.3390/microorganisms8050639.

Sobrino-López, A., and O. Martín-Belloso. 2008. "Use of Nisin and Other Bacteriocins for Preservation of Dairy Products." *International Dairy Journal* 18 (4): 329–43. doi:10.1016/j.idairyj.2007.11.009.

Stern, N. J., E. A. Svetoch, B. V. Eruslanov, V. V. Perelygin, E. V. Mitsevich, I. P. Mitsevich, V. D. Pokhilenko, V. P. Levchuk, O. E. Svetoch, and B. S. Seal. 2006. "Isolation of a *Lactobacillus salivarius* Strain and Purification of Its Bacteriocin, Which Is Inhibitory to Campylobacter Jejuni in the Chicken Gastrointestinal System." *Antimicrobial Agents and Chemotherapy* 50 (9): 3111–16. doi:10.1128/AAC.00259-06.

Uzelac, Gordana, Marija Miljkovic, Jelena Lozo, Zorica Radulovic, Natasa Tosic, and Milan Kojic. 2015. "Expression of Bacteriocin LsbB Is Dependent on a Transcription Terminator." *Microbiological Research* 179: 45–53. doi:10.1016/j.micres.2015.06.011.

Villalobos-Delgado, Luz H., G. V. Nevárez-Moorillon, Irma Caro, Emiliano J. Quinto, and Javier Mateo. 2019. "Natural Antimicrobial Agents to Improve Foods Shelf Life." *Food Quality and Shelf Life*, 125–57. doi:10.1016/b978-0-12-817190-5.00004-5.

Vimont, Allison, Benoît Fernandez, Riadh Hammami, Ahlem Ababsa, Hocine Daba, and Ismaïl Fliss. 2017. "Bacteriocin-Producing *Enterococcus faecium* LCW 44: A High Potential Probiotic Candidate from Raw Camel Milk." *Frontiers in Microbiology* 8: 865. doi:10.3389/fmicb.2017.00865.

Wencewicz, Timothy A., and Marvin J. Miller. 2018. "Sideromycins as Pathogen-Targeted Antibiotics." *Topics in Medicinal Chemistry*, 26: 151–83. doi:10.1007/7355_2017_19.

Wiebach, Vincent, Andi Mainz, Mary Ann J. Siegert, Natalia A. Jungmann, Guillaume Lesquame, Sophie Tirat, Assia Dreux-Zigha, Jozsef Aszodi, Dominique le Beller, and Roderich D. Süssmuth. 2018. "The Anti-Staphylococcal Lipolanthines Are Ribosomally Synthesized Lipopeptides Brief-Communication." *Nature Chemical Biology* 14 (7): 652–54. doi:10.1038/s41589-018-0068-6.

Yang, Shih Chun, Chih Hung Lin, Calvin T. Sung, and Jia You Fang. 2014. "Antibacterial Activities of Bacteriocins: Application in Foods and Pharmaceuticals." *Frontiers in Microbiology* 5: 241. doi:10.3389/fmicb.2014.00241.

Yildirim, Z., N. Öncül, M. Yildirim, and Karabiyikli. 2016. "Application of Lactococcin BZ and Enterocin KP against *Listeria monocytogenes* in Milk as Biopreservation Agents." *Acta Alimentaria* 45 (4): 486–92. doi:10.1556/066.2016.45.4.4.

Younes, Maged, Peter Aggett, Fernando Aguilar, Riccardo Crebelli, Birgit Dusemund, Metka Filipič, Maria Jose Frutos et al. 2017a. "Safety of Nisin (E 234) as a Food Additive in the Light of New Toxicological Data and the Proposed Extension of Use." *EFSA Journal* 15 (12): 1–16. doi:10.2903/j.efsa.2017.5063.

Zacharof, M. P., and R. W. Lovitt. 2012. "Bacteriocins Produced by Lactic Acid Bacteria a Review Article." *APCBEE Procedia* 2: 50–56. doi:10.1016/j.apcbee.2012.06.010.

Zhao, Xingchen, Ce Shi, Rizeng Meng, Zonghui Liu, Yanjun Huang, Ziwen Zhao, and Na Guo. 2016. "Effect of Nisin and Perilla Oil Combination against *Listeria monocytogenes* and *Staphylococcus aureus* in Milk." *Journal of Food Science and Technology* 53 (6): 2644–53. doi:10.1007/s13197-016-2236-6.

Zhou, Hui, Jun Fang, Yun Tian, and Xiang Yang Lu. 2014. "Mechanisms of Nisin Resistance in Gram-Positive Bacteria." *Annals of Microbiology*. 64 (2): 413–420. doi:10.1007/s13213-013-0679-9.

Zimina, Maria, Olga Babich, Alexander Prosekov, Stanislav Sukhikh, Svetlana Ivanova, Margarita Shevchenko, and Svetlana Noskova. 2020. "Overview of Global Trends in Classification, Methods of Preparation and Application of Bacteriocins." *Antibiotics* 9 (9): 1–21. doi:10.3390/antibiotics9090553.

9 Plant-based Metabolites as Source of Antimicrobial Therapeutics

Prospects and Challenges

Nagella Praveen ,[1] Ravi Shankar Thenmozhi Preetha,[1] Vishakha Pagare,[1] Jyothis Devasia,[2] Aatika Nizam,[2] Epsita Mukherjee,[3] Veerappa Lakshmaiah Vasantha,[1] and Jameel Mohammed Al-Khayri[4]*

[1]Department of Life Sciences, CHRIST (Deemed to be University), Hosur Road, Bangalore, India
[2]Department of Chemistry, CHRIST (Deemed to be University), Hosur Road, Bangalore, India
[3]Amity Institute of Biotechnology, Amity University, Noida, India
[4]Department of Agricultural Biotechnology, College of Agriculture and Food Sciences, King Faisal University, Al-Ahsa, Saudi Arabia
Corresponding author: jkhayri@kfu.edu.sa; jmkhayri@yahoo.com

CONTENTS

DOI: 10.1201/9781003268932-9

9.1 INTRODUCTION

Plants are used as traditional medicines from ancient times to today as they are the largest living storehouses of bio-chemicals and pharmaceuticals known on Earth (Abdallah, 2011). The World Checklist of Vascular Plants (WCVP) database reported in April 2021 that there are 1,383,297 plant names with 996,093 plants identified at species level, constituting 342,953 accepted vascular plant species (Govaerts et al., 2021). Around 10% of the reported vascular plants are used as medicines (Salmerón-Manzano et al., 2020). According to the MPNS, 33,443 species are recorded as being used for medicinal purpose (MNPS, 2021). Medicinal plants are those that have therapeutic properties which can pose pharmacological effect on the human or animal body (Namdeo, 2018). About 80% of the world's population depends on plant-based medicine for treatment of diseases (Okoye et al., 2014). The medicinal property of a plant is attributed to rich and diverse secondary metabolites (Allemailem, 2021). Secondary metabolites are intermediates or products of primary metabolism that are not involves directly in the growth and development of the plant (Jain et al., 2019). Plants generate secondary metabolites in response to stresses posed by biotic factors (bacteria, fungi, viruses, parasites, pests, weeds, and herbivore animals) and abiotic environmental factors (temperature, salinity, drought, UV radiation etc.) so as to adapt and survive in response to environmental stimuli during their life time (Yang et al., 2018). Secondary metabolites help the plants by inducing flowering, fruit set, and abscission. They also aid in attracting pollinators. Secondary metabolites have various applications in industries as they are sources of dyes, adhesives, oils, waxes, fragrances, and flavors. Along with all these vast applications, one of the major applications of secondary metabolites is their use in therapeutics. They possess antioxidant, antibacterial, antifungal, antiviral, and anti-inflammatory properties to mention a few (Berhardt, 2016). Khazir et al. have described the beneficial effects of secondary metabolites in cancer prevention (Khazir et al., 2014). There are many studies reported on the potential use of secondary metabolites in preventing certain chronic diseases like cardiovascular disease, neurodegenerative diseases, and type II diabetes (Berhardt, 2016). They act as antimicrobials to attract or repel microbes. Both traditional and modern medicines rely on plant secondary metabolites for their action (Teoh, 2016). Presently around two lakh different plant secondary metabolites have been isolated and identified and are used to treat infectious diseases. Ancient Indian and Chinese folk medicines have used plants extensively for the treatment of various infectious diseases (Gorlenko et al., 2020). The WHO defines infectious diseases as disease caused by pathogenic microorganisms, such as bacteria, viruses, parasites or fungi that can be spread, directly or indirectly, from one person to another (World Health Organization – Eastern Mediterranean Region). Infectious diseases are treated using antimicrobials which are defined as medicines that can prevent and treat infections in plants, animals, and humans. It includes antibiotics, antivirals, antifungals, and antiparasitics ("Antimicrobial resistance," WHO). The infectious diseases can be cured by using antibiotic drugs, but antimicrobial resistance in pathogens over time due to rapid bacterial evolution and biofilm forming ability makes the medicine no longer responsive to the infection, increasing the risk and severity of illness (Mandal et al., 2017). Antimicrobial resistance poses serious threats to human health due to inappropriate use of antibiotics. New antimicrobials are the need of the hour to combat multiple drug resistance in these pathogenic organisms ("Antimicrobial resistance," WHO). Annual death numbers worldwide due to antimicrobial resistance continue to increase and are projected to reach around ten million by 2050 (Chassagne et al., 2021). Various studies on the development of new plant-based bio-actives are increasing recently with the hope to combat antimicrobial resistance (Subramani et al., 2017). On the basis of biosynthetic origin, plant secondary metabolites can be classified broadly into three categories: phenolic compounds, terpenes, and alkaloids (Berhardt, 2016).

9.2 CLASSIFICATION OF PLANT SECONDARY METABOLITES

A mode of natural defense against microbial attacks or herbivory, plants secrete certain organic compounds which, though not involved in cell growth and development, are significant contributors to maintenance of ecological interactions. They are called secondary metabolites and are classified into specific categories on the basis of a variety of factors that govern the biochemical characteristics of secondary metabolites. Some of the factors include chemical structures including the presence of a number of sugar units, solubility in water or other organic solvents, presence or absence of nitrogen in the composition along with the variation in their biosynthetic pathways (Mera et al., 2019). The latter has been considered as the most suitable and commonly used method for classification of secondary metabolites into three specific categories, as: terpenes, alkaloids, and phenolic compounds (Figure 9.1). Having special importance in the chemical conversion processes in plants, these compounds are of utmost importance to various pharmaceutical industries and have specific utilities as gums, flavors, insecticides, and other commercially important products. As described by Czapek, secondary metabolites are the end products of nitrogen metabolism after certain "secondary modifications" like deamination which could be easily analyzed using phytochemical techniques

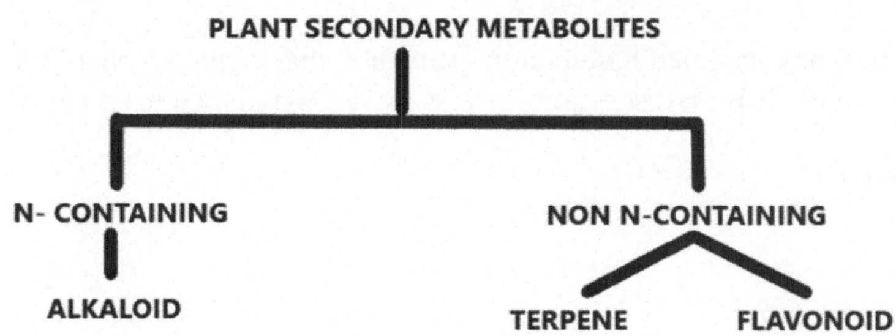

FIGURE 9.1 Most commonly accepted classification of plant secondary metabolites based on composition and variation of biochemical pathways.

like chromatography (Bourgaud et al., 2001). Thus, nitrogen plays a vital role in the classification of secondary metabolites in plants, as N-containing and non-nitrogenous substances.

These compounds are sub-classified on the basis of several other factors like number of isoprene units in case of terpenes, number of carbon atoms in the skeleton of phenolic compounds, biosynthetic origin of alkaloids into true, pseudo, and protoalkaloids and others. Apart from this, there are other criteria for classification of alkaloids, like pharmacological properties, presence or absence of heterocyclic nucleus, or distribution in other plant families. The production of secondary metabolites occurs through various pathways in the plant cells like shikimic acid pathway, malonate pathway, polyacetate pathway, and others. The functions of these secondary metabolites are more diverse than can be listed, such as generating astringent taste in them to prevent attack by herbivores. The utility also extends to the synthesis of toxic phytoalexins to prevent microbial infection.

Based on the biosynthetic pathway plant secondary metabolites are classified as:

- Phenolic compounds
- Terpenes
- Alkaloids (Berhardt, 2016).

9.2.1 Phenolic Compounds

They are the largest and most divergent group of plant secondary metabolites (Mandal et al., 2017). These compounds contain atleast an aromatic ring with one or more hydroxyl group (Berhardt, 2016; Hussein and El-Anssary, 2019). Their structure can vary from a single molecule of low molecular weight to more complex polymeric structures. Naturally occurring phenolic compounds are found to be conjugated with carbohydrate linked to one or more phenolic groups (Vuolo et al., 2019). There are more than 8,000 phenolic compounds in the plant kingdom reported (Rio et al., 2013). Phenolic compounds significantly contribute to the color, taste, and flavor to herbs and food (Hussein and El-Anssary, 2019). They have beneficial biological properties such as antioxidants, metal chelators, and free radical scavengers; they also possess various pharmacological activities like anti-inflammatory, anti-allergic, anti-carcinogenic, and antimicrobial activities. The substitution position in the aromatic ring and the length of saturated chain influences the antimicrobial activity (Mandal et al., 2017).

From a biosynthetic point of view, there are two pathways by which phenolic compounds are generated:

- Shikimate pathway – most common pathway for synthesis of phenolic acids, flavonoids and lignans.
- Phenylpropanoid pathway – common pathway for polycyclic compound synthesis like coumarins, xanthones and quinines (Vuolo et al., 2019; Berhardt, 2016).

Phenolic compounds are classified based on the structure, that is, the number and arrangement of carbon atoms (Berhardt, 2016; Hussein and El-Anssary, 2019) into two main classes:

- Non-flavonoids
- Flavonoids.

9.2.2 Non-Flavonoids

Based on the carbon skeleton and specific substituents in the basic structures, non-flavonoids are classified into sub-types:

- Simple phenols
- Phenolic acids and derivatives
- Phenones
- Phenylacetic acids and derivatives
- Hydrolysable tannins
- Stilbenes (Kougan et al., 2013).

9.2.2.1 Simple Phenols

They are compounds with one or more aromatic rings where at least one hydrogen in the ring is substituted by a hydroxyl group (Heleno et al., 2015). Simple phenol includes phenol, catechol, pyrogallol, resorcinol, phloroglucinol, and others (Kougan et al., 2013).

Catechol and pyrogallol were reported to show high antibacterial activity against *Pseudomonas putida*, *Pseudomonas pyocyanea*, and *Corynebacterium xerosis*. Catechol was also found to act as a potent antifungal agent against *Fusarium oxysporum* and *Penicillium italicum* (Kocaçalişkan et al., 2006).

9.2.2.2 Phenolic Acid and Derivatives

Phenolic acids have a benzene ring to which a carboxyl group is attached (Kougan et al., 2013). The antimicrobial activity of phenolic acids has wide application in food preservation and packaging (Sánchez-Maldonado et al., 2011).

Phenolic acids include two groups: hydroxybenzoic acids (C_6–C_1 structures) and hydroxycinnamic acids (C_6–C_3 structures) (Sánchez-Maldonado et al., 2011).

The hydroxybenzoic acid derivatives includes salicylic acid, 4-hydroxybenzoic acid, protocatechuic acid, gentisic acid, vanillic acid, syringic acid, gallic acid, ellagic acid, and hexahydroxydiphenic acid (ellagic acid dilactone) (Vuolo et al., 2019).

The hydroxycinnamic acid derivatives include cinnamic acid, coumaric acid, caffeic acid, ferulic acid, and sinapic acid. The hydroxycinnamic acids p-coumaric and ferulic acids (4-hydroxy-3-methoxycinnamic acid) are the common precursors for most of the chemically diverse phenylpropanoid derivatives having antimicrobial functions (Macoy et al., 2015).

Gallic acid is known for its wide distribution. It is found to have antibacterial, antiviral, antifungal, anti-inflammatory, antitumor, choleretic, and anti-anaphylactic properties in vitro (Harborne et al., 1998). Epigallocatechingallate, tannic acid, quercetin, and epicatechin were found to inhibit beta-lactamase and biofilm forming ability of *Staphylococcus epidermidis* and *Pseudomonas aeruginosa*, thereby acting as potent antibacterials (Mandal et al., 2017).

9.2.2.3 Phenones, Phenylacetic Acid and Derivatives

Phenones are aromatic ketones containing phenyl group attached to carbonyl group. They include acetophenone, benzophenone, and so on. Phenylacetic acid and their derivatives have a benzene ring with two side chains. Some of their derivatives include acetophenone, benzophenone, phenylacetaldehyde, and phenylethanol. They are also referred to as the C6-C2 class of phenolic compounds (Kougan et al., 2013).

9.2.2.4 Tannins

Tannins are polyphenols with large number of hydroxyl or other functional groups (Chung et al., 1998). Tannins can be hydrolysable, condensed and complex (Al Mamari, 2021). The hydrolysable tannins have a polyhydric alcohol group at the center that are esterified by gallic acid or hexahydrodiphenic acid to yield gallotannins and ellagitannins respectively. Polymers of flavan-3-ols or catechin units are termed condensed tannins, while complex tannins are complex in structure and include catechin units bound to gallotannins or ellagitannins (Chung et al., 1998; Al Mamari, 2021).

Tannic acid is a gallotannin that is found to inhibit food-borne bacteria and aquatic bacteria. Their antimicrobial property is associated with hydrolysis of ester linkage between polyols and gallic acid (Chung et al., 1998).

9.2.2.5 Stilbenes

Stilbenes are molecules containing two phenol units linked by two double bonded carbon atoms with a classical C6-C2-C6 structure with hydroxyl groups. In plants they exist as aglycones or glycosides that protect plants against bacterial, mold, or fungal infections. Examples of stilbenes include phytoalexins, resveratrol, piceatannol, pterostilbene, and so on (Watson, 2017; Al Mamari, 2021).

Resveratrol and 3,5,40 –trihydroxystilbene were found to inhibit growth of certain bacterial species like *Bacillus subtilis*, *Bacillus brevis*, *Enterobacter dissolvens*, and *Micrococcus luteus*. They act as antifungal agent against *Nematospora coryli*, *Rhizomucor miehei*, *Penicillium notatum*, and *Colletotrichum graminicola*. Their antimicrobial activity was found to act like other antibiotics like tetracycline, streptomycin, ampicillin, or kanamycin (antibacterials) and nourseothricin or hygromycin (antifungals) (Albert et al., 2011).

9.2.2.6 Flavonoids

Flavonoids are large class of small molecular secondary metabolites with phenyl benzopyran framework, that is, C6-C3-C6 carbon framework. There are different sub-types of flavonoids depending on the level of unsaturation and oxidation that include: flavonols, flavanols, flavanonols, flavones, isoflavones, flavanones, chalcones, aurones, anthocyanidins, and so on. All these sub-types have a common flavonoid structure containing a nuclear 2-phenyl-benzo-γ-pyrane with two benzene rings – Ring A and Ring B through a heterocyclic pyran or pyrone ring C. Flavonoids have potential antimicrobial activity due to their ability to act on bacterial cell wall and prevent biofilm formation. It is also being reported that plant flavonoids have the ability to revert antibiotic resistance and enhance the action of current antibiotic drugs (Górniak et al., 2019; Hussein and El-Anssary, 2019; Al Mamari, 2021). There are nearly 5,000 flavonoids reported and the number will continue to increase (Berhardt, 2016).

9.2.3 Terpenes

Terpenes are also called as isoprenoids or terpenoids due to the presence of a five-carbon isoprene unit as a building block in their structure. The isoprene unit has 5 carbon atoms and is represented as $(C_5H_8)_n$ that is present within the terpene skeleton. More than 25,000 terpenes are reported of which most have cyclic structure. The antimicrobial activity of terpenes is determined by the presence or absence of certain functional groups (Berhardt, 2016; Hussein and El-Anssary, 2019).

Terpenes are classified on the basis of the number of isoprene units present in their structure. They can be hemiterpene (n=1; contains only one isoprene unit), monoterpene (C10 i.e, two isoprene units (n=2)), sesquiterpenes (n=3; C15), diterpenes (n=4; C20), sesterpenes (n=5; C25), triterpenes (n=6; C30), and tetraterpenes (n=8; C40) (Berhardt, 2016; Hussein and El-Anssary, 2019).

9.2.3.1 Hemiterpenes

The isoprene unit itself is considered as hemiterpene. Angelic acid and isovaleric acid are examples of hemiterpenes containing oxygen derivatives (Hussein and El-Anssary, 2019).

9.2.3.2 Monoterpenes

They contain two isoprene units. Some of the examples include carvacrol, thymol, menthol, and geraniol. These monoterpenes were found to be effective antimicrobials against Gram-positive and Gram-negative bacteria. The susceptibility of multiple drug resistant *Enterobacter aerogenes* was found to be efficiently increased by geraniol. Menthol and thymol were found to effectively inhibit the growth of *S. aureus* and *E. coli.* Carvacrol and carvone were found to be efficient antimicrobials against *E. coli* and *S. typhimurium* (Mahizan et al., 2019).

α-terpinene, γ-terpinene, α-pinene, ρ-cymene, terpinen-4-ol, α-terpineol, thymol, citral, borneol, bornyl acetate, isoborneol, 1, 8-cineole, thujone, and camphor were reported to efficiently act against *Herpes simplex virus* type 1 (HSV-1) (Dunkić et al., 2010).

9.2.3.3 Sesquiterpenes

They contain three isoprene units. Their structure can be acyclic, monocyclic, or bicyclic. Farnesol, bisabolol, and caryophyllene are the examples for acyclic, monocyclic, and bicyclic sesquiterpenes respectively (Hussein and El-Anssary, 2019). Farnesol was found to show antibacterial effects against *Streptococcus mutans*, *S. sobrinus*, *S. aureus*, and *S. epidermidis* by inhibiting biofilm formation. Xanthorrhizol, a phenol sesquiterpene, was found to inhibit growth of *Mycobacterium smegmatis* (Mahizan et al., 2019; Jin et al., 2010).

9.2.3.4 Diterpenes

Diterpenes are made up of four isoprene units that can be either acyclic or macrocyclic. Diterpenes are widely used as combinational drug with antibiotics. Salvipisone and aethiopinone are diterpenes that have antibacterial and antibiofilm activities against *S. aureus*, *Enterococcus faecalis*, and *S. epidermidis* (Mahizan et al., 2019).

9.2.3.5 Triterpenes

Triterpenes are composed of six isoprene units. Oleanic acid, bonianic acid A and B, and ursolic acid are triterpenes that act effectively against *Mycobacterium tuberculosis.* Oleanic acid when combined with rifampicin, isoniazide, and ethambutol was found to act on multiple drug resistant microorganisms (Mahizan et al., 2019; Hussein and El-Anssary, 2019).

9.2.4 Alkaloids

Nitrogen-containing heterocyclic compounds are termed alkaloids. On the basis of their core chemical structures, alkaloids are classified as: acridones, aromatics, carbolines, ephedras, ergots, imidazoles, indoles, bisindoles, indolizidines, manzamines, oxindoles, quinolines, quinozolines, phenylisoquinolines, phenylethylamines, piperidines, purines, pyrrolidines, pyrrolizidines, pyrroloindoles, pyridines, and simple tetrahydroisoquinolines (Hussein and El-Anssary, 2019).

Isoquinoline alkaloids, pyridine alkaloids, indole alkaloids, and steroidal alkaloids show significant antimicrobial properties (Yan et al., 2021; Cushnie et al., 2014).

9.2.5 Isoquinoline Alkaloids

They are the largest alkaloid group that comprises of simple isoquinoline alkaloids, benzylisoquinoline alkaloids, bisbenzylisoquinoline alkaloids, aporphine alkaloids, and protoberberine alkaloids. Thalicfoetine is an isoquinoline alkaloid that can inhibit the growth of *Bacillus subtilis.* Spathullin A and B are reported to be active against *Enterobacter cloacae*, *Escherichia coli*, *Pseudomonas aeruginosa, Acinetobacter baumannii*, *Klebsiella pneumonia*, and *S. aureus* (Yan et al., 2021; Cushnie et al., 2014).

9.2.5.1 Pyridine Alkaloids

Pyridine alkaloids are derivatives of pyridine or piperidine. They are further divided into simple pyridine alkaloids, quinolizidine alkaloids, and indolizidine alkaloids. Lanatine A, cermizine C, jussiaeiine A, jussiaeiine B, 3-(4-hydroxyphenyl)-4-(3-methoxy-4-hydroxyphenyl)-3,4-dehydroquinolizidine, and (−)-N-methyl-cytisine were found to inhibit the growth of *S. aureus* and

E. coli. Pyridine alkaloids can efficiently act as broad-spectrum antimicrobials (Yan et al., 2021; Cushnie et al., 2014).

9.2.5.2 Indole Alkaloids

Indole alkaloids are the derivatives of tryptophan. Voacafricines A and B are monoterpeneindole alkaloids that can act against *S. aureus* and *Salmonella typhimurium.* Indole alkaloids can efficiently be used against drug resistant bacteria (Yan et al., 2021).

9.2.5.3 Steroidal Alkaloids

Steroidal alkaloids are nitrogen-containing derivatives, where the nitrogen atom is localized in the ring structure of natural steroids. Steroidal alkaloids have potent antibacterial activity against methicillin-sensitive *S. aureus*. Mokluangin A, B, and C steroid alkaloids were found to be active against *B. subtilis* and *E. coli* (Yan et al., 2021).

The main antibacterial mechanism of alkaloids includes inhibition of cell wall synthesis of bacteria, alteration in cell membrane permeability, inhibition of bacterial metabolism, and nucleic acid and protein synthesis inhibition (Cushnie et al., 2014b; Yan et al., 2021).

9.3 EXPLORING PLANT-BASED METABOLITES AS BIOCOMPATIBLE THERAPEUTICS

9.3.1 Antimicrobial Activity of Metal Nanoparticle Coated Plant Secondary Metabolites

Metal-nanoparticles can be made from either the complete plant or a plant extract. Plant extracts have a better bioreduction capability than microbial cultures, according to the findings. Furthermore, waste products formed by plant-mediated synthesis processes are normally environmentally friendly, whereas waste products generated by microbial synthesis methods are likely to be harmful to the environment, depending on the microbe used. The plant secondary metabolites in the extracts not only reduce metal ions, but they also stabilize the nanoparticles that have been created. Besides that, the biogenic nanoparticle manufacturing approach appears to be reproducible, environmentally benign, and safe for human therapeutic usage (Hussein and El-Anssary, 2019; Prasad et al. 2016; Srivastava et al. 2021;).

Metal-based nanomaterials serve as both a nano-carrier and a nano-drug for a variety of administration methods, as well as quick immune system detection. In the flavonoid-nanoparticle complex, hydrogen bonds and hydrophobic interactions are prominent. These interactions have also improved flavonoid's aqueous solubility at target sites. Nanoparticles shield flavonoids from GI breakdown, and the nanoparticles can then be taken up directly by epithelial cells in the small intestine, considerably increasing flavonoids' absorption and bioavailability (Sathishkumar et al., 2018).

Flavonoids have a negligible or undetectable cytotoxic effect on healthy cells. It also protects normal cells from AgNP-induced cytotoxicity and oxidative stress. As a result, flavonoids-mediated nanomaterial (FMN) is a bioactive therapeutic tool for treating a variety of disorders, as well as an immunosuppressive agent alternative. By reacting with carbonyl groups or electrons, the flavonoids in *Dalbergia spinosa* leaf extract were adsorbed onto the metal surface and showed antibacterial activity against *Bacillus subtilis*, *P. aeruginosa*, *Staphylococcus aureus*, and *E. coli* (Hussein and El-Anssary, 2019). In the treatment of methicillin-resistant *S. aureus*, baicalein works in tandem with tetracycline and lactams. It also inhibits *P. aeruginosa*'s quorum sensing and biofilm formation. Antibiofilm activity against *P. aeruginosa* was also improved by Baicalein-mediated AuNP production (Hussein and El-Anssary, 2019).

Quercetin may scavenge free radicals produced by metals, and it has been established that quercetin is primarily responsible for metal ion reduction and stabilization. The study found that bimetallic Ag-SeNPs mediated by quercetin have superior therapeutic efficacy in terms of antioxidant and antibacterial activities (Hussein and El-Anssary, 2019). According to a study, the addition of phenol and flavonoid chemicals to AgNPs increased their bioactivity. The flavonoids extracted from *Potentilla fulge* had four times the activity of phenolics, implying that flavonoids play a role in nanoparticle stability. The opportunistic pathogenic microorganisms *E. coli* and *B. subtilis* were efficiently suppressed by these AgNPs (Hussein and El-Anssary, 2019).

Silver nanoparticles made from plant extracts have been demonstrated to have antibacterial action against Gram-positive and Gram-negative bacteria, multidrug-resistant pathogens, and fungal pathogens like *Candida albicans*. The antibacterial activity of crude silver nanoparticles against *P. aeruginosa* and *E. coli* was determined to be the best in a study (Birla et al., 2009). Gallic acid coated AgNPs as active antibiotics with little toxicity against human cells were described in another investigation. AuNPs, or gold nanoparticles, have been used for diagnostic and therapeutic purposes. Gallic acid was used to make AuNPs with an average diameter of 8 nanometers. For *Escherichia coli* and *Staphylococcus aureus*, these nanoparticles are mildly hazardous. Cinnamaldehyde-coated AuNPs with an average diameter of 113 nm prevent *Candida albicans* from forming hyphae and hence reduce the organism's pathogenicity (Hussein and El-Anssary, 2019).

9.3.2 Plant Secondary Metabolites (PSMs) for the Synthesis of Polymer Thin Films

Because of the persistent problem of bacterial and fungal colonization of artificial implantable materials, as well as the declining efficacy of traditional systemic antibiotics used to treat implant-associated infections, a variety of antifouling and antibacterial techniques have been developed. Surface functionalization and the manufacture of polymer thin films are accomplished using plant secondary metabolites (PSMs). They have a specific antibacterial mechanism of action and a broad spectrum of action (Al-Jumaili et al., 2018).

Biofilm protects microorganisms from antibiotics while also strengthening their adhesion to the surface. In a process known as "quorum sensing," microorganisms communicate with one another inside a biofilm by creating chemotactic particles or pheromones (Sonawane et al. 2022). Biofilm development is mainly caused by a single bacterial species in the case of infection of medical polymers, devices, and implants. In patients with pre-existing cardiac illness, microorganisms build biofilm in the compromised endothelium of native heart valves, producing *Candida* infection and endocarditis. Biofilm formation at the tip of urinary catheters has also been reported, even after only a short duration of exposure. Gram-positive *Enterococcus faecalis*, *Candida albicans, S. aureus, S. epidermidis*, and *Streptococcus viridans*, and Gram-negative *E. coli*, *Klebsiella pneumoniae*, *Proteus mirabilis*, and *P. aeruginosa* are among the bacteria that can cause infections. The prevalence of these infections is a major issue. *S. aureus* and *S. epidermidis*, for example, create biofilms on metallic implants, such as orthopedic screws, which can lead to device failure (Al-Jumaili et al., 2018).

9.3.2.1 Terpinen-4-ol

Tea tree oil contains terpinen-4-ol, a monocyclic terpene alcohol. Cyclic terpene hydrocarbons have been found to accumulate in the cell membrane after contact with microbes. This disrupts membrane integrity, causing an increase in passive proton flux through the membrane and the loss of the proton motive force. According to a study, plasma polymerized coatings produced from terpinen-4-ol have a high potential for reducing bacterial adhesion and metabolic activity in *S. aureus* and *P. aeruginosa* (Noumi et al., 2018).

9.3.2.3 Carvone

Carvone can be found in a variety of essential oils, including caraway, spearmint, and dill, among others. With reference to microbial cells, this PSM exhibits a variety of antiproliferative actions. Detoxifying enzymes are activated by carvone. The high enzyme-inducing effect of carvone is considered to be due to the unsaturated ketone groups. The antimicrobial performance of fabricated polymer coatings made from plasma polymerization of carvone was nearly equal against Gram-negative and Gram-positive bacteria (86% reduction in *E. coli* and 84% reduction in *S. aureus*), with no cytotoxic effect on primary human endothelial cells. Microorganisms linked to the surface of plasma polymers of carvone demonstrated membrane deformation, pore formation, and membrane rupture as a result of the findings (Chan et al., 2016).

9.3.2.4 Geranium

Geranium oil (*Pelargonium graveolens*) has been shown to be effective against various bacterial infections in studies. Selected human pathogens, such as *S. aureus, P. aeruginosa*, and *E. coli*, were found to have the ability to inhibit microbial adherence and biofilm development when coated with geranium oil. Geranium polymer films had a low density, uniform covering, good adhesion, and significant physical stability in addition to their biological activity (Al-Jumaili et al., 2018).

9.4 PLANT SECONDARY METABOLITES: KEY TARGET PLAYER

Secondary metabolites in plants are not essential for regular growth or development, but they are essential for reproduction and defense against bacteria, fungi, and viruses. These items have a lot of potential to be used as medicines. They are classified into several classes, including alkaloids, phenolics, polyphenols, flavonoids, quinones, tannins, coumarins, terpenes, lectins, and polypeptides, and play a significant role as antimicrobials and antivirals. Polyphenols' antimicrobial characteristics have been utilized to generate new food preservatives to replace synthetic preservatives, as well as novel medicines for the treatment of various microbial illnesses to combat antibiotic resistance (Sathishkumar et al., 2018).

Microorganisms have evolved a number of strategies to evade the effects of antimicrobial chemicals. Antimicrobials have a specific method of action against a specific biological target, which is one of the reasons why some bacteria are resistant to various compounds. Gram-positive bacteria, for example, develop resistance to β-lactam antibiotics by changing the penicillin-binding proteins, which are transpeptidases involved in cell wall formation. Bacteria can also remove harmful chemicals that have become absorbed by a mechanism involving efflux pumps. Furthermore, pathogenic bacteria use biofilm development to protect themselves from external stressors by forming a thick and sticky extracellular matrix including DNA, proteins, and polysaccharides. Furthermore, biofilm cells divide slowly, reducing the effectiveness of antibiotic compounds that target certain biological processes (Gorlenko et al., 2020).

9.5 OVERCOMING THE BACTERIAL DRUG RESISTANCE

Due to their inherent properties, Plant-Derived Natural Products (PDNPs) are regarded attractive prospects for the creation of novel medications (Gorlenko et al., 2020):

- Plugging the Efflux Pumps
  Compounds that block efflux pump activity are critical in the fight against drug resistance. Within the bacterial cell, catechol, piperine, quercetin, and resveratrol can efficiently inhibit these pumps. Polyphenolic compounds, for example, attach directly to structural proteins in the efflux pump canal, causing conformational changes that prevent the molecule from being eliminated.
- Attenuating the Bacterial Virulence
  The antibacterial activity of plant extracts is also influenced by crucial events in the pathogenic process. Thymol reduced the synthesis of α-haemolysin and staphylococcal enterotoxins A and B in both methicillin-sensitive and methicillin-resistant *S. aureus* isolates, according to a study, which is a key virulence factor generated by *S. pneumoniae* (Ingallina et al., 2017).

- Disrupting the Biofilms Bacterial biofilms
  Antibiofilm activity of phenylpropanoids and alkaloids has been discovered against *P. aeruginosa* (Zaixiang et al., 2019), *K. pneumoniae*, and staphylococcal biofilms (Sayout et al., 2020). They have an effect on pre-formed biofilms and also inhibit new ones from forming. Disruption of intercellular communication, disruption of cell-to-cell coaggregation, suppression of cell mobility, inactivation of bacterial adhesins, and promotion of bacterium dispersal are all examples of antibiofilm actions (Gorlenko et al., 2020).

9.6 MECHANISM OF ACTION OF PLANT-BASED SECONDARY METABOLITES

The characteristics of target cells (bacterial/fungal cells) and the environment in which antimicrobial action is required to determine the mechanism of action of various PSMs. Hydrophilicity, concentration, temperature, and pH are all important environmental factors (OmojateGodstime et al., 2014). Plant metabolites' method of action is also influenced by their chemical structure and characteristics. Flavonoids like myricetin, robinetin, and epigallocatechingallate, for example, impede bacterial DNA synthesis by creating hydrogen bonds between the B ring and DNA bases. Flavonoids of a different structure, such as quercetin, may bind to the GyrB protein in *E. coli* and limit its ATPase activity (Daglia, 2012).

Inhibition of cell wall or protein synthesis, inducing leakage from cells by tampering with membrane function, interfering with DNA/RNA synthesis/function, induction of cytoplasmic constituent coagulation, and interruption of normal cell communication are all possible mechanisms of antimicrobial action for PSMs (OmojateGodstime et al., 2014).

To mediate the antibacterial effect, the following series of steps is followed: PSM interacts with the cell membrane, and then diffuses through the membrane before interacting with internal constituents and processes. Thymol (a monoterpene phenol of the aromatic p-menthane type) interacts with both outer and inner cytoplasmic cell membranes by accumulating at the lipid bilayer's polar regions. This causes the cell membrane to alternate, resulting in greater permeability and disintegration. Thymol is also implicated in the inhibition of enzymes involved in ATP generation and the downregulation of genes involved in outer membrane protein synthesis (OmojateGodstime et al., 2014).

9.6.1 Primary Mechanisms of Action

9.6.1.1 Disruption of Cytoplasmic Membrane

Secondary metabolites from plants alter the structure and stability of the phospholipid bilayer, disrupt membrane integrity, and hence increase ion permeability. As a result, the membrane electrochemical potential and the activity of membrane enzymes will vary. Certain PSMs prevent protons from crossing the membrane, causing ATP production to be disrupted. Loss of ATP causes a reduction in active transport, as well as suppression of respiration and anabolism/catabolism (OmojateGodstime et al., 2014).

PSMs have two stages of interaction with cellular membranes. Cationic amino acids, for starters, have a preference for negatively charged phospholipid groups on the surface. Second, the aliphatic fatty acids and anionic components interact with positively charged patches. Membrane instability, cytoplasmic content leakage, membrane potential loss, and membrane permeability changes are all caused by these interactions (OmojateGodstime et al., 2014).

Carvacrol interacts with the cytoplasmic membrane by inserting between acyl chains of phospholipids. It is isomeric to thymol. The membrane fluidity and permeability increase as a result of this. Increased permeability causes ion and ATP efflux, as well as a change in membrane potential and pH gradient. The impact of this molecule is amplified by its hydroxyl group, which acts as a transmembrane transporter of monovalent cations, causing the membrane potential to be disrupted (Xu et al., 2008). Eugenol, a phenylpropanoid present in many plant species, has a lytic effect on bacterial cells, with a non-specific membrane permeabilization and potassium and ATP efflux as a method of action. Eugenol is considered to attach to membrane proteins, inhibiting and altering their activity (Devi et al., 2010).

9.6.1.2 Interfering with DNA/RNA/Protein Synthesis

Allicin's thiosulfonate moiety (found in *Allium sativum*) rapidly interacts with intracellular enzymes' free SH groups. The action of allicin inhibits RNA synthesis significantly, but DNA and protein synthesis are less affected. By oxidative breakdown of disulfide bonds, allylisothiocyanate inhibits enzymes and causes protein changes. Membrane disruption caused by allylisothiocyanate causes cellular metabolite leakage in *E. coli* and *Salmonella* sp. Flavonoids disrupt both cytoplasmic membrane function and DNA synthesis, making them one of the most active plant chemicals. Intercalation with DNA is also a mechanism of action for highly aromatic quaternary alkaloids like berberine and harmane (Saravanan et al., 2010). Apigenin and quercetin, as well as a number of other flavonoids, have been discovered to suppress the activities of DNA gyrase and hydroxyacyl-acyl carrier protein dehydratase. Quercetin decreases the ATPase activity of *E. coli* DNA gyrase by binding to the GyrB subunit. Quercetin also improves the permeability of the inner membrane and reduces membrane potential dissipation (OmojateGodstime et al., 2014). PSMs from *M. oleifera* work against bacteria by inhibiting enzymes (sortase inhibition) and interfering with DNA replication. Similarly, pterygospermin works by inhibiting the transaminase enzyme and disrupting cell membranes (OmojateGodstime et al., 2014). Tannins are polyphenols that have a strong ability to inhibit bacterial cell proliferation by inhibiting microbial metabolism enzymes like the proteolytic macerating enzymes. Gallic acids and other polyphenols work by binding to the bacterial dihydrofolate reductase (DHFR) enzyme. It also binds to bacterial DNA and inhibits the supercoiling activity of *E. coli* bacterial gyrase by binding to the ATP binding site of gyrase B, causing topoisomerase IV enzyme-mediated DNA breakage and bacterial growth standstill (OmojateGodstime et al., 2014).

Cinnamaldehyde exerts antimicrobial activity through three mechanisms: sub-inhibitory doses disrupt enzymes involved in cytokinesis, greater concentrations inhibit the enzyme ATPase, and deadly quantities cause cell membrane disturbance (OmojateGodstime et al., 2014).

9.6.1.3 Interrupting Communication between Bacterial Cells

Quorum sensing (QS) involves the creation and detection of diffusible signal molecules that regulate bacterial expression of specific genes, the products of which alter the local host environment to promote pathogen invasion and persistence. QS is used by many pathogenic bacteria to control their pathogenicity. The ideal QS inhibitor would be a low-molecular-mass chemical that lowers the expression of QS-controlled genes while remaining resistant to the host organism's metabolic and disposal processes. Secondary metabolites from plants appear to be particularly promising in this regard. Sprout, chamomile, carrot, garlic, habanero (*Capsicum chinensis* Jacq.), propolis, water lily, and yellow pepper extracts were observed to suppress *P. aeruginosa* QS. Garlic extract inhibits QS in a concentration-dependent manner, with a structure–activity relationship that suggests a competitive binding mechanism. It significantly lowers the biofilm of *P. aeruginosa* (Bjarnsholt et al., 2005).

9.6.1.4 Synergistic Approaches to Enhance Activity

The use of a single medicine in monotherapy leads to an increase in the number of multidrug-resistant bacteria. Due to their synergistic therapeutic effect, mixtures of bioactive chemicals in botanical medications are often touted to be preferable to mono-substances. Antimicrobial synergy is based on the idea that when two or more antibiotics are used together, they can improve efficacy, reduce toxicity or undesirable side effects, increase bioavailability, lower the dose, and slow the spread of antimicrobial resistance. In the case of the malaria parasite (*Plasmodium* sp.), artemisinin combination therapy is indicated since it may help to reduce resistance. The combination of amoxicillin (a lactam antibiotic) with clavulanicacid is a well-known commercial antibiotic. Clavulanic acid binds to lactamase-producing bacteria and protects amoxicillin from lactamase attack, allowing amoxicillin to have a broader spectrum of activity. PSMs help to increase the absorption of conventional medicines by inhibiting efflux (Radulovic et al., 2013).

9.7 GREEN SYNTHESIS OF PLANT-BASED METABOLITES AS ANTIMICROBIAL ACTIVITY

Green synthesis route is a safe, biocompatible, environment-friendly approach to synthesize various nanoparticles which can be carried out using different biological resources such as fungi, algae, bacteria, and plants. Various plant parts have been employed successfully in the synthesis of nanoparticles, as phytochemicals present in the plant extract act as a stabilization and reducing agent (Roy et al., 2019). In comparison with the plant extract synthesis methods, microorganism-assisted synthesis of nanoparticles is slow, and has limited number of size and shape outcomes. A straightforward synthesis, easy scale up, less contamination chances and production of stable nanoparticles gives the plant-based methods an upper hand among the aforementioned methods. In general, an extract prepared from the plant is mixed with the metal source under prescribed conditions. During the process, biological reduction and nucleation takes place. As a result, stabilized and capped nanoparticles are formed (Vanlalveni et al., 2021) (Figure 9.2).

In the twentieth century, death related to microbial infections had arisen to one in every two cases in developing nations, where the majority were connected to respiratory infections, diarrhoeal diseases, malaria, measles, and others (Lederberg et al., 2003). The United States alone suffers from 2.8 million infections with more than 35,000 deaths each year. These drastic numbers are ghastly and demand an urgent attention to all possible remedies including antibiotic preparations. Out of many other sources, our autotrophs can be a good source of antimicrobial agents, thereby benefitting the mankind on an industrial scale. "Green" synthesis on a large scale refers to the usage of technologies in an eco-friendly manner using plants and their parts to develop organic compounds with therapeutic applications as antimicrobial agents, coupled with synthesis of nanoparticles (Prasad 2014; Joshi et al. 2018). The research and development for investigation of these producers have been in development for some time and processes adopted for their synthesis are majorly aimed at being pollution-free and in the absence of any malicious intermediates. The matters worsen due to the increased antibiotic resistance and lack of effective therapeutic strategies. Not only a good effect, the antibiotics should also have worldwide availability for the entire population, especially in developing countries. For centuries plants have been studied as the source of several potent antimicrobial components that can act against the various infectious agents that are becoming ubiquitous every day. Treatment of infectious conditions of the skin like acne have long been treated using essential oils from tee trees (*Melaleuca alternifolia*), while infections of the urinary tract have found their remedy in usage of bearberry (*Arctostaphylos uva-ursi* (L.) Spreng) and cranberry (*Vaccinium macrocarpon* Ait.) (Carson, Mee, and Riley 2002; Radulović et al. 2010). Statistics reveals that almost 30% of the higher plants have been regarded as

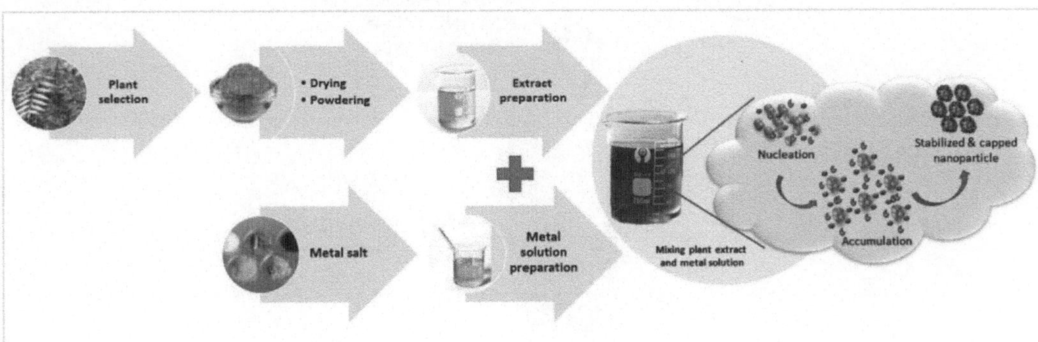

FIGURE 9.2 Schematic representation of green synthesis of metal-nanoparticles.

"medicinal" owing to the varied applications of plants in the therapeutic field (Leonti et al., 2009). Another interesting fact that gains attention is that even today, about two-thirds of the world's population still relies on traditional agents and plant-derived remedies due to economic reasons and other unaffordable high prices of allopathic substances. This increases the reason for development of higher phytomedicines and medical formulations on an industrial scale, to provide affordability to all sections of the population. Examples of leaf extracts of *Cymbopogon citrates* and *Ximenia americana* have already proved to possess antimicrobial effects against *E. coli* and *P. aeruginosa* due to the presence of few phenolic compounds such as cyclohexane-1-3,5-trione, 2-phenyl-1,4-benzopyrone, methyl guanidine, and 3-meyhylheptyl acetate (Hassan et al., 2021). A higher responsibility of developing the therapeutics lies in plants alone because of their abundant reserves, leading to a wider scope, as well as as the fact that they are mostly free from adverse side effects after human consumption.

The plant-based metabolites synthesized to be consumed by the human body cannot be taken in directly due to factors like polar nature and their huge size, which makes them difficult to cross barriers like blood and brain (BBB), the gastrointestinal tract and the mucosa and other endothelial lining of blood vessels (Ahmad et al., 2021). Thus, it is necessary to couple these metabolites along with nanocarriers that enable them to be consumed based on the effective stability of the gastrointestinal tract. To further dive into the preparation of plant- based medicinal therapeutics, the techniques used for the same need to be non-invasive, anti-pollution based and commercialized at cheaper rates. An interesting method for the green synthesis of silver nanoparticles from sunflower (*Helianthus annuus*) using its secondary metabolites like flavonoids, terpenoids, amino acids and amides has been described by Thakore et al., (2014) (Figure 9.3). It is well known that sunflower oil has capacity to be effective against microbes like *Bacillus subtilis*, *E. coli*, *Staphylococcus aureus*, and *C. albicans* and thus can be treated against urinary tract infections. Tabassum and Vidyasagar in 2014 have reported that out of a group of edible medicinal oils, sunflower seed oil has exhibited maximum antimicrobial efficacy with MIC values 0.62-40 mg/mL against a number of bacteria and fungi, indicating their efficacy against skin diseases as well. Owing to the presence of a good number of phenolics like α-tocopherol, reducing the property of sunflower oil has been utilized in the preparation of silver nanoparticles. When mixed in the ratio of 1:1, sunflower oil and petroleum ether, in the presence of alcoholic $AgNO_3$ solution, could lead to the formation of yellow-colored silver nanoparticles in a domestic microwave oven under certain environmental conditions. It was

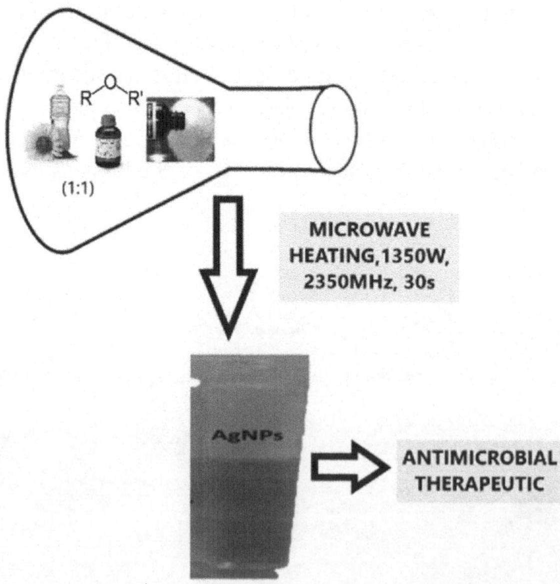

FIGURE 9.3 Green synthesis of phenolic mediated silver nanoparticle using sunflower oil.

further stored at a 4°C refrigerator for the homogenization of $AgNO_3$ with sunflower oil. Further, these nanoparticles were characterized using optical property measurement with UV-vis spectrophotometer followed by recording of FT-IR spectra and particle size measurement using TEM.

Another antibacterial activity was exhibited against *E. coli*, *P. aeruginosa*, *S. aureus*, and *B. subtilis* by the palladium nanoparticles synthesized on coupling with metabolites obtained from *Filicium decipiens* (Sharmila et al., 2017). In this case, potential antimicrobial activity was observed against both Gram-positive and -negative microorganisms owing to the secondary metabolite saponin extracted from leaves of the plant. This was also the first saponin exhibiting antifungal properties against *Aspergillus flavus* extracted from *Filicium decipiens*, using gel filtration and ion exchange chromatographic techniques. On dissolving into 0.15 M NaCl, the saponin extract was analyzed using ion exchange chromatography, followed by other filtration and elution techniques to finally activate the antifungal activity with minimum inhibitory concentration of 7.5 μg/mL and IC50 5.8 μg/mL. There are a number of other phytometabolites coupled with nanomaterials for the activity as antimicrobial agents using greener and safer technologies (Table 9.1). Similarly, there have been a number of strategies adopted for the synthesis of phytometabolites to be used as therapeutic agents.

TABLE 9.1
Few Metabolites Utilized in the Green Synthesis of Nanoparticles

Plant species	Nanoparticles	Metabolites Identified in the green synthesis	Reference
Abutilon indicum	Au	Flavonoids, phenolic compounds	Mata et al., 2016
Acalypha indica	Ag	Quercetin, plant pigment	Krishnaraj et al., 2010
Achyranthes aspera	Ag	Polyols	Elumalai et al., 2016
Aegle marmelos	Ag	Tannin	Rao and Paria, 2013
Agathosmabetulina	ZnO	Hydroxyl group	Thema et al., 2015
Aloe barbadensis	ZnO	Phenol, amines, alcohol groups	Ali et al., 2016
Alternanthera sessilis	Ag	Tannins, carbohydrates, proteins, ascorbic acid	Niraimathi et al., 2013
Anacardium occidentale	Ag	Proteins, polyols	Mukunthan and Balaji, 2012
Anisochilus carnosus	ZnO	Phenol, carboxylic acid	Anbuvannan et al., 2015
Annona squamosa	Ag	Alkaloids, glycoside, saponins, tannins, phenolic, carbohydrates	Vivek et al., 2012
Argemone mexicana	Au	Phosphorous compounds	Varun et al., 2015
Averrhoa bilimbi	Au	Phenols, tertiary amides	Isaac et al., 2013
Azadirachta indica	Ag	Flavanoids, terpenoids	Ahmed et al., 2016
Azadirachta indica	ZnO	Flavonoids, phenolic acid, terpenoids, protein	Elumalai and Velmurugan, 2015
Boswellia serrata	Ag	Proteins	Kora et al., 2012
Butea monosperma	Au	Polyphenols	Patra et al., 2015
Cacumen platycladi	Pt	Flavonoids, proteins	Zheng et al., 2013
Calotropis procera	ZnO	Hydroxyl groups, aldehydes, amines, ketones, carboxylic acids	Gawade et al., 2017
Carica papaya	Ag	Catechnis, hydroxyflavones	Banala et al., 2015
Cassia auriculata	Au	Hydroxyl group	Venkatachalam et al., 2013
Cassia fistula	ZnO	Flavonoids, polyphenols	Suresh et al., 2015
Cinnamomum zeylanicum	Au	Proteins	Smitha et al., 2009
Citrus maxima	Au	Flavonoids, terpenes, vitamins	Yu et al., 2016
Citrus sinensis	Ag	Flavonoids, ascorbic acid, volatile oils	Kaviya et al., 2011
Coptis chinensis	ZnO	Alcohol, carboxylic acid, alkyl halide, alkynes	Nagajyothi et al., 2014
Couroupita guianensis	Au	Phenol group	Sathishkumar et al., 2016
Cymbopogon citratus	Cu	Polyphenols, proteins	Brumaugh et al., 2014
Decalepis hamiltonii	Ag	Polyols, phenols	Rashmi and Sanjay, 2017

(continued)

TABLE 9.1 (Continued)
Few Metabolites Utilized in the Green Synthesis of Nanoparticles

Plant species	Nanoparticles	Metabolites Identified in the green synthesis	Reference
Delonix regia	Pd	Polyphenols	Dauthal and Mukhopadhyay, 2013
Desmodium triflorum	Ag	Ascorbic acid	Ahmad et al., 2011
Dioscorea bulbifera	Ag	Diosgenin, ascorbic acid	Ghosh et al., 2012
Dioscorea bulbifera	Ag	Flavonoids, polyphenols	Ghosh et al., 2012
Dioscorea bulbifera	Pt-Pd	Hydroxyl group of polyphenolic compounds	Ghosh et al., 2015
Diospyros kaki	Pt	Terpenoids	Song et al., 2010
Elaeis guineensis	Au	Phenolic, carboxylic, amines	Ahmad et al., 2018
Eucalyptus	Ag	Alcohol, phenols, alkyl aldehyde	Liu et al., 2018
Euphorbia granulate	Pd	Hydroxyflavones, phenolics	Nasrollahzadeh and Sajadi, 2016
Ficus benghalensis	Au	Proteins	Francis et al., 2014
Galaxaura elongata	Au	Glutamic acid, hexadecanoic acid, oleic acid, 11-eicosenoic acid, stearic acid, gallic acid, epigallocatechin, catechin, epicatechin gallate	Abdel-Raouf et al., 2017
Galenia africana	Au	Flavonoids, terpenoids, phenolic compounds and/or carbohydrates	Elbagory et al., 2017
Gardenia jasminoides	Ag	Rutin, gallic acid, chlorogenic acid	Fenfen et al., 2014
Glycyrrhiza glabra	Ag	Flavonoids, thiamine and terpenoids	Dinesh et al., 2012
Gymnema sylvestre	Au	Proteins, polypeptides	Arunachalam et al., 2014
Helianthus annuus	Ag	Flavonoids, proteins, amino acids, amides terpenoids	Thakore et al., 2014
Hibiscus rosa- sinensis	Ag	Carboxylate ion groups	Nayak et al., 2015
Hibiscus rosa- sinensis	Au	Flavonoids	Yasmin et al., 2014
Hippophae rhamnoides	Pd	Flavonoids	Nasrollahzadeh et al., 2016
Hybanthus enneaspermus	Ag	Several bioactive phytochemical compounds	Suman et al., 2016
Hygrophila spinosa	Au	Hydroxyl group	Koperuncholan, 2015
Hypoxis hemerocallidea	Au	Flavonoids, terpenoids, phenolic compounds and/or carbohydrates	Elbagory et al., 2017
Ipomoea carnea	Au	Polysaccharides, protein	Abbasi et al., 2015
Iresine herbstii	Ag	Phenolic compound	Dipankar and Murugan, 2012
Jatropha	Au	Proteins	Kanchi et al., 2018
Jatropha curcas	Ag	Cyclic peptides (curcacycline A and curcacycline B)	Bar et al., 2009
Lansium domesticum	Au	Carboxylic acid	Shankar et al., 2014
Lantana camara	Ag	Flavonoids, glycosides and carbohydrates	Ajitha et al., 2015
Lawsonia inermis	Cu	Phenolic compounds	Cheirmadurai et al., 2014
Leonuri herba	Ag	Hydroxyl, polyphenols groups	Im et al., 2012
Lippia citriodora	Ag	Verbascoside, isoverbascoside, chrysoeriol-7-O-diglucoronide, luteonin-7-O-diglucoronide	Cruz et al., 2010
Lonicera japonica	Ag	Phenolic and hydroxyl groups of chlorogenic acid	Zhou and Tang, 2018
Mangifera indica	Ag	Ketone, aldehydes, hydroxyl, carboxyl groups	Yang and Li, 2013
Mangifera indica	Au	Flavonoids, terpenoids, thiamine	Philip, 2010
Mentha piperita	Ag	Alkaloids, flavones, steroids, polysaccharides, amino acids, oximes, proteins, menthol	Mubarak et al., 2011
Mentha piperita	Au	Menthol	Mubarak et al., 2011
Mimusop selengi	Ag	Polyphenols	Kumar et al., 2014
Mirabilis jalapa	Au	Polysaccharides	Vankar and Bajpai, 2010

TABLE 9.1 (Continued)
Few Metabolites Utilized in the Green Synthesis of Nanoparticles

Plant species	Nanoparticles	Metabolites Identified in the green synthesis	Reference
Morinda citrifolia	Au	Proteins	Suman et al., 2014
Morinda pubescens	Ag	Catechins, hydroxyflavones	Kumar et al., 2015
Nigella arvensis	Ag	Flavonoids, alkaloids	Chahardoli et al., 2017
Nigella arvensis	Au	Flavonoids, phenolic compounds	Chahardoli et al., 2018
Ocimum sanctum	Ag	Flavonoid, proteins, gallic acid, terpenoids	Ramteke et al., 2013
Ocimum sanctum	Ag	Quercetin	Jain and Mehata, 2017
Olea europaea	Au	Proteins	Khalil et al., 2012
Panax ginseng	Au	Polysaccharides, phenolic compounds	Pérez et al., 2017
Parthenium hysterophorus	ZnO	Phosphorus compound, secondary sulfonamide, monosubstituted alkyne	Rajiv et al., 2013
Phoenix dactylifera	Au	Hydroxyl group	Zayed and Eisa, 2014
Piper betle	Ag	Proteins	Lagashetty, 2015
Coleus aromaticus	Ag	Flavonoids	Vanaja and Annadurai, 2013
Platycodon grandiflorum	Au	Hydroxyl group	Choi et al., 2018
Plumeria rubra	Ag	Proteins	Patil et al., 2012
Punica granatum	Cu	Flavonoids, alkaloids, polyphenols	Nazar et al., 2018
Rosa canina	ZnO	Phenolic and carboxylic acids	Jafarirad et al., 2016
Rosa rugosa	Ag	Carboxylate, amine groups	Dubey et al., 2010
Rosmarinus officinalis	Ag	Polyphenols	Ghaedi et al., 2015
Salix alba	Au	Proteins, metabolites having functional groups of amines, alcohols, ketones, aldehydes, carboxylic acids (salicin)	Islam et al., 2019
Sesbania grandiflora	Au	Flavonoids, polyphenols	Das and Velusamy, 2014
Sesuvium portulacastrum	Ag	Flavones, proteins, terpenoids	Nabikhan et al., 2010
Solanum xanthocarpum	Ag	Alkaloids, phenolics, sugars	Sengottaiyan et al., 2016
Sterculia acuminata	Au	Polyphenols	Bogireddy et al., 2015
Suaeda monoica	Au	Flavonoids, terpenoids, soluble proteins	Rajathi et al., 2014
Syzygium cumini	Ag	Flavonoids	Prasad and Swamy, 2013
Syzygium cumini	Ag	Polyphenols	Kumar et al., 2010
Syzygium jambos	Au	Saccharides, phenolics	Dutta et al., 2017
Tagetes sp. and *Rosa* sp.	Cd	Alcoholic, amide, C–C, –OCH3 groups (tannins, flavonoids, alkaloids and carotenoids)	Hajra et al., 2016
Tamarindus indica	Au	Phenolic compounds	Kaur and Bhullar, 2016
Terminalia arjuna	Au	Hydrolyzable tannins	Gopinath et al., 2013
Terminalia arjuna	Au	Polyphenols	Kumar et al., 2013
Terminalia catappa	Au	Hydroxyl group of phenols	Ankamwar, 2010
Trianthema decandra	Ag	Catechins, hydroxyflavones	Geethalakshmi and Sharada, 2013
Trianthema decandra	Ag	Saponin	Geethalakshmi and Sharada, 2013
Trianthema decandra	Au	Saponin	Geethalakshmi and Sharada, 2013
Trifolium pratense	ZnO	Hydroxyl group	Dobrucka and Długaszewska, 2016
Vitis vinifera	Au	Flavonoids	Ismail et al., 2014
Withania somnifera	Ag	Catechin p-coumaric acid, luteolin-7-glucoside, withanolides	Marslin et al., 2015
Zingiber officinale	Ag	Flavonoid, alkaloids	Velmurugan et al., 2014
Zostera noltii	Au	Flavone sulfates	Zarzuela et al., 2018

9.8 MECHANISM OF ANTIMICROBIAL ACTIVITY

In history, there are instances and examples of metals being used as antimicrobial agents. Various cultures, namely, Persians, Indians, Greek, Egyptians, and Romans, have used silver and copper metals to preserve and store food materials, also to disinfect water. The small size and larger surface area of metals have shown significantly efficient activity against pathogenic microbes. With the advent of areas such as nanotechnology, several studies have led to the discovery of more effective nanoparticles (Nisar et al., 2019).

Although the mechanism of nanoparticle action against microbes is still under investigation, some insights based on experimental observations led to proposed possibilities. NPs have the ability to attach to the surface of the microbial cell wall due to charge on the surface. NPs can penetrate and disturb the structure of cell membrane, and the formation of 'pits' observed on treated bacterial cell surface suggest cell membrane damage (Shaikh et al., 2019, Dakal et al., 2016). Electron spin resonance spectroscopy studies suggested the formation of free radicals at the contact point of nanoparticle with the cell membrane. Hence, these free radicals can damage the cell membrane and make it porous which eventually leads to the death of the cells. Reactive oxygen species can also be involved in cell membrane damage (Aziz et al. 2014, 2015, 2016, 2019). In other cases, the nanoparticle gets attached on the surface of the cell or they penetrate through the membrane and disrupt the functions of the cell by interfering with different cellular components, such as the DNA present in the nucleus, proteins with essential functions, and organelles such as mitochondria, ribosomes, lysosomes, Golgi apparatus, and electron transport channels, and so on, as depicted in Figure 9.4 (Hemlata et al., 2020; Sánchez-López et al., 2020).

The majority of nanomaterials described in recent studies have antibacterial activity attributable to at least one of the mechanisms described above. Some of the well-documented nanoparticles with antimicrobial activity and their probable mechanism of actions are listed in Table 9.2 (Nisar et al., 2019; Anees Ahmad et al., 2020; Nguyen et al., 2018; Gudkov et al., 2021).

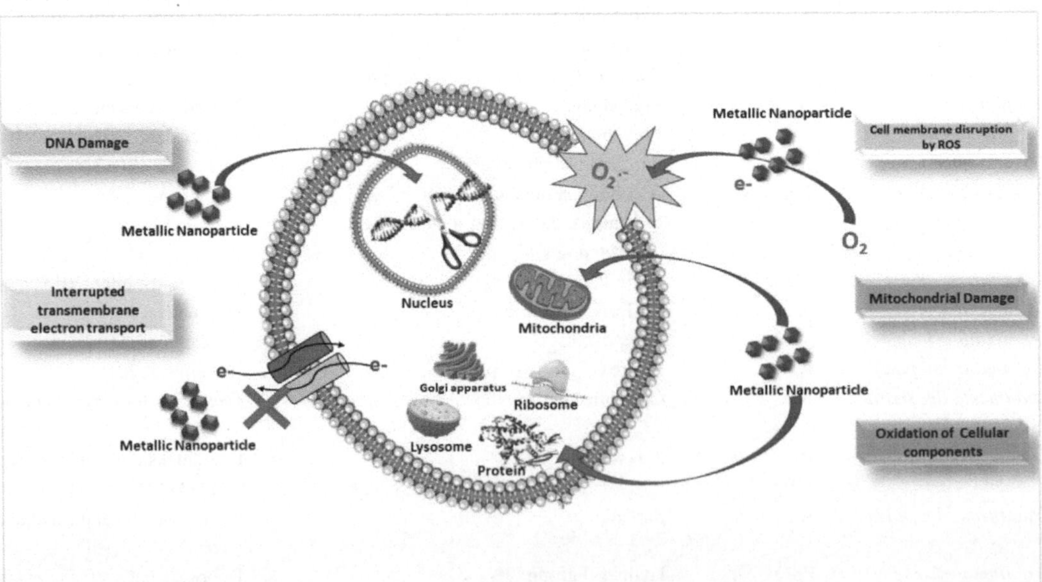

FIGURE 9.4 Schematic representation of different mechanisms of antimicrobial activity by nanoparticles.

TABLE 9.2
Mechanism of Antimicrobial Activity by Various Metal-Nanoparticles

NPs	Mechanism of antimicrobial action
Copper	CuNPs rupture the bacterial cell wall by releasing Cu ions that bind negatively charged cell, transformation of helical structure of DNA by interacting between and within nucleic acids
Gold	Heavy electrostatic attraction, accumulation at cell surfaces, and interaction with cell membrane
Iron	Generation of ROS, Damages cell membrane and DNA
Magnesium	Oxidative stress, Ca^{2+} ion concentrations, membrane damage, and quorum sensing
Silver	Interferes with cell membrane, damages DNA and electron transport
Titanium	Damages cell membranes and releases reactive oxygen species (ROS)
Zinc	Disrupts the cell membrane, accumulates inside the cell and produces toxic H_2O_2

9.9 WHY DOES GREEN SYNTHESIZED METAL NANOPARTICLE SHOW BETTER ANTIMICROBIAL ACTIVITY?

Different reports show that smaller-sized nanoparticles enhance antimicrobial properties due to the increase in surface area to volume ratio. Green synthesized nanoparticles, being capped with different phytochemicals, slow down the leaching process and thus a constant and uniform concentration of the ions is delivered in order to fight the microorganisms. In other words, a controlled release mechanism takes place. Similarly, agglomeration is a common behavior observed in the nanoparticle suspended in biological medium. This phenomenon appeared to reduce antimicrobial effectiveness. Green synthesized nanoparticles on the other hand are stabilized and capped with various phytochemicals which drastically reduces the possibility of agglomeration by directly influencing the antimicrobial activity. There are unexplored factors that can positively affect the antimicrobial activities, such as using multiple mechanism simultaneously, overcoming the existing antibiotic resistance mechanisms, and so on (Li et al., 2015; Yang et al., 2021; Wang et al., 2017).

9.10 GREEN SYNTHESIZED NANOPARTICLE AS ANTIMICROBIAL AGENTS

Silver nanoparticles (AgNPs) are one of the extensively studied nanomaterials (Guilger-Casagrande and de Lima, 2019). AgNPs with high surface area to volume ratio are of prime interest due to high antimicrobial actions against various microorganisms, such as bacteria, fungi, and viruses. It also offers potential effects against multidrug-resistant strains of bacteria such as ampicillin-resistant *E. coli*, erythromycin-resistant *Streptococcus pyogenes*, methicillin-resistant *Staphylococcus aureus* (MRSA), and so on. These biocidal actions of silver nanoparticles lead to several physical and chemical routes for the synthesis of AgNPs, but these methods present drawbacks, including use of hazardous precursor chemicals (such as sodium borohydride, potassium bitartrate, methoxypolyethylene glycol, and hydrazine), use of toxic solvents (such as sodium dodecyl benzyl sulphate and polyvinyl pyrrolidone), and so on (Shaikh et al., 2019; Dakal et al., 2016; Liu et al., 2020). With the advancement of knowledge, the synthesis of silver nanoparticles took alternative eco-friendly routes. A variety of biological agents (plant extracts, bacterial and fungal derived compounds) are used to convert metal salts to nanoparticles. However, the slow rate of nanoparticle synthesis using microbes makes the plant extract method the preferred choice due to its simplicity, efficiency, and viability. Plants possess the ability to accumulate and detoxify heavy metals. Generally, plants (extract to be precise) are used as a bio reductant in the process of synthesizing silver nanoparticles. Terpenoids, flavones, ketones, aldehydes, amides, carboxylic acids, carbohydrates, proteins, and vitamins are majorly involved in the process (Guilger-Casagrande and de Lima, 2019; Hamouda et al., 2019).

TABLE 9.3
Green Synthesis of Silver Nanoparticles and its Application Against Various Microbes

Sl. No	Extract source	Method of synthesis/ Conditions	Nanoparticle size (nm)	Antimicrobial activity	Reference
1	*Azadirachta indica*	Room temperature, dark condition	64–68nm	*E. coli*	Ulaeto et al., 2019
2	*Azadirachta indica*	Room temperature	<40nm	*P. nitroreducens, Aspergillus unguis - NII 08123*	Roy et al., 2017
3	*Caesalpinia digyna*	30ºC, dark	11–45nm	*B. subtilis, E. coli, S. typhi*	Niloy et al., 2020
4	Lichens	mechanochemical approach	39–112nm	*E. coli, S. aureus.*	Baláž et al., 2020
5	*Lotus lalambensis*	Room temperature, dark	6–26nm	*C. albicans*	Abdallah and Ali, 2021
6	*Lysilomaa capulcensis*	Room temperature	1.2–62nm	*E. coli, S. aureus, P. aeruginosa, C. albicans.*	Garibo et al., 2020
7	*Mangifera indica* seed	Room temperature, dark for 24 hrs.	26–27nm	*S. aureus, C. rubrum, B. subtilis, B. cereus, P. aeruginosa, K. pneumoniae, E. coli, S. typhimurium, C. albicans*	Donga and Chanda, 2021
8	*Mentha longifolia*	70ºC	10nm	*K. pneumoniae, S. aureus, B. subtilis*	Rauf et al., 2021
9	*Ornithogalum narbonense L.*	Room temperature	8–9nm	*E. coli, P. aeruginosa, E. faecalis, S. aureus, B. cereus*	Ozturk et al., 2022
10	Pu'er tea	Room temperature	4–5nm	*E. coli, K. pneumoniae, S. typhimurium, S. enteritidis*	Loo et al., 2018
11	*Tagetes erecta*	Room temperature for 24 hrs. in dark.	10–90nm	*B. cereus, S. aureus*	Padalia et al., 2015
12	*Vitex Negundo L.*	Room Temperature stirring, 48hrs.	10–30nm	*E. coli, S. aureus*	Zargar et al., 2011

Various plants and different plant parts are used in the preparation of the plant extract. The extract is then used in the synthesis of AgNP by the reduction of the precursor, generally $AgNO_3$. Care needs to be taken during the process due to the light sensitivity of the precursor salt. Numerous methods are reported in the literature. A few of them are enlisted in the Table 9.3. Plant extract preparation has been achieved by plants from different families and plant parts such as root, stem, leaf, bark, flowers, and so on.

Apart from silver nanoparticles, there are various biosynthesized metal-nanoparticles that can be used as antimicrobial agents. These metals have been used as an antimicrobial agent for centuries in several countries. Among the large pool of the nanoparticles synthesized, a few of them stand out for their advantages such as low toxicity, high biocompatibility with human body, and a large surface area to volume ratio. Gold (Au), Copper (Cu), Zinc (Zn), Iron (Fe), Titanium (Ti), and Magnesium (Mg) are known for their antimicrobial activity with good biocompatibility (Kumari et al., 2021; Singh et al., 2020; Zhang et al., 2020).

Gold nanoparticles lead this group with various tunable properties with a suitable biocompatibility. *Myristica fragans* fruit extract, *Nigella sativa* plant extract, *Terminalia bellirica* dry fruit pericarp, *Malva Verticillata* leaf extract, and Brazilian red propolis are a few among the reported plant sources used in the synthesis of the nanoparticles (Castillo-Henríquez et al., 2020), (Botteon et al., 2021). In a similar manner, biosynthesized copper nanoparticles also provide extraordinary biological activity against microbes. Copper nanoparticles can also be synthesized using various plant extracts including tea leaf, *Murraya koenigii* leaves, *Eclipta prostrata* leaves, tomato, *Punica granatum* peel extract, *Aegle marmelos* leaf, and so on (Jayarambabu et al., 2020; Wu et al., 2020). Biosynthesized ZnO NPs on the other hand show strong antimicrobial activity against clinical pathogens compared to standard drugs, suggesting that plant-based synthesis of NPs can be an excellent strategy to develop versatile and eco-friendly antimicrobial products, which could be used as an alternative method against the drug resistant microbes in the worldwide emergence of drug resistance (Bhuyan et al. 2015; Naseer et al., 2021; Gomathi and Suhana, 2021; Gupta et al., 2018).

Several studies were carried out for green synthesis of iron-based nanoparticles from various plants parts like fruit extract of *Cynometra ramiflora*, rind of *Persea americana*, seeds extract of *Punica granatum*, and flower extract of *Avicennia marina*. Most of the reports suggest good antimicrobial activities against different pathogens (Bhuiyan et al., 2020; Arakha et al., 2015; Üstün et al., 2022). TiO_2-Nanoparticles (TiO_2 NPs) are well documented for their safe, stable, nontoxic, high surface activity (Jassal et al. 2022). TiO_2 NPs have found use treating different pathogenic diseases. Various plants and their parts have been reported for the biosynthesis of TiO_2NPs, including *Acanthophyllum laxiusculum* (roots), *Aloe barbadensis* (leaves), *Annona squamosa* (peel), *Calotropis gigantean*, and so on (Ahmad et al., 2020; Al-Shabib et al., 2020).

Even though magnesium (Mg) is an essential mineral element for organisms, and is nontoxic, the magnesium oxide nanoparticles (MgNPs) can be employed as an antimicrobial agent. Magnesium oxide nanoparticles display extraordinary biological applications. They are used in ointments for the treatment of heart burn, wounds, and bone regeneration. Furthermore, these nanoparticles have displayed excellent toxic effects against pathogens; the profound biocompatibility makes this nanoparticle very suitable for antimicrobial activity against human pathogens (Amina et al., 2020; Khan et al., 2020).

9.11 IN SILICO STUDIES FOR SCREENING OF POTENTIAL METABOLITES AS THERAPEUTIC AGENTS

The "in silico" study of any experiment includes the usage of silicon in general terms, referring to the experiments in biological systems while conducting a computer model-based understanding of cellular behavior. Any successful study on green technology involves complete isolation, identification and further confirmation of the presence of any metabolite in the extract that has the potential of a therapeutic effect. As stated above, there have been many efforts to produce such potential phytometabolites along with their subsequent screening procedures to ensure the presence of the compound using advanced technologies like chromatography. For the screening of phytometabolites, identification of potential compounds for the antimicrobial therapy becomes a necessity along with understanding their various adverse side effects too.

One of the screening processes included phytometabolite testing from the *Betula alnoides* bark where ethyl alcohol extract was majorly the solvent used for the isolation of lupeol. Lupeol is a triterpene of lupine type and has been known for its various antimicrobial activities isolated from commonly found vegetables and species like *Tamarindus indica*, *Celastrus paniculatus*, *Allanblackia monticola*, *Himatanthus sucuuba*, *Zanthoxylum riedelianum*, and others. Khan et al. (2021) reported the usage of HPTLC (High Performance Thin Layer Chromatography) densitometric chromatogram with n-hexane as the solvent which yielded an extraction potential of 3.45% w/w with a resolved peak at R_f 0.61. Specific microbes that the triterpene was effective against included *B. subtilis*, *P. aeruginosa*, and *S. aureus* as bacterial organisms and *Epidermophyton floccosum* and *Candida albicans* as fungus against the standard drugs, ciprofloxacin and fluconazole, respectively.

A member of the family Rubiaceae, *Randia spinosa* has prominent secondary metabolites like alkaloids, terpenoids, and tannins that possess several antimicrobial activities against organisms like *E. coli* and *S. aureus* that are known for infections like those of the urinary tract (UTI), travelers' diarrhea, impetigo, folliculitis, as well as diseases like pneumonia, sepsis, neonatal meningitis, and several others (Sridhar et al., 2014). Thus, discovering the potent antimicrobial effects of the ethanolic extract of *R. spinosa* becomes a necessity and is generally done using various phytochemical assessments and other chromatographic techniques (Agre et al., 2021). The usage of thin layer chromatography has revealed that an R_f of 0.217, 0.434 for ethyl acetate extract and 0.75 for petroleum ether was sufficient for exhibiting a positive zone of inhibition for the fruits of this species. Another important criterion that needs to be researched in this category refers to the antimicrobial resistance in microorganisms and phyto-antimicrobials that have been sufficient to combat this issue as proven by various *in silico* studies (Tiwari et al., 2021). Synthetic biology has helped in either reconstitution of the metabolic pathway, or insertion of gene cluster or introducing it completely into a different plant system. The various approaches under this branch are metabolic engineering, genome editing, and systems and computational biology using plant models. The approaches for using plants as the basis for the phytochemical production ranges from designing and using computers and DNA assembly where simulation, genomics, molecular docking using structure–activity relationship studies are done (Hammami and Fliss, 2010) to various silencing mechanisms using RNA interference (RNAi) (Guo et al., 2016). Apart from these, various spectroscopic technologies like UV or NMR can be adopted for the characterization and isolation of phytometabolites. Rhizomes of the *Alpinia galanga* plant have been home to two new phytometabolites as characterized using spectroscopic assays like FTIR, COSY, C-NMR, ^{1}H-NMR, DEPT, UV and Mass spectrometry (Sharma et al., 2021). In a similar way, few phytometabolites can also be utilized as biomarkers for quality control, isolation and identification of respective plants, especially when they are at an existential threat.

9.12 PLANT GENOME EDITING USING CRISPR-CAS 9

Genome editing, in recent times, has been advanced to the usage of various tools like Zinc Finger nucleases (ZFN), Transcription Activator-Like Effector Nucleases (TALENS), or most commonly CRISPR-Cas 9 that enables the insertion or modification of any plant genome in a desirable manner (Dangi et al., 2018; Sarma et al. 2021). In 2017, de la Fuente-Núñez and Lu devised an innovative tool using CRIPSR-Cas 9 strategies to remove those microbes which were specifically resistant to antibiotics through biofilm production. It is not only microbes, but also this system is emerging as a revolutionary strategy in plant-based systems as hosts too. Lowder et al. in 2015 devised the molecular toolbox using the same which consisted of information about the reagents, especially about the plant DNA assembly, and that has proved successful in plants like *Arabidopsis thaliana*, *Oryza sativa*, and *Nicotiana benthamiana*. Apart from this, CRISPR-Cas 9 has also been used in the mutagenesis of rosmarinic acid synthase gene (RAS) and maintenance of phenol biosynthetic pathway in *Salvia miltiorrhiza* (Zhou et al., 2018). In the engineered hairy roots, there was a decreased production of phenol and RAS, thereby altering the amount of metabolite production. The plant *Glycine max* also has been genetically modified using CRISPR-Cas 9, leading to a higher content of secondary metabolite isoflavone, ultimately resistant to soyabean mosaic virus (Zhang et al., 2020). The pYLCRISPR-Cas9 system has also helped in the elevation of metabolite γ-aminobutyric acid (GABA) in *Solanum lycopersicum* by metabolic engineering, which, when fermented, can be used as an antioxidant along with antimicrobials in pharmaceutical compounds. *V. parahaemolyticus*, *S. typhimurium*, *B. cereus*, and *S. aureus* have shown inhibition in growth when GABA was added in their growth medium (Mau et al., 2012). There are innumerable other phytometabolites that have been screened from plant models after their genetic modification, thereby elevating their content and increasing the application in the antimicrobial field, including the antimicrobial resistant compounds (Table 9.4). There is always a wider scope in this study that aims to explore newer strategies and understanding into the production of secondary metabolites in the usage as antimicrobial and antibacterial therapy.

TABLE 9.4
Comparative *in silico* Molecular Docking Studies of Various Phyto-metabolites and Their Effects Against Selected Microorganisms

Microorganism	Plant species	Metabolite	Metabolite structure	Docking scores (Kcal/mol)	Reference
	Enantia chlorantha	Canadine	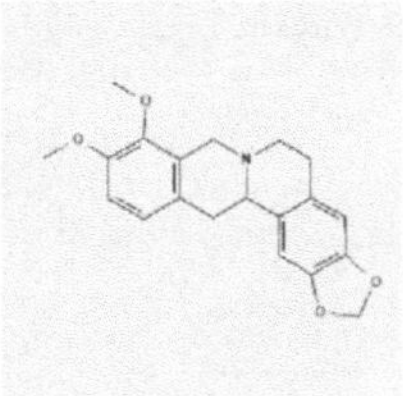	-6.18	Abike et al., 2020
Candida albicans	*Enantia chlorantha*	Pseudocolumbamine	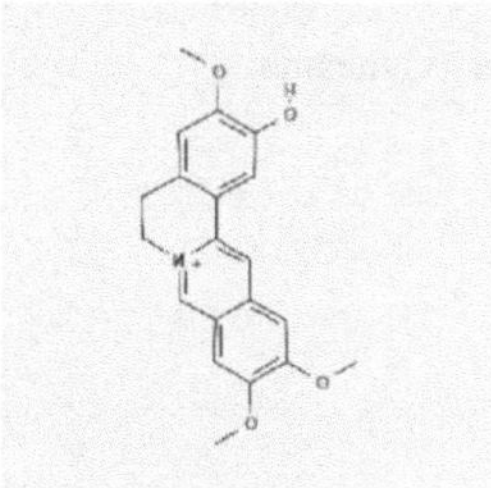	-8.006	Abike et al., 2020
Multidrug-resistant (MDR) uropathogenic *Escherichia coli*	*Terminalia chebula*	Gallotannin	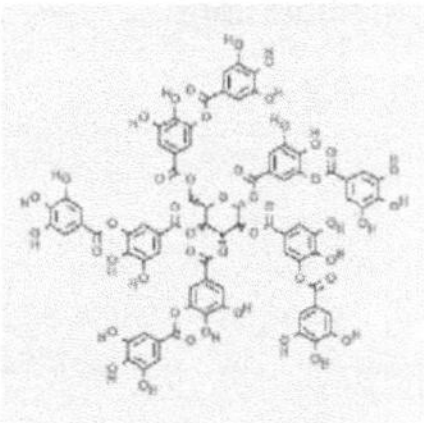		Bag and Chattopadhyay, 2014
Mycobacterium smegmatis	*Allium cepa*	Quercetin	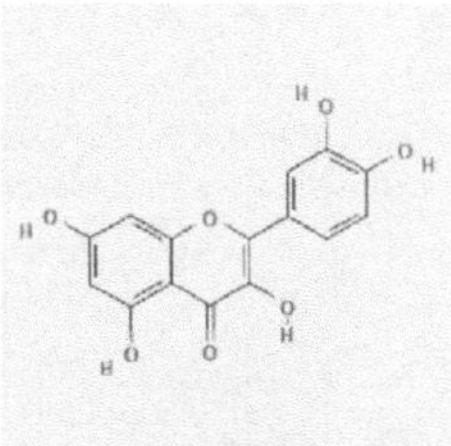	-4.80	Sharma et al., 2019
Phytophtohora capsici	*Chromolaena odorata*	Tianshic acid	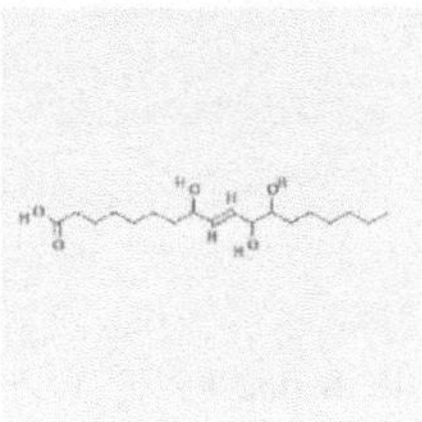	-9.305	David et al., 2019

(continued)

TABLE 9.4 (Continued)
Comparative *in silico* Molecular Docking Studies of Various Phyto-metabolites and Their Effects Against Selected Microorganisms

Microorganism	Plant species	Metabolite	Metabolite structure	Docking scores (Kcal/mol)	Reference
Plasmodium falciparum	*Cannabis sativa*	Isovitexin	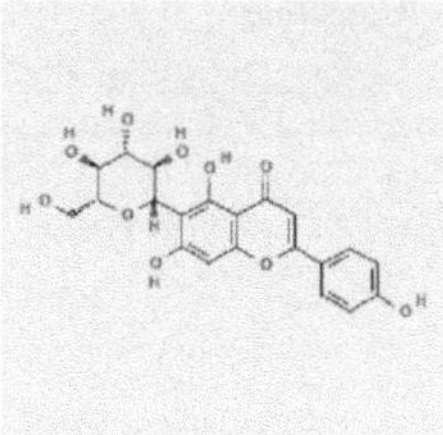	-11.485	David et al., 2018
Plasmodium falciparum	*Cannabis sativa*	Vitexin	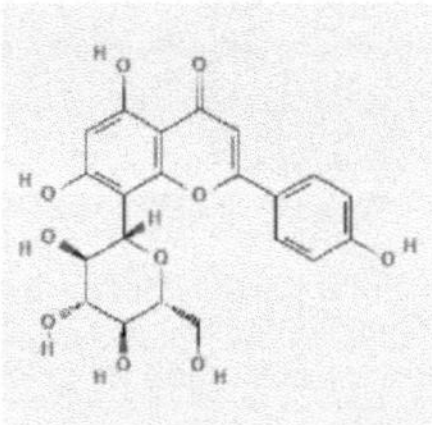	-10.601	David et al., 2018
Pseudomonas aeruginosa	*Chromolaena odorata*	Medioresinol	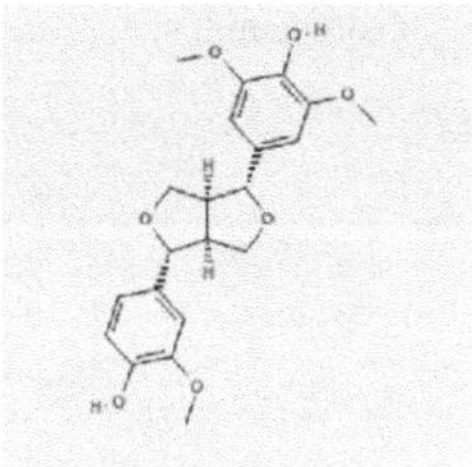	-6.121	David et al., 2019
Pseudomonas aeruginosa (MexAB-OprM)	*Holarrhena antidysenterica*	Conessine	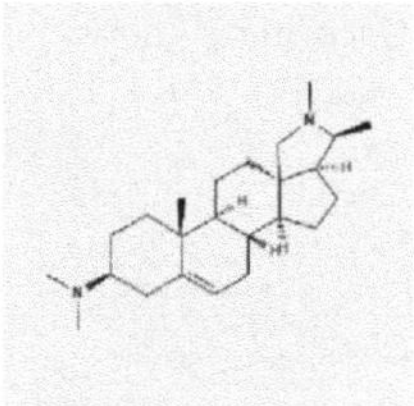	-8.75	Siriyong et al., 2017
Salmonella spp.	*Vaccinium sect. Cyanococcus*	Gallic acid	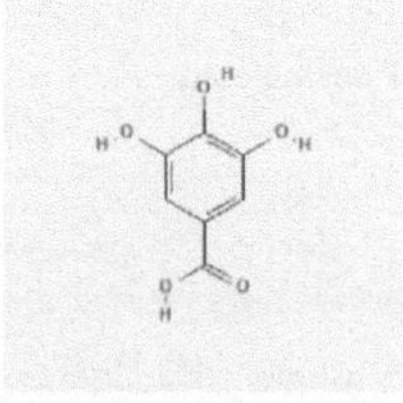	-8.4	Nohynek et al., 2006
Salmonella typhi	*Enantia chlorantha*	Berberine-1	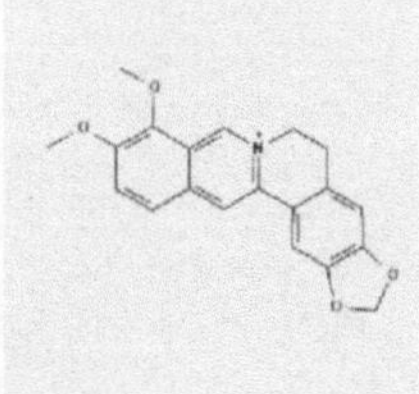	-0.467	Abike et al., 2020

TABLE 9.4 (Continued)
Comparative *in silico* Molecular Docking Studies of Various Phyto-metabolites and Their Effects Against Selected Microorganisms

Microorganism	Plant species	Metabolite	Metabolite structure	Docking scores (Kcal/mol)	Reference
Salmonella typhi	*Enantia chlorantha*	Atherosperminine	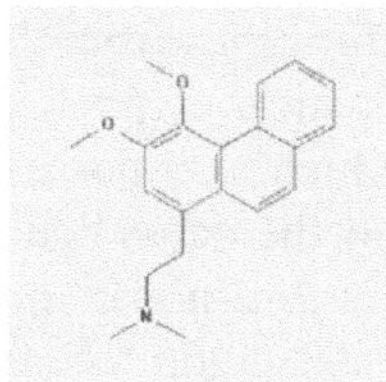	-1.144	Abike et al., 2020
Staphylococcus aureus	*Enantia chlorantha*	Levofloxacin	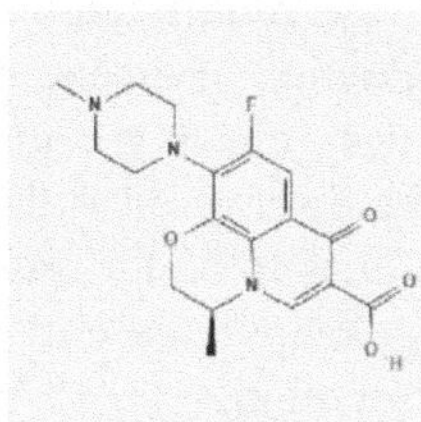	-3.436	Abike et al., 2020
Staphylococcus aureus	*Enantia chlorantha*	Palmatine	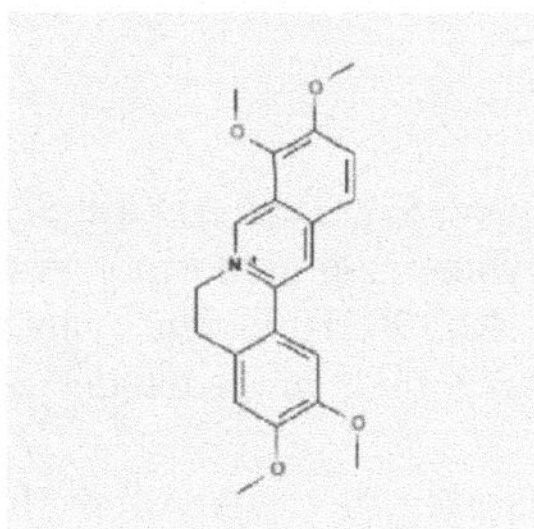	-3.287	Abike et al., 2020
Streptococcus pyogenes	*Chromolaena odorata*	Lycopsamine	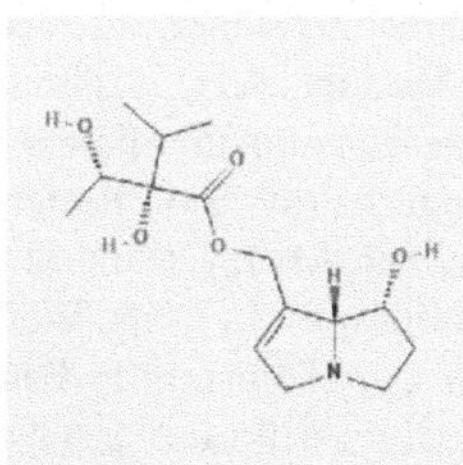	-5.196	David et al., 2019

9.13 CONCLUSION AND FUTURE PROSPECTS

With increasing statistics in the rate of microbial infections daily, it is the need of the hour to develop more sources of antimicrobials from as many renewable sources as possible. Green synthesis for the production of phytometabolites in order to develop antimicrobials using high-throughput technologies has enabled the discovery of roads leading to the procurement of secondary metabolites. For the process, plant extracts were initially subjected to specific conditions along with respective inorganic compounds for the development of nanoparticles and their subsequent applications in the antimicrobial world. But due to the problems of confinement of these metabolite compounds in specific species of plants only, the question of threat to biodiversity exists. Therefore, the stress is being laid on the isolation of specific molecules responsible for the antimicrobial property rather than using the entire plant extract alone. This strategy can be adopted either through various sophisticated chromatographic techniques or characterization using spectrophotometric assays. The inevitable importance of nanoparticles is being discovered in fields ranging from biosensing, bio-imaging, and biomolecular recognition to drug delivery. Thus, green synthesis of nanoparticles is a simplified approach using eco-friendly and nontoxic procedures in antimicrobial production using secondary metabolites in plants. Along with this, another alternative to the conservation of biodiversity and protection against extinction is by the infection of plants using a few species of endophytic fungi. These are found to possibly obtain genes from their host and mimic the entire metabolic pathway in plants, leading to the reproduction of plant metabolites like polyketides, alkaloids, phenols, quinones, steroids, and others which also possess high antimicrobial value. The method being a natural one is less threatening to the ecosystem, and more sustainable, while further molecular cloning of genes in plants to develop higher quantity of phytometabolites is artificial and thus requires extensive planning and designing techniques. Optimization of plant metabolic pathways and tailoring plant hosts are always required to be updated, to allow wider routes for plant genetic engineering needed for the production of essential phytometabolites and discover their potential as antibiotics. The avenues for newer, cheaper, and faster production of antimicrobials through secondary metabolites are yet to be discovered with the help of systems biology and ever-increasing genomic data of new plants suffices to be the basis of pathways being channelized for better optimization of metabolic pathways keeping ease of commercialization by production at an industrial scale and at cost effective rates in mind.

REFERENCES

Abbasi, T., Anuradha, J., Ganaie, S. U., and Abbasi, S. A. (2015). Gainful utilization of the highly intransigent weed ipomoea in the synthesis of gold nanoparticles. J. King Saud Univ. Sci., 27(1), 15–22.

Abdallah, B. M., and Ali, E. M. (2021). Green synthesis of silver nanoparticles using the Lotus lalambensis aqueous leaf extract and their anti-candidal activity against oral candidiasis. ACS Omega, 6(12), 8151–8162.

Abdallah, E. M. (2011). Plants: An alternative source for antimicrobials. J. Appl. Pharm. Sci., 1(6), 16–20.

Abdel-Raouf, N., Al-Enazi, N. M., and Ibraheem, I. B. (2017). Green biosynthesis of gold nanoparticles using Galaxharactengata and characterization of their antibacterial activity. Arab. J. Chem., 10, S3029–S3039.

Abike, T. O., Osuntokun, O. T., Modupe, A. O., Adenike, A. F., and Atinuke, A. R. (2020). Antimicrobial efficacy, secondary metabolite constituents, ligand docking of Enantia chlorantha on selected multidrug resistance bacteria and fungi. J. Adv. Biol. Biotechnol. 23(6), 17–32.

Agre, K. G., Das, R. J., and Umekar, M. J. (2021). Phytochemical screening, thin layer chromatography and antimicrobial study of Randia spinosa fruits. World J. Pharm. Res., 10(8), 1000–1008.

Ahmad, N., Sharma, S., Singh, V. N., Shamsi, S. F., Fatma, A., and Mehta, B. R. (2011). Biosynthesis of silver nanoparticles from Desmodium triflorum: a novel approach towards weed utilization. Biotechnol. Res. Int., 1–8.

Ahmad, R., Srivastava, S., Ghosh, S., and Khare, S. K. (2021). Phytochemical delivery through nanocarriers: A review. Colloids Surf. B, 197, 111389.

Ahmad, S. A., Das, S. S., Khatoon, A., Ansari, M. T., Afzal, M., Hasnain, M. S., and Nayak, A. K. (2020). Bactericidal activity of silver nanoparticles: A mechanistic review. Mater. Sci. Technol., 3, 756–769.

Ahmad, T., Bustam, M. A., Irfan, M., Moniruzzaman, M., Asghar, H. M. A., and Bhattacharjee, S. (2018). Green synthesis of stabilized spherical shaped gold nanoparticles using novel aqueous Elaeis guineensis (oil palm) leaves extract. J. Mol. Struct., 1159, 167–173.

Ahmed, S., Saifullah, Ahmad, M., Swami, B. L., and Ikram, S. (2016). Green synthesis of silver nanoparticles using Azadirachta indica aqueous leaf extract. J. Radiat. Res. Appl., 9(1), 1–7.

Ajitha, B., Reddy, Y. A. K., Shameer, S., Rajesh, K. M., Suneetha, Y., and Reddy, P. S. (2015). Lantana camara leaf extract mediated silver nanoparticles: antibacterial, green catalyst. J. Photochem. Photobiol. B: Biol., 149, 84–92.

Al- Mamari, H. H., (2021). Phenolic compounds: Classification, chemistry, and updated techniques of analysis and synthesis. In Al- Mamari, H. H (Ed.), Phenolic Compounds – Chemistry, synthesis, diversity, non-conventional industrial, pharmaceutical and therapeutic applications. IntechOpen.

Albert, S., Horbach, R., Deising, H. B., Siewert, B., and Csuk, R. (2011). Synthesis and antimicrobial activitIf (E) stilbene derivatives. Bioorg. Med. Chem., 19(17), 5155–5166.

Ali, K., Dwivedi, S., Azam, A., Saquib, Q., Al-Said, M. S., Alkhedhairy, A. A., and Musarrat, J. (2016). Aloe vera extract functionalized zinc oxide nanoparticles as nanoantibiotics against multi-drug resistant clinical bacterial isolates. J. Colloid Interface Sci., 472, 145–156.

Al-Jumaili, A., Kumar, A., Bazaka, K., and Jacob, M. V. (2018). Plant secondary metabolite-derived polymers: A potential approach to develop antimicrobial films. Polym. J., 10(5), 515.

Allemailem, K. S. (2021). Antimicrobial potential of naturally occurring bioactive secondary metabolites. J. Pharm. Bioallied Sci., 13(2), 155.

Al-Shabib, N. A., Husain, F. M., Qais, F. A., Ahmad, N., Khan, A., Alyousef, A. A., Arshad, M., Noor, S., Khan, J. M., Alam, P., Albalawi, T. H., and Shahzad, S. A. (2020). Phyto-mediated synthesis of porous titanium dioxide nanoparticles from Withania somnifera root extract: Broad-spectrum attenuation of biofilm and cytotoxic properties against HepG2 cell lines. Front. Microbiol., 11, 1680.

Amina, M., Al Musayeib, N. M., Alarfaj, N. A., El-Tohamy, M. F., Oraby, H. F., Al Hamoud, G. A., Bukhari, S. I., and Moubayed, N. M. (2020). Biogenic green synthesis of MgO nanoparticles using Saussurea costus biomasses for a comprehensive detection of their antimicrobial, cytotoxicity against MCF-7 breast cancer cells and photocatalysis potentials. PLoS One, 15(8), e0237567.

Anbuvannan, M., Ramesh, M., Viruthagiri, G., Shanmugam, N., and Kannadasan, N. (2015). Anisochilus carnosus leaf extract mediated synthesis of zinc oxide nanoparticles for antibacterial and photocatalytic activities. Mater. Sci. Semicond. Process., 39, 621–628.

Ankamwar, B. (2010). Biosynthesis of gold nanoparticles (green-gold) using leaf extract of Terminalia catappa. New J. Chem., 7(4), 1334–1339.

Arakha, M., Pal, S., Samantarrai, D., Panigrahi, T. K., Mallick, B. C., Pramanik, K., Mallick, B., and Jha, S. (2015). Antimicrobial activity of iron oxide nanoparticle upon modulation of nanoparticle-bacteria interface. Sci. Rep., 5(1), 1–12.

Arunachalam, K. D., Arun, L. B., Annamalai, S. K., and Arunachalam, A. M. (2014). Biofunctionalized gold nanoparticles synthesis from Gymnema sylvestre and its preliminary anticancer activity. Int. J. Pharm. Pharm. Sci, 6(4), 423–430.

Aziz, N., Faraz, M., Pandey, R., Sakir, M., Fatma, T., Varma, A., Barman, I., Prasad, R. (2015). Facile algae-derived route to biogenic silver nanoparticles: Synthesis, antibacterial and photocatalytic properties. Langmuir, 31, 11605–11612. DOI: 10.1021/acs.langmuir.5b03081

Aziz, N., Faraz, M., Sherwani, M. A., Fatma, T., Prasad, R. (2019). Illuminating the anticancerous efficacy of a new fungal chassis for silver nanoparticle synthesis. Front Chem, 7, 65. DOI: 10.3389/fchem.2019.00065

Aziz, N., Fatma, T., Varma, A., Prasad, R. (2014). Biogenic synthesis of silver nanoparticles using Scenedesmus abundans and evaluation of their antibacterial activity. Journal of Nanoparticles, Article ID 689419, 2014. http://dx.doi.org/10.1155/2014/689419

Aziz, N., Pandey, R., Barman, I., Prasad, R. (2016). Leveraging the attributes of Mucor hiemalis-derived silver nanoparticles for a synergistic broad-spectrum antimicrobial platform. Front Microbiol, 7, 1984. DOI: 10.3389/fmicb.2016.01984

Bag, A., and Chattopadhyay, R. R. (2014). Efflux-pump inhibitory activity of a gallotannin from Terminalia chebula fruit against multidrug-resistant uropathogenic Escherichia coli. Nat. Prod. Res., 28(16), 1280–1283.

Baláž, M., Goga, M., Hegedüs, M., Daneu, N., Kováčová, M., Tkáčiková, L. U., L'udmila Balážová and Bačkor, M. (2020). Biomechanochemical solid-state synthesis of silver nanoparticles with antibacterial activity using lichens. ACS Sustain. Chem. Eng., 8(37), 13945–13955.

Banala, R. R., Nagati, V. B., and Karnati, P. R. (2015). Green synthesis and characterization of Carica papaya leaf extract coated silver nanoparticles through X-ray diffraction, electron microscopy and evaluation of bactericidal properties. Saudi J. Biol. Sci, 22(5), 637–644.

Bar, H., Bhui, D. K., Sahoo, G. P., Sarkar, P., De, S. P., and Misra, A. (2009). Green synthesis of silver nanoparticles using latex of Jatropha curcas. Colloids Surf. A: Physicochem. Eng. Asp., 339(1–3), 134–139.

Bhuiyan, M. S. H., Miah, M. Y., Paul, S. C., Aka, T. D., Saha, O., Rahaman, M. M., Sharif, M. J. I., Habib, O., and Ashaduzzaman, M. (2020). Green synthesis of iron oxide nanoparticle using Carica papaya leaf extract: Application for photocatalytic degradation of remazol yellow RR dye and antibacterial activity. Heliyon, 6(8), e04603.

Bhuyan, T., Mishra, K., Khanuja, M., Prasad, R., Varma, A. (2015). Biosynthesis of zinc oxide nanoparticles from *Azadirachta indica* for antibacterial and photocatalytic applications. Mater Sci Semicond Process, 32, 55–61.

Birla, S. S., Tiwari, V. V., Gade, A. K., Ingle, A. P., Yadav, A. P., and Rai, M. K. (2009). Fabrication of silver nanoparticles by Phoma glomerata and its combined effect against Escherichia coli, Pseudomonas aeruginosa and Staphylococcus aureus. Lett. Appl. Microbiol., 48(2), 173–179.

Bjarnsholt, T., Jensen, P. Ø., Rasmussen, T. B., Christophersen, L., Calum, H., Hentzer, M., Hougen, H. P., Rygaard, J., Moser, C., Eber, L., Høiby, N., and Givskov, M. (2005). Garlic blocks quorum sensing and promotes rapid clearing of pulmonary Pseudomonas aeruginosa infections. Microbiology, 151(12), 3873–3880.

Bogireddy, N. K. R., Anand, K. K. H., and Mandal, B. K. (2015). Gold nanoparticles—synthesis by Sterculia acuminata extract and its catalytic efficiency in alleviating different organic dyes. J. Mol. Liq., 211, 868–875.

Bourgaud, F., Gravot, A., Milesi, S., and Gontier, E. (2001). Production of plant secondary metabolites: A historical perspective. Plant Sci., 161(5), 839–851.

Brumbaugh, A. D., Cohen, K. A., and St. Angelo, S. K. (2014). Ultrasmall copper nanoparticles synthesized with a plant tea reducing agent. ACS ACS Sustainable Chem. Eng., 2(8), 1933–1939.

Carson, C. F., Mee, B. J., and Riley, T. V. (2002). Mechanism of action of Melaleuca alternifolia (tea tree) oil on Staphylococcus aureus determined by time-kill, lysis, leakage, and salt tolerance assays and electron microscopy. Antimicrob. Agents Chemother., 46(6), 1914–1920.

Castillo-Henríquez, L., Alfaro-Aguilar, K., Ugalde-Álvarez, J., Vega-Fernández, L., Montes de Oca-Vásquez, G., and Vega-Baudrit, J. R. (2020). Green synthesis of gold and silver nanoparticles from plant extracts and their possible applications as antimicrobial agents in the agricultural area. Nanomaterials, 10(9), 1763.

Chahardoli, A., Karimi, N., and Fattahi, A. (2017). Biosynthesis, characterization, antimicrobial and cytotoxic effects of silver nanoparticles using Nigella arvensis seed extract. Iran. J. Pharm. Sci.: IJPR, 16(3), 1167.

Chahardoli, A., Karimi, N., Sadeghi, F., and Fattahi, A. (2018). Green approach for synthesis of gold nanoparticles from Nigella arvensis leaf extract and evaluation of their antibacterial, antioxidant, cytotoxicity and catalytic activities. Artif. Cells Nanomed. Biotechnol. 46(3), 579–588.

Chan, Y. W., Siow, K. S., Ng, P. Y., Gires, U., and Majlis, B. Y. (2016). Plasma polymerized carvone as an antibacterial and biocompatible coating. Mater. Sci. Eng. C., 68, 861–871.

Chassagne, F., Samarakoon, T., Porras, G., Lyles, J. T., Dettweiler, M., Marquez, L., Akram M. Salam, Shabih, S., Farrokhi, D. R., and Quave, C. L. (2021). A systematic review of plants with antibacterial activities: A taxonomic and phylogenetic perspective. Front. Pharmacol., 11, 2069.

Cheirmadurai, K., Biswas, S., Murali, R., and Thanikaivelan, P. (2014). Green synthesis of copper nanoparticles and conducting nanobiocomposites using plant and animal sources.RSC Adv., 4(37), 19507–19511.

Choi, Y., Kang, S., Cha, S. H., Kim, H. S., Song, K., Lee, Y. J., Kim, K., Kim, Y. S., Cho, S., and Park, Y. (2018). Platycodon saponins from Platycodi radix (Platycodon grandiflorum) for the green synthesis of gold and silver nanoparticles. Nanoscale Res. Lett., 13(1), 1–10.

Chung, K. T., Wong, T. Y., Wei, C. I., Huang, Y. W., and Lin, Y. (1998). Review of the tannins and human health. Crit. Rev. Food Sci. Nutr., 38, 421–464.

Cruz, D., Falé, P. L., Mourato, A., Vaz, P. D., Serralheiro, M. L., and Lino, A. R. L. (2010). Preparation and physicochemical characterization of Ag nanoparticles biosynthesized by Lippia citriodora (Lemon Verbena). Colloids Surf., B, 81(1), 67–73.

Cushnie, T. T., Cushnie, B., and Lamb, A. J. (2014). Alkaloids: An overview of their antibacterial, antibiotic-enhancing and antivirulence activities. Int. J. Antimicrob. Agents, 44(5), 377–386.

Daglia, M. (2012). Polyphenols as antimicrobial agents. Curr. Opin. Biotechnol., 23(2), 174–181.

Dakal, T. C., Kumar, A., Majumdar, R. S., and Yadav, V. (2016). Mechanistic basis of antimicrobial actions of silver nanoparticles. Front. Microbiol. 7, 1831.

Dangi, A. K., Sinha, R., Dwivedi, S., Gupta, S. K., and Shukla, P. (2018). Cell line techniques and gene editing tools for antibody production: A review. Front. Pharmacol., 9, 630.

Das, J., and Velusamy, P. (2014). Catalytic reduction of methylene blue using biogenic gold nanoparticles from Sesbania grandiflora L. J Taiwan Inst Chem Eng. 45, 2280–2285.

Dauthal, P., and Mukhopadhyay, M. (2013). Biosynthesis of palladium nanoparticles using Delonix regia leaf extract and its catalytic activity for nitro-aromatics hydrogenation. Ind. Eng. Chem. Res., 52(51), 18131–18139.

David, T. I., Adelakun, N. S., Omotuyi, O. I., Metibemu, D. S., Ekun, O. E., Inyang, O. K., Adewumi, B., Enejoh, O. A., Owolabi, R. T., and Oribamise, E. I. (2018). Molecular docking analysis of phyto-constituents from Cannabis sativa with pfDHFR. Bioinformation, 14(9), 574.

David, T. I., Okonkwo, D. C., Adelakun, N. S., and Omotuyi, O. I. (2019). Automated QSAR and docking study of phyto-ligands of Chromolaena odorata as inhibitor of bacteria beta-lactamase enzyme. J. Biol. Eng., 6(2), 32–42.

de la Fuente-Núñez, C., and Lu, T. K. (2017). CRISPR-Cas9 technology: Applications in genome engineering, development of sequence-specific antimicrobials, and future prospects. Integr. Biol., 9(2), 109–122.

Devi, K. P., Nisha, S. A., Sakthivel, R., and Pandian, S. K. (2010). Eugenol (an essential oil of clove) acts as an antibacterial agent against Salmonella typhi by disrupting the cellular membrane. J. Ethnopharmacol., 130(1), 107–115.

Dinesh, S., Karthikeyan, S., and Arumugam, P. (2012). Biosynthesis of silver nanoparticles from Glycyrrhiza glabra root extract. Arch Appl Sci Res, 4(1), 178–187.

Dipankar, C., and Murugan, S. (2012). The green synthesis, characterization and evaluation of the biological activities of silver nanoparticles synthesized from Iresine herbstii leaf aqueous extracts. Colloids Surf., B, 98, 112–119.

Dobrucka, R., and Długaszewska, J. (2016). Biosynthesis and antibacterial activity of ZnO nanoparticles using Trifolium pratense flower extract. Saudi J Biol Sci., 23, 517–523.

Donga, S., and Chanda, S. (2021). Facile green synthesis of silver nanoparticles using Mangifera indica seed aqueous extract and its antimicrobial, antioxidant and cytotoxic potential (3-in-1 system). Artif. Cells Nanomed. Biotechnol., 49(1), 292–302.

Dubey, S. P., Lahtinen, M., and Sillanpää, M. (2010). Green synthesis and characterizations of silver and gold nanoparticles using leaf extract of Rosa rugosa. Colloids Surf., A, 364(1–3), 34–41.

Dunkić, V., Bezić, N., Vuko, E., and Cukrov, D. (2010). Antiphytoviral activity of Satureja montana L. haracteregata (Host) PW Ball essential oil and phenol compounds on CMV and TMV. Molecules, 15(10), 6713–6721.

Dutta, P. P., Bordoloi, M., Gogoi, K., Roy, S., Narzary, B., Bhattacharyya, D.I, ... and Mazumder, B. (2017). Antimalarial silver and gold nanoparticles: Green synthesis, characterization and in vitro study. Biomed. Pharmacother., 91, 567–580.

Elbagory, A. M., Meyer, M., Cupido, C. N., and Hussein, A. A. (2017). Inhibition of bacteria associated with wound infection by biocompatible green synthesized gold nanoparticles from South African plant extracts. Nanomaterials, 7(12), 417.

Elumalai, D., Kaleena, P. K., Ashok, K., Suresh, A., and Hemavathi, M. (2016). Green synthesis of silver nanoparticle using Achyranthes aspera and its larvicidal activity against three major mosquito vectors. Eng. Agric. Environ. Food, 9(1), 1–8.

Elumalai, K., and Velmurugan, S. (2015). Green synthesis, characterization and antimicrobial activities of zinc oxide nanoparticles from the leaf extract of Azadirachta indica (L.). Appl. Surf. Sci., 345, 329–336.

Enwa, F. O. (2014). A mini review on the microbiochemical properties of sauerkraut. Afr. J. Sci. Technol. Innov. Dev., 3(1), 15–16.

Fenfen, L. Ü., Yixian, G. A. O., Huang, J., Daohua, S., and Qingbiao, L. I. (2014). Roles of biomolecules in the biosynthesis of silver nanoparticles: Case of Gardenia jasminoides extract. Chin. J. Chem. Eng., 22(6), 706–712.

Francis, G., Thombre, R., Parekh, F., and Leksminarayan, P. (2014). Bioinspired synthesis of gold nanoparticles using Ficus benghalensis (Indian Banyan) leaf extract. Chem Sci Trans, 3(1), 470–474.

Garibo, D., Borbón-Nuñez, H. A., de León, J. N. D., García Mendoza, E., Estrada, I., Toledano-Magaña, Y., Tiznado, H., Ovalle-Marroquin, M., Soto-Ramos, A. G., Blanco, A., Rodríguez, A. J., Romo, O. A., Chávez-Almazán, L. A., and Susarrey-Arce, A. (2020). Green synthesis of silver nanoparticles using Lysiloma acapulcensis exhibit high-antimicrobial activity. Sci. Rep., 10(1), 1–11.

Gawade, V. V., Gavade, N. L., Shinde, H. M., Babar, S. B., Kadam, A. N., and Garadkar, K. M. (2017). Green synthesis of ZnO nanoparticles by using Calotropis procera leaves for the photodegradation of methyl orange. J. Mater. Sci.: Mater. Electron., 28(18), 14033–14039.

Geethalakshmi, R., and Sarada, D. V. L. (2013). Characterization and antimicrobial activity of gold and silver nanoparticles synthesized using saponin isolated from Trianthema decandra L. Ind. Crops Prod., 51, 107–115.

Ghaedi, M., Yousefinejad, M., Safarpoor, M., Khafri, H. Z., and Purkait, M. K. (2015). Rosmarinus officinalis leaf extract mediated green synthesis of silver nanoparticles and investigation of its antimicrobial properties. J. Ind. Eng. Chem., 31, 167–172.

Ghosh, S., Nitnavare, R., Dewle, A., Tomar, G. B., Chippalkatti, R., More, P., Kitture, R., Kale, S., Bellare, J., and Chopade, B. A. (2015). Novel platinum-palladium bimetallic nanoparticles synthesized by Dioscorea bulbifera: anticancer and antioxidant activities. Int. J. Nanomed., 10, 7477.

Ghosh, S., Patil, S., Ahire, M., Kitture, R., Kale, S., Pardesi, K., Cameotra, S. S., Bellare, J., Dhavale, D. D., Jabgunde, A., and Chopade, B. A. (2012). Synthesis of silver nanoparticles using Dioscorea bulbifera tuber extract and evaluation of its synergistic potential in combination with antimicrobial agents. Int. J. Nanomed., 7, 483.

Gomathi, R., and Suhana, H. (2021). Green synthesis, characterization and antimicrobial activity of zinc oxide nanoparticles using Artemisia pallens plant extract. Inorg. Nano-Metal Chem., 51(12), 1663–1672.

Gopinath, K., Venkatesh, K. S., Ilangovan, R., Sankaranarayanan, K., and Arumugam, A. (2013). Green synthesis of gold nanoparticles from leaf extract of Terminalia arjuna, for the enhanced mitotic cell division and pollen germination activity. Industrial Crops and Products, 50, 737–742.

Gorlenko, C. L., Kiselev, H. Y., Budanova, E. V., Zamyatnin, A. A., and Ikryannikova, L. N. (2020). Plant secondary metabolites in the battle of drugs and drug-resistant bacteria: New heroes or worse clones of antibiotics? Antibiotics, 9(4), 170.

Górniak, I., Bartoszewski, R., and Króliczewski, J. (2019). Comprehensive review of antimicrobial activities of plant flavonoids. Phytochem. Rev., 18(1), 241–272.

Govaerts, R., Nic Lughadha, E., Black, N., Turner, R., and Paton, A. (2021). The World checklist of vascular plants, a continuously updated resource for exploring global plant diversity. Sci. Data, 8(1), 1–10.

Gudkov, S. V., Burmistrov, D. E., Serov, D. A., Rebezov, M. B., Semenova, A. A., and Lisitsyn, A. B. (2021). Do iron oxide nanoparticles have significant antibacterial properties? Antibiotics, 10(7), 884.

Guilger-Casagrande, M., and Lima, R. D. (2019). Synthesis of silver nanoparticles mediated by fungi: A review. Front. Bioeng. Biotechnol., 7, 287.

Guo, Q., Liu, Q., A Smith, N., Liang, G., and Wang, M. B. (2016). RNA silencing in plants: Mechanisms, technologies and applications in horticultural crops. Curr Genomics, 17(6), 476–489.

Gupta, M., Tomar, R.S., Kaushik, S., Mishra, R.K. and Sharma, D. (2018). Effective Antimicrobial Activity of Green ZnO Nano Particles of Catharanthus roseus. *Front. Microbiol.* 9:2030. doi: 10.3389/fmicb.2018.02030

Gupta, C. L., Tiwari, R. K., and Cytryn, E. (2020). Platforms for elucidating antibiotic resistance in single genomes and complex metagenomes. Environ. Int., 138, 105667.

Hajra, A., Dutta, S., and Mondal, N. K. (2016). Mosquito larvicidal activity of cadmium nanoparticles synthesized from petal extracts of marigold (Tagetes sp.) and rose (Rosa sp.) flower. J Parasit Dis., 40(4), 1519–1527.

Hammami, R., and Fliss, I. (2010). Current trends in antimicrobial agent research: Chemo- and bioinformatics approaches. Drug Discovery Today, 15(13–14), 540–546.

Hamouda, R. A., Hussein, M. H., Abo-Elmagd, R. A., and Bawazir, S. S. (2019). Synthesis and biological characterization of silver nanoparticles derived from the cyanobacterium Oscillatoria limnetica. Sci. reports, 9(1), 1–17.

Harborne, J. B., and Williams, C. A. (1998). Anthocyanins and other flavonoids. Nat. Prod. Rep., 15(6), 631–652.

Hassan, A. U., Madu, A. H., Ozojiofor, U. O., Galadanci, A. H., Mato, I. B., and Jafaru, R. Antimicrobial Activities of Cymbopogon citratus and Ximenia americana leaf extracts against some selected bacterial and yeast clinical isolates. Asian J. Biochem. Genet. Mol. Biol. 9(1): 1–10.

Heleno, S. A., Martins, A., Queiroz, M. J. R., and Ferreira, I. C. (2015). Bioactivity of phenolic acids: Metabolites versus parent compounds: A review. Food Chem., 173, 501–513.

Hemlata, P. R. M., Singh, A. P., and Tejavath, K. K. (2020). Biosynthesis of silver nanoparticles using Cucumis prophetarum aqueous leaf extract and their antibacterial and antiproliferative activity against cancer cell lines. ACS omega, 5(10), 5520.

Hussein, R. A., and El-Anssary, A. A. (2019). Plants secondary metabolites: The key drivers of the pharmacological actions of medicinal plants. Herb. Med. J. 1, 13.

Im, A. R., Han, L., Kim, E. R., Kim, J., Kim, Y. S., and Park, Y. (2012). Enhanced antibacterial activities of Leonuri herba extracts containing silver nanoparticles. Phytother. Res., 26(8), 1249–1255.

Ingallina, C., Costa, P. M., Ghirga, F., Klippstein, R., Wang, J. T., Berardozzi, S., Hodgins, N., Infante, P., Pollard, S. M., Botta, B., and Al-Jamal, K. T. (2017). Polymeric glabrescione B nanocapsules for passive targeting of Hedgehog-dependent tumor therapy in vitro. Nanomedicine, 12(7), 711–728.

Isaac, R. S., Sakthivel, G., and Murthy, C. H. (2013). Green synthesis of gold and silver nanoparticles using Averrhoa bilimbi fruit extract. J. Nanotechnol., 2013.

Islam, N. U., Jalil, K., Shahid, M., Rauf, A., Muhammad, N., Khan, A., Shah, M. R., and Khan, M. A. (2019). Green synthesis and biological activities of gold nanoparticles functionalized with Salix alba. Arabian J. Chem., 12(8), 2914–2925.

Ismail, E. H., Khalil, M. M., Al Seif, F. A., El-Magdoub, F., Bent, A. N., Rahman, A., and Al, U. S. D. (2014). Biosynthesis of gold nanoparticles using extract of grape (Vitis vinifera) leaves and seeds. Prog Nanotechnol Nanomater, 3, 1–12.

Jafarirad, S., Mehrabi, M., Divband, B., and Kosari-Nasab, M. (2016). Biofabrication of zinc oxide nanoparticles using fruit extract of Rosa canina and their toxic potential against bacteria: A mechanistic approach. Adv. Mater. Sci. Eng.: C, 59, 296–302.

Jain, C., Khatana, S., and Vijayvergia, R. (2019). Bioactivity of secondary metabolites of various plants: A review. Int. J. Pharm. Sci. Res, 10(2), 494–504.

Jain, S., and Mehata, M. S. (2017). Medicinal plant leaf extract and pure flavonoid mediated green synthesis of silver nanoparticles and their enhanced antibacterial property. Sci. Rep., 7(1), 1–13.

Jassal, P. S., Kaur, D., Prasad, R., Singh, J. (2022). Green synthesis of titanium dioxide nanoparticles: Development and applications. Journal of Agriculture and Food Research, 10, 100361. https://doi.org/10.1016/j.jafr.2022.100361

Jayarambabu, N., Akshaykranth, A., Rao, T. V., Rao, K. V., and Kumar, R. R. (2020). Green synthesis of Cu nanoparticles using Curcuma longa extract and their application in antimicrobial activity. Mater. Lett., 259, 126813.

Jin, J., Guo, N., Zhang, J., Ding, Y., Tang, X., Liang, J., Li, L., Deng, X., and Yu, L. (2010). The synergy of honokiol and fluconazole against clinical isolates of azole resistant Candida albicans. Lett. Appl. Microbiol., 51(3), 351–357.

Joshi, N., Jain, N., Pathak, A., Singh, J., Prasad, R., Upadhyaya, C. P. (2018). Biosynthesis of silver nanoparticles using *Carissa carandas berries* and its potential antibacterial activities. J Sol-Gel Sci Techn, 86(3), 682–689. https://doi.org/10.1007/s10971-018-4666-2

Kanchi, S., Kumar, G., Lo, A. Y., Tseng, C. M., Chen, S. K., Lin, C. Y., and Chin, T. S. (2018). Exploitation of de-oiled jatropha waste for gold nanoparticles synthesis: a green approach. Arabian J. Chem., 11(2), 247–255.

Kaur, M., and Bhullar, G. K. (2016). Partial characterization of tamarind (Tamarindus indica L.) kernel starch oxidized at different levels of sodium hypochlorite. Int. J. Food Prop., 19(3), 605–617.

Kaviya, S., Santhanalakshmi, J., Viswanathan, B., Muthumary, J., and Srinivasan, K. (2011). Biosynthesis of silver nanoparticles using Citrus sinensis peel extract and its antibacterial activity. Spectrochim. Acta A Mol. Biomol. Spectrosc., 79(3), 594–598.

Khalil, M. M., Ismail, E. H., and El-Magdoub, F. (2012). Biosynthesis of Au nanoparticles using olive leaf extract: 1st nano updates. Arabian J. Chem., 5(4), 431–437.

Khan, S., Singh, H., Mishra, A. K., and Khan, N. A. (2021) Quantification of Lupeol as secondary metabolite by HPTLC technique and assessment of antimicrobial potential of ethyl acetate extract of Betula alnoides bark. Orient. J. Chem., 37(2), 426–432.

Khazir, J., Mir, B. A., Pilcher, L., and Riley, D. L. (2014). Role of plants in anticancer drug discovery. Phytochem. Lett., 7, 173–181.

Kocaçalışkan, I., Talan, I., and Terzi, I. (2006). Antimicrobial activity of catechol and pyrogallol as allelochemicals. Z. Naturforsch. C, 61(9–10), 639–642.

Koperuncholan, M. (2015). Bioreduction of chloroauric acid (HAuCl4) for the synthesis of gold nanoparticles (GNPs): A special empathies of pharmacological activity. Int. J. Phytopharm, 5(4), 72–80.

Kora, A. J., Sashidhar, R. B., and Arunachalam, J. (2012). Aqueous extract of gum olibanum (Boswellia serrata): A reductant and stabilizer for the biosynthesis of antibacterial silver nanoparticles. Process Biochem., 47(10), 1516–1520.

Kougan, G. B., Tabopda, T., Kuete, V., and Verpoorte, R. (2013). Simple phenols, phenolic acids, and related esters from the medicinal plants of Africa. J. Med. Plants Res. 225–249.

Krishnaraj, C., Jagan, E. G., Rajasekar, S., Selvakumar, P., Kalaichelvan, P. T., and Mohan, N. (2010). Synthesis of silver nanoparticles using Acalypha indica leaf extracts and its antibacterial activity against water borne pathogens. Colloids Surf., B, 76(1), 50–56.

Kumar, H. A. K., Mandal, B. K., Kumar, K. M., babu Maddinedi, S., Kumar, T. S., Madhiyazhagan, P., and Ghosh, A. R. (2014). Antimicrobial and antioxidant activities of Mimusops elengi seed extract mediated isotropic silver nanoparticles. Spectrochim. Acta, Part A, 130, 13–18.

Kumar, K. M., Mandal, B. K., Kumar, H. A. K., and Maddinedi, S. B. (2013). Green synthesis of size controllable gold nanoparticles. Spectrochim. Acta, Part A: Molecular and Biomolecular Spectroscopy, 116, 539–545.

Kumar, K. R., Nattuthurai, N., Gopinath, P., and Mariappan, T. (2015). Synthesis of eco-friendly silver nanoparticles from Morinda tinctoria leaf extract and its larvicidal activity against Culex quinquefasciatus. J. Parasitol. Res., 114(2), 411–417.

Kumar, V., Yadav, S. C., and Yadav, S. K. (2010). Syzygium cumini leaf and seed extract mediated biosynthesis of silver nanoparticles and their characterization. J. Chem. Technol. Biotechnol., 85(10), 1301–1309.

Kumari, M., Sharma, N., Manchanda, R., Gupta, N., Syed, A., Bahkali, A. H., and Nimesh, S. (2021). PGMD/curcumin nanoparticles for the treatment of breast cancer. Sci. Rep., 11(1), 1–17.

Kwiecien, J. M., and Berhardt, L. V. (2016). Post-traumatic inflammation of the white matter; new disease for anti-inflammatory drugs. Adv. Exp. Med. Biol., 94, 149–157.

Lagashetty, A. N. (2015). Green synthesis and characterization of silver nanoparticles using piper betel leaf extract. Bull. Adv. Sci. Res, 1(5), 136–138.

Lederberg, J., Hamburg, M. A., and Smolinski, M. S. (eds.). (2003). Microbial threats to health: Emergence, detection, and response. National Academies Press.

Leonti, M., Casu, L., Sanna, F., and Bonsignore, L. (2009). A comparison of medicinal plant use in Sardinia and Sicily—De Materia Medica revisited? J. Ethnopharmacol., 121(2), 255–267.

Li, H., Chen, Q., Zhao, J.,and Urmila, K. (2015). Enhancing the antimicrobial activity of natural extraction using the synthetic ultrasmall metal nanoparticles. Scientific reports, 5(1), 1–13.doi: 10.1038/srep11033

Liu, T., Baek, D. R., Kim, J. S., Joo, S. W., and Lim, J. K. (2020). Green synthesis of silver nanoparticles with size distribution depending on reducing species in glycerol at ambient pH and temperatures. ACS omega, 5(26), 16246–16254.

Liu, Y., Jin, X., and Chen, Z. (2018). The formation of iron nanoparticles by Eucalyptus leaf extract and used to remove Cr (VI). Sci. Total Environ., 627, 470–479.

Loo, Y. Y., Rukayadi, Y., Nor-Khaizura, M. A. R., Kuan, C. H., Chieng, B. W., Nishibuchi, M., and Radu, S. (2018). In vitro antimicrobial activity of green synthesized silver nanoparticles against selected Gram-negative foodborne pathogens. Front. Microbiol., 9, 1555.

Lowder, L. G., Zhang, D., Baltes, N. J., Paul III, J. W., Tang, X., Zheng, X., Voytas, D. F., Hsieh, T. F., Zhang, Y., and Qi, Y. (2015). A CRISPR/Cas9 toolbox for multiplexed plant genome editing and transcriptional regulation. J. Plant Physiol., 169(2), 971–985.

Macoy, D. M., Kim, W. Y., Lee, S. Y., and Kim, M. G. (2015). Biosynthesis, physiology, and functions of hydroxycinnamic acid amides in plants. Plant Biotechnol. Rep., 9(5), 269–278.

Mahizan, N. A., Yang, S. K., Moo, C. L., Song, A. A. L., Chong, C. M., Chong, C. W., Abushelaibi, A., Lim, S. E. and Lai, K. S. (2019). Terpene derivatives as a potential agent against antimicrobial resistance (AMR) pathogens. Molecules, 24(14), 2631.

Mandal, S. M., Dias, R. O., and Franco, O. L. (2017). Phenolic compounds in antimicrobial therapy. J. Med. Food, 20(10), 1031–1038.

Marslin, G., Selvakesavan, R. K., Franklin, G., Sarmento, B., and Dias, A. C. (2015). Antimicrobial activity of cream incorporated with silver nanoparticles biosynthesized from Withania somnifera. Int. J. Nanomed., 10, 5955.

Mata, R., Nakkala, J. R., and Sadras, S. R. (2016). Polyphenol stabilized colloidal gold nanoparticles from Abutilon indicum leaf extract induce apoptosis in HT-29 colon cancer cells. Colloids Surf., B, 143, 499–510.

Mau, J. L., Chiou, S. Y., Hsu, C. A., Tsai, H. L., and Lin, S. D. (2012). Antimutagenic and antimicrobial activities of γ-aminobutyric acid (Gaba) tea extract. Int. Conf. Nutr. Food Sci. 39, 178–182.

Mera, G., Francesca, I., Falconí, D. E. G., and Córdova, V. M. (2019). Secondary metabolites in plants: Main classes, phytochemical analysis and pharmacological activities. Bionatura 4 (4): 1000–1009.

Mubarak Ali, D., Thajuddin, N., Jeganathan, K., and Gunasekaran, M. (2011). Plant extract mediated synthesis of silver and gold nanoparticles and its antibacterial activity against clinically isolated pathogens. Colloids Surf., B: Biointerfaces, 85(2), 360–365.

Mukunthan, K. S., and Balaji, S. (2012). Cashew apple juice (Anacardium occidentale L.) speeds up the synthesis of silver nanoparticles. Int. J. Green Nanotechnol., 4(2), 71–79.

Nabikhan, A., Kandasamy, K., Raj, A., and Alikunhi, N. M. (2010). Synthesis of antimicrobial silver nanoparticles by callus and leaf extracts from saltmarsh plant, Sesuvium portulacastrum L. Colloids Surf., B: Biointerfaces, 79(2), 488–493.

Nagajyothi, P. C., Sreekanth, T. V. M., Tettey, C. O., Jun, Y. I., and Mook, S. H. (2014). Characterization, antibacterial, antioxidant, and cytotoxic activities of ZnO nanoparticles using Coptidis Rhizoma. Bioorg. Med. Chem. Lett., 24(17), 4298–4303.

Namdeo, A. G. (2018). Cultivation of medicinal and aromatic plants. In Natural products and drug discovery edited by Namdeo, A. G (pp. 525–553). Elsevier.

Naseer, M., Ramadan, R., Xing, J., and Samak, N. A. (2021). Facile green synthesis of copper oxide nanoparticles for the eradication of multidrug resistant Klebsiella pneumonia and Helicobacter pylori biofilms. International Biodeterioration and Biodegradation, 159, 105201.

Nasrollahzadeh, M., and Sajadi, S. M. (2016). Pd nanoparticles synthesized in situ with the use of Euphorbia granulate leaf extract: catalytic properties of the resulting particles. J. Colloid Interface Sci., 462, 243–251.

Nayak, D., Ashe, S., Rauta, P. R., and Nayak, B. (2015). Biosyntheharacterizationation and antimicrobial activity of silver nanoparticles using Hibiscus rosa-sinensis petals extracts. IET Nanobiotechnol., 9(5), 288–293.

Nazar, N., Bibi, I., Kamal, S., Iqbal, M., Nouren, S., Jilani, K., Umair, M., and Ata, S. (2018). Cu nanoparticles synthesis using biological molecule of P. granatum seeds extract as reducing and capping agent: Growth mechanism and photo-catalytic activity. Int. J. Biol. Macromol., 106, 1203–1210.

Nguyen, N. Y. T., Grelling, N., Wetteland, C. L., Rosario, R., and Liu, H. (2018). Antimicrobial activities and mechanisms of magnesium oxide nanoparticles (nMgO) against pathogenic bacteria, yeasts, and biofilms. Sci. Reports, 8(1), 1–23.

Niloy, M. S., Hossain, M. M., Takikawa, M., Shakil, M. S., Polash, S. A., Mahmud, K. M., Uddin, F., Alam, M., Shubhra, R. D., Shawan, M. M. A. K., Saha, T., Takeoka, S., Hasan, M. A., and Sarker, S. R. (2020). Synthesis of biogenic silver nanoparticles using Caesalpinia digyna and investigation of their antimicrobial activity and in vivo biocompatibility. ACS Appl. Bio Mater., 3(11), 7722–7733.

Niraimathi, K. L., Sudha, V., Lavanya, R., and Brindha, P. (2013). Biosynthesis of silver nanoparticles using Alternanthera sessilis (Linn.) extract and their antimicrobial, antioxidant activities. Colloids Surf., B: Biointerfaces, 102, 288–291.

Nisar, P., Ali, N., Rahman, L., Ali, M., and Shinwari, Z. K. (2019). Antimicrobial activities of biologically synthesized metal nanoparticles: An insight into the mechanism of action. JBIC J. Biol. Inorg. Chem., 24(7), 929–941.

Nohynek, L. J., Alakomi, H. L., Kähkönen, M. P., Heinonen, M., Helander, I. M., Oksman-Caldentey, K. M., and Puupponen-Pimiä, R. H. (2006). Berry phenolics: antimicrobial properties and mechanisms of action against severe human pathogens. Nutr. Cancer, 54(1), 18–32.

Noumi, E., Merghni, A., M Alreshidi, M., Haddad, O., Akmadar, G., De Martino, Mastouri, M. L., Ceylan, O., Snoussi, M., Al-sieni, A., and De Feo, V. (2018). Chromobacterium violaceum and Pseudomonas aeruginosa PAO1: Models for evaluating anti-quorum sensing activity of Melaleuca alternifolia essential oil and its main compoharacterpinen-4-ol. Molecules, 23(10), 2672.

Okoye, T. C., Uzor, P. F., Onyeto, C. A., and Okereke, E. K. (2014). Safe African medicinal plants for clinical studies. In V. Kuete, ed., Toxicological Survey of African Medicinal Plants (pp. 535–555). Elsevier.

OmojateGodstime, C., Enwa Felix, O., Jewo Augustina, O., & Eze Christopher, O. (2014). Mechanisms of antimicrobial actions of phytochemicals against enteric pathogens–a review. J Pharm Chem Biol Sci, 2(2), 77–85.

Ozturk, D., Ozguven, A., Yonten, V., and Ertas, M. (2022). Green synthesis, characterization and antimicrobial activity of silver nanoparticles using Ornithogalum narbonense L. Inorg. Nano-Metal Chem., 52(3), 329–341.

Padalia, H., Moteriya, P., and Chanda, S. (2015). Green synthesis of silver nanoparticles from marigold flower and its synergistic antimicrobial potential. Arabian J. Chem., 8(5), 732–741.

Patil, C. D., Patil, S. V., Borase, H. P., Salunke, B. K., and Salunkhe, R. B. (2012). Larvicidal activity of silver nanoparticles synthesized using Plumeria rubra plant latex against Aedes aegypti and Anopheles stephensi. J. Parasitol. Res., 110(5), 1815–1822.

Patra, S., Mukherjee, S., Barui, A. K., Ganguly, A., Sreedhar, B., and Patra, C. R. (2015). Green synthesis, characterization of gold and silver nanoparticles and their potential application for cancer therapeutics. Mater. Sci. Eng., C, 53, 298–309.

Pérez, Z. E. J., Ramya Mathiyalagan, J. M., Kim, Y. J., Kang, H. M., Abbai, R., Seo, K. H., Wang, D., Soshnikova, V., and Yang, D. C. (2017). Ginseng-berry-mediated gold and silver nanoparticle synthesis and evaluation of their in vitro antioxidant, antimicrobial, and cytotoxicity effects on human dermal fibroblast and murine melanoma skin cell lines. Int. J. Nanomed., 12, 709.

Philip, D. (2010). Rapid green synthesis of spherical gold nanoparticles using Mangifera indica leaf. Spectrochim. Acta, Part A, 77(4), 807–810.

Pivetta, T. P., Botteon, C. E., Ribeiro, P. A., Marcato, P. D., and Raposo, M. (2021). Nanoparticle systems for cancer phototherapy: An overview. Nanomaterials, 11(11), 3132.

Prasad, R. (2014). Synthesis of silver nanoparticles in photosynthetic plants. Journal of Nanoparticles, Article ID 963961, 2014. http://dx.doi.org/10.1155/2014/963961

Prasad, R., and Swamy, V. S. (2013). Antibacterial activity of silver nanoparticles synthesized by bark extract of Syzygium cumini. J. Nanopart. Res., 2013.

Prasad, R., Pandey, R., Barman, I. (2016). Engineering tailored nanoparticles with microbes: quo vadis. WIREs Nanomed Nanobiotechnol, 8, 316-330. DOI: 10.1002/wnan.1363

Radulovic, N. S., Blagojevic, P. D., Stojanovic-Radic, Z. Z., and Stojanovic, N. M. (2013). Antimicrobial plant metabolites: Structural diversity and mechanism of action. Curr. Med. Chem., 20(7), 932–952.

Radulović, N., Blagojević, P., and Palić, R. (2010). Comparative study of the leaf volatiles of Arctostaphylos uva-ursi (L.) Spreng. and Vaccinium vitis-idaea L. (Ericaceae). Molecules, 15(9), 6168–6185.

Rajathi, F. A. A., Arumugam, R., Saravanan, S., and Anantharaman, P. (2014). Phytofabrication of gold nanoparticles assisted by leaves of Suaeda monoica and its free radical scavenging property. J. Photochem. Photobiol. B: Biol., 135, 75–80.

Rajiv, P., Rajeshwari, S., and Venckatesh, R. (2013). Bio-fabrication of zinc oxide nanoparticles using leaf extract of Parthenium hysterophorus L. and its size-dependent antifungal activity against plant fungal pathogens. Spectrochim. Acta, Part A: Molecular and Biomolecular Spectroscopy, 112, 384–387.

Ramteke, C., Chakrabarti, T., Sarangi, B. K., and Pandey, R. A. (2013). Synthesis of silver nanoparticles from the aqueous extract of leaves of Ocimum sanctum for enhanced antibacterial activity. J. Chem., 2013.

Rao, K. J., and Paria, S. (2013). Green synthesis of silver nanoparticles from aqueous Aegle marmelos leaf extract. Mater. Res. Bull., 48(2), 628–634.

Rashmi, V., and Sanjay, K. R. (2017). Green syntheharacterizationation and bioactivity of plant-mediated silver nanoparticles using Decalepis hamiltonii root extract. IET Nanobiotechnol., 11(3), 247–254.

Rauf, A., Ahmad, T., Khan, A., Maryam, Uddin, G., Ahmad, B., Nasser, Y., Bawazeer, S., Riaz, N., Malikovna, B. K., Almarhoon, Z. M., and Al-Harrasi, A. (2021). Green synthesis and biomedicinal applications of silver and gold nanoparticles functionalized with methanolic extract of Mentha longifolia. Artif. Cells Nanomed. Biotechnol., 49(1), 194–203.

Rio, D., Rodriguez-Mateos, A., Spencer, J. P., Tognolini, M., Borges, G., and Crozier, A. (2013). Dietary (poly) phenolics in human health: structures, bioavailability, and evidence of protective effects against chronic diseases. Antioxid. Redox Signaling, 18(14), 1818–1892.

Roy, A., Bulut, O., Some, S., Mandal, A. K., and Yilmaz, M. D. (2019). Green synthesis of silver nanoparticles: Biomolecule-nanoparticle organizations targeting antimicrobial activity. RSC Adv., 9(5), 2673–2702.

Roy, P., Das, B., Mohanty, A., and Mohapatra, S. (2017). Green synthesis of silver nanoparticles using Azadirachta indica leaf extract and its antimicrobial study. Appl. Nanosci., 7(8), 843–850.

Salmerón-Manzano, E., Garrido-Cardenas, J. A., and Manzano-Agugliaro, F. (2020). Worldwide research trends on medicinal plants. Int. J. Environ. Res. Public Health, 17(10), 3376.

Sánchez-López, E., Gomes, D., Esteruelas, G., Bonilla, L., Lopez-Machado, A. L., Galindo, R., Cano, A., Espina, M., Ettcheto, M., Camins, A., Silva, A. M., Durazzo, A., Santini, A., Garcia, M. L., and Souto, E. B. (2020). Metal-based nanoparticles as antimicrobial agents: an overview. Nanomaterials, 10(2), 292.

Sánchez-Maldonado, A. F., Schieber, A., and Gänzle, M. G. (2011). Structure–function relationships of the antibacterial activity of phenolic acids and their metabolism by lactic acid bacteria. J. Appl. Microbiol., 111(5), 1176–1184.

Saravanan, P., Ramya, V., Sridhar, H., Balamurugan, V., and Umamaheswari, S. (2010). Antibacterial activity of Allium sativum L. on pathogenic bacterial strains. Glob. Vet., 4(5), 519–522.

Sarma, H., Forid, N., Prasad, R., Prasad, M. N. V., Ma., L. Q., Rinklebe, J. (2021). Enhancing phytoremediation of hazardous metal(loid)s using genome engineering CRISPR–Cas9 technology. Journal of Hazardous Materials, 414, 125493. https://doi.org/10.1016/j.jhazmat.2021.125493

Sathishkumar, G., Jha, P. K., Vignesh, V., Rajkuberan, C., Jeyaraj, M., Selvakumar, M., Jha, R., and Sivaramakrishnan, S. (2016). Cannonball fruit (Couroupita guianensis, Aubl.) extract mediated synthesis of gold nanoparticles and evaluation of its antioxidant activity. J. Mol. Liq., 215, 229–236.

Sathishkumar, P., Gu, F. L., Zhan, Q., Palvannan, T., and Yusoff, A. R. M. (2018). Flavonoids mediated "Green"nanomaterials: A novel nanomedicine system to treat various diseases – Current trends and future perspective. Mater. Lett., 210, 26–30.

Sayout, A., Ouarhach, A., Dilagui, I., Soraa, N., and Romane, A. (2020). Antibacterial activity and chemical composition of essential oil from Lavandula tenuisecta Coss. ex Ball. an endemic species from Morocco. Eur. J. Integr. Med., 33, 101017.

Sengottaiyan, A., Mythili, R., Selvankumar, T., Aravinthan, A., Kamala-Kannan, S., Manoharan, K., Thiyagarajan, P., Govarthanan, M., and Kim, J. H. (2016). Green synthesis of silver nanoparticles using Solanum indicum L. and their antibacterial, splenocyte cytotoxic potentials. Res. Chem. Intermed., 42(4), 3095–3103.

Shaikh, S., Nazam, N., Rizvi, S. M. D., Ahmad, K., Baig, M. H., Lee, E. J., and Choi, I. (2019). Mechanistic insights into the antimicrobial actions of metallic nanoparticles and their implications for multidrug resistance. Int. J. Mol. Sci., 20(10), 2468.

Shankar, S., Jaiswal, L., Aparna, R. S. L., and Prasad, R. G. S. V. (2014). Synthesis, characterization, in vitro biocompatibility, and antimicrobial activity of gold, silver and gold silver alloy nanoparticles prepared from Lansium domesticum fruit peel extract. Mater. Lett., 137, 75–78.

Sharma, D., Rani, R., Chaturvedi, M., Rohilla, P., and Yadav, J. P. (2019). In silico and in vitro approach of Allium cepa and isolated quercetin against MDR bacterial strains and Mycobacterium smegmatis. S. Afr. J. Bot., 124, 29–35.

Sharma, P. K., Fuloria, S., Ali, M., Singh, A., Kushwaha, S. P., Sharma, V. K., Vetriselvan, S., and Fuloria, N. K. (2021). Isolation of new phytometabolites from Alpinia galanga wild rhizomes. Pak. J. Pharm. Sci., 34(4).

Sharmila, G., Fathima, M. F., Haries, S., Geetha, S., Kumar, N. M., and Muthukumaran, C. (2017). Green synthesis, characterization and antibacterial efficacy of palladium nanoparticles synthesized using Filicium decipiens leaf extract. J. Mol. Struct., 1138, 35–40.

Singh, B., Yang, S., Krishna, A., and Sridhar, S. (2020). Nanoparticle formulations of poly (ADP-ribose) polymerase inhibitors for cancer therapy. Front. Chem., 8, 594619.

Siriyong, T., Srimanote, P., Chusri, S., Yingyongnarongkul, B. E., Suaisom, C., Tipmanee, V., and Voravuthikunchai, S. P. (2017). Conessine as a novel inhibitor of multidrug efflux pump systems in Pseudomonas aeruginosa. BMC Complement Altern. Med., 17(1), 1–7.

Smitha, S. L., Philip, D., and Gopchandran, K. G. (2009). Green synthesis of gold nanoparticles using Cinnamomum zeylanicum leaf broth. Spectrochim. Acta, Part A: Molecular and Biomolecular Spectroscopy, 74(3), 735–739.

Sonawane, J. M., Rai, A. K., Sharma, M., Tripathi, M., Prasad, R. (2022) Microbial biofilms: Recent advances and progress in environmental bioremediation. Science of the Total Environment, 824, 153843, https://doi.org/10.1016/j.scitotenv.2022.153843

Song, J. Y., Kwon, E. Y., and Kim, B. S. (2010). Biological synthesis of platinum nanoparticles using Diopyros kaki leaf extract. Bioprocess. Biosyst. Eng., 33(1), 159–164.

Sridhar, V., Mamatha, P., and Prasad, S.H. (2018). Pharmacognostical, phytochemical screening and evaluation of antioxidant activity of ethanol and aqueous extracts of Randia spinosa leaves. *Int J Pharm Sci & Res.*, 9(11), 4854–4858. doi: 10.13040/IJPSR.0975-8232.9

Srivastava, S., Usmani, Z., Atanasov, A. G., Singh, V. K., Singh, N. P., Abdel-Azeem, A. M., Prasad, R., Gupta, G., Sharma, M., Bhargava, A. (2021) Biological nanofactories: Using living forms for metal nanoparticle synthesis. Mini-Reviews in Medicinal Chemistry, 21(2), 245–265

Subramani, R., Narayanasamy, M., and Feussner, K. D. (2017). Plant-derived antimicrobials to fight against multi-drug-resistant human pathogens. 3 Biotech, 7(3), 1–15.

Suman, T. Y., Rajasree, S. R., Jayaseelan, C., Mary, R. R., Gayathri, S., Aranganathan, L., and Remya, R. R. (2016). GC-MS analysis of bioactive components and biosynthesis of silver nanoparticles using Hybanthus enneaspermus at room temperature evaluation of their stability and its larvicidal activity. Environ. Sci. Pollut. Res., 23(3), 2705–2714.

Suman, T. Y., Rajasree, S. R., Ramkumar, R., Rajthilak, C., and Perumal, P. (2014). The green synthesis of gold nanoparticles using an aqueous root extract of Morinda citrifolia L. Spectrochim. Acta, Part A: Molecular and Biomolecular Spectroscopy, 118, 11–16.

Suresh, D., Nethravathi, P. C., Rajanaika, H., Nagabhushana, H., and Sharma, S. C. (2015). Green synthesis of multifunctional zinc oxide (ZnO) nanoparticles using Cassia fistula plant extract and their photodegradative, antioxidant and antibacterial activities. Mater. Sci. Semicond. Process., 31, 446–454.

Tabassum, N., and Vidyasagar, G. M. (2014). In vitro antimicrobial activity of edible oils against human pathogens causing skin infections. Int. J. Pharm. Sci. Res. (IJPSR), 5(10), 4493–4498.

Teoh, E. S. (2016). Medicinal orchids of Asia (Vol. 16, No. 4). Cham: Springer.

Thakore, S., Rathore, P. S., Jadeja, R. N., Thounaojam, M., and Devkar, R. V. (2014). Sunflower oil mediated biomimetic synthesis and cytotoxicity of monodisperse hexagonal silver nanoparticles. J. Mater. Sci. Eng: C, 44, 209–215.

Thema, F. T., Manikandan, E., Dhlamini, M. S., and Maaza, M. J. M. L. (2015). Green synthesis of ZnO nanoparticles via Agathosma betulina natural extract. Mater. Lett., 161, 124–127.

Tiwari, P., Khare, T., Shriram, V., Bae, H., and Kumar, V. (2021). Plant synthetic biology for producing potent phyto-antimicrobials to combat antimicrobial resistance. Biotechnol. Adv., 48, 107729.

Ulaeto, S. B., Mathew, G. M., Pancrecious, J. K., Nair, J. B., Rajan, T. P. D., Maiti, K. K., and Pai, B. C. (2019). Biogenic Ag nanoparticles from neem extract: Their structural evaluation and antimicrobial effects against Pseudomonas nitroreducens and Aspergillus unguis (NII 08123). ACS Biomater. Sci. Eng., 6(1), 235–245.

Üstün, E., Önbaş, S. C., Çelik, S. K., Ayvaz, M. Ç., and Şahin, N. (2022). Green synthesis of iron oxide nanoparticles by using Ficus carica leaf extract and its antioxidant activity. Biointerface Res. Appl. Chem, 2021, 2108–2116.

Vanaja, M., and Annadurai, G. (2013). Coleus aromaticus leaf extract mediated synthesis of silver nanoparticles and its bactericidal activity. Appl. Nanosci., 3(3), 217–223.

Vankar, Padma S., and Dhara Bajpai. 2010. Preparation of gold nanoparticles from Mirabilis jalapa flowers. Indian J. Biochem. Biophys., 47 (3): 157–160.

Vanlalveni, C., Lallianrawna, S., Biswas, A., Selvaraj, M., Changmai, B., and Rokhum, S. L. (2021). Green synthesis of silver nanoparticles using plant extracts and their antimicrobial activities: A review of recent literature. RSC Advances, 11(5), 2804–2837.

Varun, S., Sellappa, S., RafiqKhan, M., and Vijayakumar, S. (2015). Green synthesis of gold nanoparticles using Argeharacteicana L. leaf extract and its characterization. Int J Pharm Sci Rev Res, 32, 42–44.

Velmurugan, P., Anbalagan, K., Manosathyadevan, M., Lee, K. J., Cho, M., Lee, S. M., Cho, M., Lee, S. M., Park, J. H., Oh, S. G., Bang, K. S., and Oh, B. T. (2014). Green synthesis of silver and gold nanoparticles using Zingiber officinale root extract and antibacterial activity of silver nanoparticles against food pathogens. Bioprocess. Biosyst. Eng., 37(10), 1935–1943.

Venkatachalam, M., Govindaraju, K., Sadiq, A. M., Tamilselvan, S., Kumar, V. G., and Singaravelu, G. (2013). Functionalization of gold nanoparticles as antidiabetic nanomaterial. Spectrochim. Acta, Part A: Molecular and Biomolecular Spectroscopy, 116, 331–338.

Vivek, R., Thangam, R., Muthuchelian, K., Gunasekaran, P., Kaveri, K., and Kannan, S. (2012). Green biosynthesis of silver nanoparticles from Annona squamosa leaf extract and its in vitro cytotoxic effect on MCF-7 cells. Process Biochem., 47(12), 2405–2410.

Vuolo, M. M., Lima, V. S., and Junior, M. R. M. (2019). Phenolic compounds: Structure, classification, and antioxidant power. In Bioactive compounds edited by Vuolo, M. M., Lima, V. S., and Junior, M. R. M (33–50). Woodhead Publishing.

Wang, L., Hu, C., and Shao, L. (2017). The antimicrobial activity of nanoparticles: Present situation and prospects for the future. Int. J. Nanomed., 12, 1227.

Watson, R. (2017). Nutrition and functional foods for healthy aging. Academic Press.

Wu, S., Rajeshkumar, S., Madasamy, M., and Mahendran, V. (2020). Green synthesis of copper nanoparticles using Cissus vitiginea and its antioxidant and antibacterial activity against urinary tract infection pathogens. Artif. Cells Nanomed. Biotechnol., 48(1), 1153–1158.

Xu, J., Zhou, F., Ji, B. P., Pei, R. S., and Xu, N. (2008). The antibacterial mechanism of carvacrol and thymol against Escherichia coli. Lett. Appl. Microbiol., 47(3), 174–179.

Yan, Y., Li, X., Zhang, C., Lv, L., Gao, B., and Li, M. (2021). Research progress on antibacterial activities and mechanisms of natural alkaloids: A review. J. Antibiot., 10(3), 318.

Yang, L., Wen, K. S., Ruan, X., Zhao, Y. X., Wei, F., and Wang, Q. (2018). Response of plant secondary metabolites to environmental factors. Molecules, 23(4), 762.

Yang, L., Wen, K. S., Ruan, X., Zhao, Y. X., Wei, F., and Wang, Q. (2018). Response of plant secondary metabolites to environmental factors. Molecules, 23(4), 762.

Yang, N., and Li, W. H. (2013). Mango peel extract mediated novel route for synthesis of silver nanoparticles and antibacterial application of silver nanoparticles loaded onto non-woven fabrics. Ind. Crops Prod., 48, 81–88.

Yang, XiuYi, Etelka Chung, Ian Johnston, Guogang Ren, and Yuen-Ki Cheong. 2021. Exploitation of Antimicrobial Nanoparticles and Their Applications in Biomedical Engineering. Applied Sciences, 11(10), 4520. https://doi.org/10.3390/app11104520

Yasmin, A., Ramesh, K., and Rajeshkumar, S. (2014). Optimization and stabilization of gold nanoparticles by using herbal plant extract with microwave heating. Nano Convergence, 1(1), 1–7.

Yu, J., Xu, D., Guan, H. N., Wang, C., and Huang, L. K. (2016). Facile one-step green synthesis of gold nanoparticles using Citrus maxima aqueous extracts and its catalytic activity. Mater. Lett., 166, 110–112.

Zai-xiang, C. H. I., Shi-hua, J. I. A. N. G., Hui, B. A. I., Gui-qiong, L. I., Xiang, S. U. N., Bei, W. A. N. G., and Xian-ju, L. O. N. G. (2019). Conditions of low temperature and high humidity in winter and spring easily lead to red cartridge kiwifruit canker disease in Guizhou. Chin. J. Agrometeorol., 40(09), 591.

Zargar, M., Hamid, A. A., Bakar, F. A., Shamsudin, M. N., Shameli, K., Jahanshiri, F., and Farahani, F. (2011). Green synthesis and antibacterial effect of silver nanoparticles using Vitex negundo L. Molecules, 16(8), 6667–6676.

Zarzuela, R., Luna, M. J., Gil, M. L. A., Ortega, M. J., Palacios-Santander, J. M., Naranjo-Rodríguez, I., Delgado, J. J., and Cubillana-Aguilera, L. M. (2018). Analytical determination of the reducing and stabilization agents present in different Zostera noltii extracts used for the biosynthesis of gold nanoparticles. J. Photochem. Photobiol., B: Biology, 179, 32–38.

Zayed, M. F., and Eisa, W. H. (2014). Phoenix dactylifera L. leaf extract phytosynthesized gold nanoparticles; controlled synthesis and catalytic activity. Spectrochim. Acta, Part A: Molecular and Biomolecular Spectroscopy, 121, 238–244.

Zhang, P., Du, H., Wang, J., Pu, Y., Yang, C., Yan, R., Yang, H., Cheng, H., and Yu, D. (2020). Multiplex CRISPR/Cas9-mediated metabolic engineering increases soya bean isoflavone content and resistance to soya bean mosaic virus. Plant Biotechnol. J., 18(6), 1384–1395.

Zheng, B., Kong, T., Jing, X., Odoom-Wubah, T., Li, X., Sun, D., Lu, F., Zheng, Y., Huang, J., and Li, Q. (2013). Plant-mediated synthesis of platinum nanoparticles and its bioreductive mechanism. J. Colloid Interface Sci., 396, 138–145.

Zhou, Y., and Tang, R. C. (2018). Facile and eco-friendly fabrication of AgNPs coated silk for antibacterial and antioxidant textiles using honeysuckle extract. J. Photochem. Photobiol., B: Biology, 178, 463–471.

Zhou, Z., Tan, H., Li, Q., Chen, J., Gao, S., Wang, Y., Chen, W., and Zhang, L. (2018). CRISPR/Cas9-mediated efficient targeted mutagenesis of RAS in Salvia miltiorrhiza. Phytochemistry, 148, 63–70.

10 New Alternatives of Treatment Against Intestinal Parasite Infection

Lissethe Palomo-Ligas[1], José Roberto Vargas-Villanueva[1], Mariana Garza-Ontiveros[1], Filiberto Gutiérrez-Gutiérrez[2], Rocio Guadalupe Castillo-Godina[1], Lizeth Guadalupe Campos-Muzquiz[1], Adriana Carolina Flores-Gallegos[1], Raúl Rodríguez Herrera[1] and Sendar Daniel Nery-Flores[1]*

[1]Departamento de Investigación en Alimentos, Facultad de Ciencias Químicas, Universidad Autónoma de Coahuila, Blvd. V. Carranza s/n. Col. República Oriente 25280 Saltillo, Coahuila, México

[2]Departamento de Química, Centro Universitario de Ciencias Exactas e Ingenierías, Universidad de Guadalajara, Blvd. Gral. Marcelino García Barragán 1421, Olímpica, 44430 Guadalajara, Jalisco, México

*Corresponding author: sendar.nery@uadec.edu.mx

CONTENTS

DOI: 10.1201/9781003268932-10

10.1 INTRODUCTION

Intestinal parasite infections are mainly ubiquitous, and their transmission is a growing problem in tropical underdeveloped countries as well as in returning travelers from economically developed temperate countries. Although the infection is normally asymptomatic, untreated infections are associated with acute watery diarrhea, relapsing or persistent symptoms generally unresponsive to the usual antibiotic treatment of traveler's diarrhea (Barbosa et al., 2018). According to World Health Organization, diarrheal diseases ranked 5 in the top 10 causes of death in low-income countries (World Health Organization, 2020). Therefore, underestimation of these pathologies can lead to health emergencies.

Protozoan pathogen prevalence is 3% to 7% worldwide, and 20 % to 30% in developing countries with poor sanitation and low socioeconomic levels. High-risk population includes children, exposed travelers, as well as immunocompromised individuals (Aschale et al., 2022; Njoku et al., 2022). On the other hand, helminthic parasites are mainly prevalent in developing countries with tropical and subtropical climates; low socioeconomic background, as well as poor sanitation and personal hygiene, are also of importance for the prevalence of the parasites. Children and pregnant women are a high-risk population for helminthic infections (de Lima Corvino and Shawn, 2021; Ghodeif, 2022; Viswanath Avinash et al., 2021).

Protozoan pathogens normally encountered in intestinal infections are *Giardia lamblia*, *Entamoeba histolytica*, *Cyclospora cayetanensis*, and *Cryptosporidium parvum* (Hemphill et al., 2019). For example, *G. lamblia* causes a variety of gastrointestinal diseases commonly labeled as giardiasis, and symptoms include nausea, weight loss, malabsorption, diarrhea, and vomiting, among others. Its transmission is by fecal-oral contamination, thus its prevalence is higher among poor hygiene and poor sanitization populations (Vivancos et al., 2018). In addition, *E. histolytica* causes amebiasis, which spreads through the fecal-oral route, and symptoms include bloody diarrhea, intestinal ulcers, weight loss, fever, and peritonitis; unlike *G. lamblia*, *E. histolytica* can disseminate and spread to the liver, lungs, and brain (Vique-Sánchez et al., 2021). Also, *C. cayetanensis* is the causal agent of cyclosporiasis which is characterized by watery diarrhea with anorexia, mild to severe nausea, abdominal cramps, and constipation, vomiting, and fever. The main transmission route is the fecal-oral, causing a self-limited infection, however, in immunosuppressed patients causes severe diarrhea (Alkorta et al., 2022). Conversely, *C. parvum* is a coccidia genus that causes cryptosporidiosis, and the risk population includes children, pregnant women, and patients being treated with immunosuppressing drugs; its main symptom is severe, watery diarrhea that could be lethal. It can be transmitted via domestic and livestock animals, the fecal-oral route, and contaminated food and water (Ren et al., 2022). Common intestinal helminthic parasites, also known as geohelminths, are *Ancylostoma duodenale*, *Necator americanus*, *Trichuris trichiura*, and *Ascaris lumbricoides* (Lim-Leroy and Chua, 2020). Common symptoms are ground itch, which may follow skin penetration with intense vesicular, pruritic, and tortuous lesions caused by migration of *A. duodenale* and *N. americanus* larvae; they can also cause intestinal tract chronic infection and suck their host blood which, in most cases, leads to iron deficiency anemia, though the infections are commonly asymptomatic (Jourdan et al., 2018).

T. trichiura is a roundworm that causes trichuriasis in humans; the worm is acquired via fecal-oral transmission, by ingesting embryonated eggs. Children have appeared to be vulnerable to trichuriasis infection and poor sanitary conditions are associated with a heavy disease burden. Symptoms include abdominal pain, and discomfort, painful stool passage, and mucus discharge.

Most cases are asymptomatic since the clinical infection is parasite burden dependent (Viswanath et al., 2021).

A. lumbricoides is a large nematode transmitted through the soil and is the causal agent of ascariasis. Infected patients can be asymptomatic, presenting only malnutrition and growth retardation as long-term manifestations; bloating, abdominal pain, anorexia, nausea, and intermittent diarrhea commonly occur in symptomatic infections. Once the number of larvae passing through the lungs is significant, Loeffler syndrome may manifest (pneumonitis and eosinophilia) (de Lima Corvino D. and Horrall Shawn, 2021). Other pathogens include *Taenia solium*, *Eimeria tenella*, *Ascaris suum*, *Trichinella spiralis*, *Gymnophalloides seoi*, and *Ancylostoma ceylanicum*, which have zoonotic potential.

Common drugs for parasitosis treatment include members of the 5-nitroximidazole family like metronidazole and tinidazole, nitrotazoles like nitazoxanide, nitrofurans such as furazolidone, and benzimidazoles like mebendazole and albendazole (Movahedi et al. 2017).

Metronidazole is widely used as a treatment against intestinal parasites such as *G. lamblia* and *E. histolytica*, as well as most Gram-positive and Gram-negative anaerobic bacteria (Krakovka et al., 2022). Due to its low cost and general availability, it remains the first-choice treatment against intestinal parasitic infections. However, adverse effects of metronidazole usage include encephalopathy, peripheral neuropathy, convulsive seizures, aseptic meningitis, and cases of severe hepatotoxicity or acute hepatic failure (Roy et al., 2016).

Albendazole is used as an antiparasitic agent, and it has better efficiency than mebendazole, highly due to its great tissue penetration into cysts. Few and mild side effects have been reported such as upset gastrointestinal tract, dizziness, rashes, lethargy, and, at higher dosages, liver toxicity, allergic reactions, and rarely severe myelosuppression (Chai et al., 2021).

Mebendazole is a broad-spectrum anthelmintic used as treatment of intestinal nematode infections (Chai et al., 2021), and common side effects include loss of appetite, nausea, diarrhea, vomiting, tinnitus, and high liver enzymes; convulsions and urticaria may manifest in some patients (Thakur and Patel, 2021). Tinidazole side effects include metallic taste, rash, nausea, vomiting, and anorexia (Divakaruni et al., 2019).

Nitazoxamide is a commonly used drug against nematodes and protozoan parasites such as *Ascaris*, *Trichuris*, *Ancylostoma*, *Cryptosporidium*, *Giardia*, and *Entamoeba*. Common side effects include abdominal pain, headaches, nausea, diarrhea, dizziness, general fatigue, and vomiting (Bethesda, 2020).

Furazolidone is used as an antigiardial drug, not commonly used due to its large dose volume (400 mg three times a day for 7–10 days) and high price. During treatment, side effects such as nausea, vomiting, headache, urticaria, hypersensitivity reactions and hypotension may manifest (Vivancos et al., 2018).

For any given pathogen, drug effectiveness is the most important criteria for its use as treatment. Susceptibility to the drug of interest is often measured in pathogen cultures to measure the drug effect experimentally; however, drug resistance is an emerging issue, as first-choice drugs used against different parasitosis are showing an increasing decline in cure rates. Results of a study in 2008 showed that a single dose of albendazole and mebendazole cure rates against *A. lumbricoides* were of 88% and 95%. With the same doses, cure rates against *T. trichiura* were 28% and 36%, and against hookworm infection of 72% and 15% respectively (Keiser and Utzinger, 2008). Failed therapeutic success has also been observed for all standard treatments for giardiasis, compromised of metronidazole, tinidazole, furazolidone, and nitazoxanide (Loderstädt and Frickmann, 2021). As intestinal parasites have less susceptibility to standard treatments, new and innocuous treatment alternatives are very much needed to help the rising battle against drug resisting intestinal parasites, and the infections they bring forth before they reach the resistant strain levels seen in several bacteria genera.

10.2 BIOACTIVE COMPOUNDS

Natural sources and their derivatives represent a growing and safe alternative for obtaining bioactive compounds. They have a wide range of metabolites with potential therapeutic value (Davoodi and Abbasi-Maleki, 2018). Many medicinal plants, agricultural and agro-industrial residues, plant extracts, essential oils, and isolated molecules have been tested against intestinal parasites. Phytochemicals, such as alkaloid, flavonoid, terpenoids, and tannins are the main families probed with antiparasitic effects *in vitro* and *in vivo*. The utilization of different tissues (flowers, seeds, leaves, green stems, woody stems, and roots), the methodology of extraction, solvent election, sample treatment, seasonal variability, and so on, are also important for isolation of phytochemicals and their activities (Ghadage et al., 2017; Guzowska et al., 2022; Piotrowicz et al., 2021). Several studies described the potential effects of bioactive compounds against intestinal parasites. In a review about medicinal plants and their effects for antigiardial purposes, the authors mentioned 1585 papers related to this topic and found that the most widely employed plants against *Giardia* belong to the *Lamiaceae* (30.0%), *Asteraceae* (13.5%), and *Apiaceae* (10.5%) families. Bioactive compounds from these plant families are aqueous extracts (30.0%), essential oils (25.4%), and hydroalcoholic and methanolic extracts (10.5%) (Alnomasy et al., 2021). On the other hand, an *in vitro* test showed that *Origanum vulgare* (Lamiaceae) hydroalcoholic extract had a time-dose effect on *G. lamblia* cysts; they attributed the antiprotozoal activity to phenolic compounds, for example naringenin, kaempferol, and quercetin, among others (Davoodi and Abbasi-Maleki, 2018).

In addition, the evaluation of six different extracts and two essential oils from *Ageratum conyzoides* on *G. lamblia* trophozoites was performed; LW-P (white-purple flowered) and FP (Purple flowered) crude extracts and essential oils, showed the highest anti-Giardia activity. Structural alterations observed in TEM including those in flagella, ventral discs, nuclei, cellular vacuoles, shapes, and chromatin (Pintong et al., 2020). Another example is the evaluation of three different plant extracts from pomegranate, rhubarb (rhein), and maple syrup against *E. histolytica* trophozoite growth, being Rhein and maple syrup, the extracts that showed the highest inhibitory activity (Hunt et al., 2018). A pomegranate peel extract with punicalin, punicalagin, galloyl-dihexahydroxydiphenoyl-hexoside, luteolin, and ellagic acid also affects the rich microtubule cytoskeleton of *G. lamblia* trophozoites *in vitro* causing alterations on tubulin distribution, adhesion capacity reduction, and morphology affectations (Palomo-Ligas et al., 2022). In addition, experiments *in vivo* demonstrated that pomegranate alcoholic extracts are effective for the reduction of *Giardia* cysts shedding and are helpful in prevention and restoration of intestinal cells (Al-Megrin, 2017; El-Kady et al., 2021). Other investigations have showed the potential of methanolic extracts from medicinal plants against *E. histolytica* trophozoites. The testing of 32 extracts revealed that *Lippia graveolens Kunth* and *Ruta chalepensis* Pers. have the highest antiprotozoal activity (91.54% and 90.50% growth inhibition, respectively); the authors identified carvacrol and chalepensin as the agents responsible of the cytotoxic effects (Quintanilla-Licea et al., 2014). In addition, the evaluation of 80 plant species used in Mexican traditional medicine for diarrhea treatment as potential antiprotozoal agents against *E. histolytica* and *G. lamblia* was performed. The authors isolated 150 compounds from these samples. Out of all compounds, they found that epicatechin, kaempferol, eupomatenid 1, geranin A, incomptine A, and tiliroside were the most potent (Calzada and Bautista, 2020). In addition, Drinić et al. (2019) tested 14 compounds from three plant families (*Meliaceae, Rutaceae,* and *Asteraceae*) that belong to different chemical classes (flavagline, furoquinoline, acridone, quinolinone, quinazoline, sulphur-containing amide, and coumarin). The sulphur-containing amide, methylgerambullin showed the highest activity against both parasites, showing EC_{50} against *E. histolytica* after 24 h and 48 h of 14.5 μM (6.08 μg/mL) and 17.4 μM (7.33 μg/mL), respectively and the EC_{50} against *G. lamblia* after 24 h and 48 h were 14.6 μM (6.14 μg/mL) and 36.5 μM (15.34 μg/mL), respectively (Tuvshintulga et al., 2019). On the other hand, the amoebicidal potential of curcumin in *E. histolytica* trophozoites was observed: the highest effects were with 300 μM at 24 h, presenting a 65.5% of growth inhibition, and only 28.8% trophozoites being viable, SEM also showed cell disintegration (Rangel-Castañeda et al. 2018). Curcumin's anti-Giardia potential has also been demonstrated, the authors found that the highest inhibitory concentrations of curcumin (3 and 15 μM) disrupted the cytoskeletal structures of *G. lamblia* trophozoites (ventral disk and flagella; membrane was also affected). The compound altered distribution of tubulin on ventral disk and flagella as well as clear reduction of tubulin expression. They hypothesize that curcumin binds at the interface of the tubulin dimer close to the vinblastine binding site (Gutiérrez-Gutiérrez et al. 2017).

As mentioned before, organic plant waste and by-products from the food industry represent a rich source of bioactive compounds. The *in vitro* inhibitory activity of 42 plant-derived by-products against *C. parvum* were tested, being an ethanolic extract obtained from olive (*Olea europaea*) pomace, the most effective against the parasite, showing a MIC100 of 250-500 μg/mL (Teichmann et al., 2016). On another side, bioactive compounds from different plants (polyphenols from blueberry and pomegranate, phenolic compounds from cinnamon, flavonoids and sulfide compounds from onion, allicin from garlic, mangiferin from mango, oleuropein from olive from pomace and carvacrol from oregano) have been associated with crypto sporicidal effects against *C. parvum* and *C. hominis*; this is very important because represents a potential alternative to nitazoxanide, the only approved medicine against this parasite (Almoradie et al., 2018).

Eimeria spp is a parasite that affects human and chickens. An interesting study tested the anticoccidial effects of plant extracts on *E. tenella* infection in broiler chickens. These authors found that *Nectaroscordum tripedale* extract showed the best results according to a decreased intensity of bloody diarrhea and oocysts count per bird (Habibi et al., 2016). Recently, a processed sugarcane extract (Polygain™) was tested for anti-Eimeria activity, the results showed a reduced number of viable sporozoites of several species of *Eimeria in vitro*. The biological effects are due to polysaccharides and phenolic compounds (Daneshmand et al., 2021). Other studies *in vitro* and *in vivo* confirmed the effect of *Fructus Meliae toosendan* extracts on unsporulated oocysts, the high dose of the extract strongly inhibited sporulation of oocyst, presenting a LC_{50} value of 245.83mg/mL (Yong et al., 2020). A combination of organic acid also has been tested against *Eimeria*; the mixture of maltodextrin, sodium chloride, citric acid, sodium citrate, silica, malic acid, citrus extract, and olive extract inhibit *Eimeria* infection in MDBK and CLEC213 epithelial cells *in vitro*, and improved the immune response, by reducing inflammatory oxidative stress, and increasing antioxidant enzymes *in vivo*. Also, early inclusion of the natural antimicrobial mixture in broiler drinking water was effective in reducing coccidian infection (Balta et al., 2021). The prophylactic focus of herbal powder "Shi Ying Zi" in chickens infected with *Eimeria tenella* was tested. The mixture contains *Conidium monnieri* (L.) Cuss and *Taraxacum mongolicum* Hand.-Mazz. Results showed that treated chickens with "Shi Ying Zi" had a lower mortality rate than non-treated chickens (Song et al., 2020). In addition, the use of leaves of *Rumex nervosus* not only prevents the infection against *E. tenella*, but also could enhance growth performance indices of infected broiler chicken and elevate the meat quality. This supplementation is a promising alternative against commercial anticoccidials with benefits to consumer health (Qaid et al., 2021).

In relation with human hookworm, a study evaluated *in vitro* anthelmintic activity of *Dichapetalum filicaule* Breteler extracts and compounds against this helminth. The bioactive compounds from *D. filicaule* were obtained from root extracts using petroleum ether, chloroform-acetone, and methanol. These authors identified three triterpenoids (dichapetalin X, dichapetalin A, and glycerol monostearate) responsible of the inhibitory effects against *N. americanus*, they also exhibited egg-hatching inhibition activities with IC_{50} values of 162.4, 523.2, and 306.0 mg/mL (Chama et al., 2016). Plant bioactive compounds and extracts have been tested against other nematodes: an assay was performed evaluating the ovicidal and larvicidal effects of different *Andrographis paniculata* leaf extract obtained with ethanol, methanol, ethyl acetate, and petroleum ether against *A. duodenale*. These authors mentioned that the ethanol and methanol extracts showed the highest inhibition of egg hatching while ethyl acetate extract provoked the highest larval motility; the compound andrographolide is one of the components responsible for these effects (Banerjee et al., 2019). A model nematode used to study *A. lumbricoides* is *A. suum* which is a pig infecting parasite causing significant economic losses. The test of 29 ethanolic extracts from Africa and the Caribbean plants found that *Clausena anisata*, *Zanthoxylum zanthoxyloides*, and *Punica granatum* extracts showed the best *A. suum* control; the authors proposed that the effect could be due to presence of tannins (ellagitannins and gallic acid), flavonoids, terpenes, and alkaloids (Williams et al., 2016). Another strategy for parasite infection control is the combination of phytochemicals with reference drugs. In this sense, the combination of carvone and abamectin was assessed *in vitro* and *in vivo* in lambs naturally infected with resistant gastrointestinal nematodes. The co-administration of the natural compound prolonged abamectin absorption in lamb infected with *Haemonchus contortus*, and increased the efficacy of the anthelmintic from 94.9 to 99.8%. Additionally, the potential targets of the pharmacological combination were identified *in silico*, these results are promising tools for the designing of successful strategies for parasite control (Miró et al., 2020).

For tapeworm, activity of the crude aqueous extract from the medicinal herb *Artemisia absinthium* against *Hymenolepis nana* demonstrated that the extract induced worm paralysis, death, and ultrastructural alterations, such as tegumental damage, lipid accumulation, and destruction of the nephridial canal and the intrauterine egg *in vitro*. Also, *in vivo* induced significant reductions in egg counts per gram of faces and the worm burden. These results are comparable to praziquantel, the drug of choice for *H. nana* infection (Beshay, 2018).

10.3 REPURPOSING OF EXISTING DRUGS

Currently, infections caused by intestinal parasites are among the leading causes of morbidity and mortality. As mentioned, there is a growing increase in strains of parasites resistant to conventional treatments, which makes them a public health problem, either due to recurrent reinfections, limitation in treatment options, and/or economic burden that this represents.

To solve this problem, different alternatives have been proposed as sources of obtaining new drugs. A promising alternative due to its speed, effectiveness, and low cost, is the repositioning of drugs or also known as redirection and redefinition of drugs, which consists of finding new uses for clinically approved drugs that have pharmacological, toxicological, and bioavailability data and a known security profile (Mehndiratta et al., 2016; Serafin and Hörner, 2018). The most important advantage of this alternative source for obtaining drugs is that, compared to the development of innovative drugs, the time and costs are substantially lower (Oprea and Overington 2015; Ashburn and Thor 2004; Serafin and Hörner 2018). In the case of intestinal parasites that cause diseases in humans, the use of drug repositioning as an alternative for drug development is especially interesting, since the disease models are relatively simple in addition to keeping a relatively high homology with species on which multiple drug developments have been made. Below, some examples of evaluation drugs for a second use against intestinal parasites are mentioned.

10.3.1 Beta-blockers

Promisingly, propranolol, a weak beta-blocker used clinically against hypertension, has been shown to have an anti-*Giardia* effect with an IC_{50} of 180 μM, and it has also been shown to be useful against *Giardia* resistant to metronidazole alone, or in combination with metronidazole (Popović et al. 1991; Popović and Milović 1990). In the case of *E. histolytica*, it has been shown that propranolol inhibits parasite growth with an IC_{50} of 400 μM, reducing its viability by 90%; in addition, it reduces liver damage, which is induced by infection of the parasite, and decreases γ-glutamyl transpeptidase, glutathione oxidized and attenuates NF-κB and IL-1β (Aldaba-Muruato et al., 2017; Muñoz et al., 1991).

10.3.2 Proton Pump Inhibitors

Proton pump inhibitors, drugs that irreversibly block ATPases H+/K+ of gastric parietal cells used in the treatment of gastroesophageal reflux and peptic ulcer, have demonstrated potential as antiparasitic agents in *Giardia*: omeprazole (9.55 nM), pantoprazole (1.57 nM), lansoprazole (7.31 nM), and rabeprazole (1.81 nM) inhibit parasite growth (Pérez-Villanueva et al., 2011). Furthermore, some of these inhibitors are capable of specifically inhibiting the triosephosphate isomerase enzyme, a limiting enzyme in the pathway of glycolysis (Reyes-Vivas et al., 2014). Recombinant protein activity assays and targeted mutations demonstrated that omeprazole interacts with allosteric Cys222 in *Giardia* (García-Torres et al., 2016), in terms of ultrastructural effects; omeprazole generates profuse cytoplasmic vacuolization and the emergence of lamellar structures suggestive of autophagy (López-Velázquez et al., 2019). Furthermore, computational prediction studies against the *Giardia* arginine deaminase enzyme showed that rabeprazole can inhibit this enzyme activity (Fernández-Lainez et al., 2021). In the case of *E. histolytica*, it has also been shown that some proton pump inhibitors such as omeprazole, rabeprazole, lansoprazole, and pantoprazole are capable of inhibiting growth in trophozoites with an IC_{50} of 49.22, 34.66, 2.37, and 0.26 nM, respectively (Pérez-Villanueva et al., 2011).

10.3.3 Anti-obesity

Within this group of drugs, the anti-*Giardia* effect of tetrahydrolipstatin, a potent inhibitor of gastric and pancreatic lipases, used in the treatment of obesity, has been demonstrated. Tetrahydrolipstatin shows activity against *Giardia* with an IC_{50} of 2.8 µM and against strains resistant to metronidazole (IC_{50} 6.2 µM), causing blisters on the dorsal surface and tips of trophozoites flagella (Hahn et al., 2013). In the case of *Cryptosporidium*, atorvastatin, a statin used to lower blood cholesterol levels and in the prevention of cardiovascular diseases, has shown its ability to reduce the number of shed oocysts, after a dose of 20 mg/kg, 40 mg/kg, and the combination of both with 1000 mg/kg of nitazoxanide, with a decrease of 53.7, 67.2, 70.1 and 77.5 %, respectively (Madbouly Taha et al., 2017). On the other hand, it has also been shown that mevastatin, a hypolipidemic agent, showed an anti-*Cryptosporidium* effect with an IC_{50} of 5.1 µM (Bessoff et al., 2013).

10.3.4 Antimalarials

It has been shown that multiple drugs initially used as antimalarial agents have an antiparasitic effect against other parasites, an example of this is quinacrine, which is already used clinically to treat giardiasis, inhibiting its growth in a dose of 0.38 µg/mL and affecting its adhesion capacity (Favennec et al. 1992). In addition, quinacrine achieved a clinical cure of giardiasis, in 81% and parasitological in 100% in clinical trials (Neumayr et al., 2021). On the other hand, chloroquine inhibits the growth of *Giardia* by 45.3% at a dose of 100 µg/mL. It was also found that chloroquine treatment decreases the level of the CWP1 protein, increases the number of cytosolic vesicles, and inhibits cyst formation (Wu et al., 2021). Furthermore, in a randomized, controlled, and open clinical trial, chloroquine for five days of treatment showed a cure frequency equal to that of metronidazole (Karademir et al., 2016). Some other antimalarials evaluated that inhibited the growth of *Giardia* in doses of 100 µg/mL are mepacrine (81.5 %), primaquine (39.6 %), quinine (74.8 %), and mefloquine (94.4 %), the latter notoriously affecting its adhesion (Crouch, Seow, and Thong 1986; Favennec et al. 1992). In the case of *Cryptosporidium*, mefloquine has been shown to prophylactically decrease the number of oocysts by 77% and 100% in combination with nitazoxanide (El-Wakil et al., 2021). On the other hand, in trophozoites of *E. histolytica*, mefloquine has an IC_{50} of 1.1 µM, being more active than metronidazole (5 µM), while, in cysts, mefloquine maintains its activity unlike metronidazole (Nagaraja and Ankri, 2019).

10.3.5 Inhibitors of Ethanol Metabolism

During the search for active compounds against variant *Giardia* surface proteins, it was shown that disulfiram, an inhibitor of acetaldehyde dehydrogenase, an enzyme involved in the ethanol metabolism, which is used for the treatment of alcohol dependence, inhibits the growth of *Giardia* with an IC_{50} of 0.9 μM *in vitro* and 1.23 μM *in vivo*, even demonstrated greater activity in resistant strains than non-resistant ones (Galkin et al., 2014; Nash and Rice, 1998). Furthermore, disulfiram has been shown to interact with *Giardia* carbamate kinase, reversibly inhibiting it (IC_{50} of 0.58 μM), and interacting with Cys242 (Galkin et al., 2014). Similarly, disulfiram has been shown to specifically inhibit *Giardia* triosephosphate isomerase enzyme with an IC_{50} of 2 μM (Ni et al., 2017). As with *Giardia*, the anti-*Cryptosporidium* effect of disulfiram has been demonstrated, and this can reduce the number of oocysts after oral administration, inhibiting the enzyme inosine 5′-monophosphate dehydrogenase, an essential enzyme to produce guanine nucleotides (Sarwono et al., 2019). On the other hand, the effect of disulfiram was evaluated in a mouse model that simulated human amoebic colitis, which showed that disulfiram in combination with zinc was very effective in eliminating *E. histolytica* (Ghosh, Farr, et al., 2020; Shirley et al., 2021). While the effects of disulfiram on eggshell morphology in *T. muris* were evaluated *in vitro* and *in vivo*. Daily disulfiram treatment of mice with 25 days of infection with *T. muris* for 26 days resulted in production of malformed eggs by adult female worms at doses of 5.0 or 7.5 mg/kg/day of disulfiram. In addition to parasites incubated with 4 and 8 μg/mL of disulfiram, release malformed eggs in the culture medium after 30 hours in culture, which do not cause an infection in mice (Hill and Fetterer, 1997).

10.3.6 Anti-rheumatic Agents

A molecule that has recently arose great interest is auranofin, an organic gold compound used for treatment of rheumatoid arthritis, which can inhibit the *Giardia* enzyme thioredoxin reductase by inhibiting its growth (Debnath et al., 2013; Tejman-Yarden et al., 2013). A recent clinical trial designed to gauge the relevance of auranofin as a treatment for giardiasis and amebiasis has shown that this drug is safe when administered at 6 mg/d for 7 days (Capparelli et al., 2016). On the other hand, through computational prediction studies directed against the *Giardia* arginine deaminase enzyme, it was shown that aurothiomalate, a gold compound used for its anti-rheumatic immunosuppressive effects, can inhibit this enzyme activity, without affecting the survival of trophozoites (Fernández-Lainez et al., 2021). In the case of *T. crassiceps*, the effect of auranofin on the viability and respiration of cysticercoid in culture was evaluated, respiratory activity had decreased by 50 % in just 4 hours with 10 μM and at 12 h, and no survivors or respiration were detected. The IC_{50} was calculated at 3.8 μM, and, in addition, auranofin inhibits the activity of the thioredoxin glutathione reductase enzyme from *T. crassiceps* with an IC_{50} of 0.6 μM (Martínez-González et al., 2010); the same happens with *T. solium* where auranofin inhibits the thioredoxin glutathione reductase enzyme with an IC_{50} of 3.25 nM (Plancarte and Nava, 2015). While, during a high-throughput automated screening of anti-amoebic drugs, it was discovered that auranofin, as in *Giardia*, also inhibits the thioredoxin reductase enzyme (Debnath et al 2012, Debnath, Ndao, and Reed 2013). Therefore, auranofin increases parasite sensitivity toward death mediated by reactive oxygen species (ROS) (Andrade and Reed 2015; Debnath et al. 2012; Debnath, Ndao, and Reed 2013). Finally, auranofin also inhibits the growth of *Cryptosporidium* with an IC_{50} of 2 μM (Debnath et al., 2013), and for isolates of *B. hominis* ST1-NUH9, ST4-WR1, and ST7-B with an IC_{50} of 62.2, 122, and 113 ng/mL, respectively (Yason et al., 2018).

10.3.7 Anti-Parkinson

A virtual scan based on the structure of drugs approved by the FDA as possible inhibitors of the triosephosphate isomerase enzyme from *Giardia* was recently carried out to identify compounds with antiprotozoal activity; the results show that tolcapone, a benzophenone capable of reversibly inhibiting the catechol-O-methyl-transferase enzyme and used for Parkinson's disease treatment, is capable of inhibiting *Giardia* growth with an IC_{50} of 0.05 µg/mL (Juárez-Saldivar et al., 2021).

10.3.8 Anti-cancer

One of the groups of drugs most investigated as anti-parasite are anti-cancer drugs: within these, 5-fluorouracil, an antimetabolite used for cancer treatment, is capable of inhibiting *Giardia* growth by 83% at a dose of 200 µM, and also inhibits the uridine phosphorylase enzyme (Jiménez et al., 1989; Reaume et al., 2013). On the other hand, etoposide, a topoisomerase II inhibitor, is capable of inhibiting *Giardia* growth with an IC_{50} of 400 µM, in addition to inhibiting the expression of the cyst wall proteins (CWP) gene and the formation of cysts (B.-C. Lin et al., 2013). It is worth highlighting the group of anti-cancer agents that inhibit mitosis: vincristine reduces the viability of *Giardia* by 83.38% in just 9 h, affecting adhesion capacity and morphology at doses of 60 µg/mL (Sandhu et al., 2004), while demecolcine inhibits the growth of *Giardia* by 90% in doses of 5 µM, the same as colchicine with an IC_{50} of 200 µM affecting parasite morphology and inhibiting cytokinesis. Another of these drugs is podophyllotoxin, which inhibits *Giardia* growth, affects its adhesion capacity, and notably alters its morphology and cell viability, by interacting with tubulin (Gutiérrez-Gutiérrez et al. 2019; Gutiérrez-Gutiérrez et al. 2017). Imatinib, a protein tyrosine kinase inhibitor, has recently been shown to be capable of affecting *Giardia* growth with an IC_{50} of 3.46 µg/mL, and computer data suggested that it is capable of binding to the *Giardia* triosephosphate isomerase enzyme (Juárez-Saldivar et al., 2021). In the case of *Cryptosporidium*, in the same way, multiple anti-cancer drugs have been tested against the parasite; among these, anti-mitotic drugs such as docetaxel with an IC_{50} of 0.087 µM, paclitaxel of 0.099 µM, podophyllotoxin of 0.138 µM, colchicine of 0.187 µM also stand out. The DNA intercalants daunorubicin and doxorubicin have been shown to inhibit *Cryptosporidium* growth with an IC_{50} of 0.787 and 1.092 µM, respectively (Guo et al., 2018). On the other hand, it has also been shown that the antimetabolites floxuridine, 6-azauridine, carmofur, and 5-fluorouracil inhibit the growth of *Cryptosporidium* with an IC_{50} of 0.0063, 3.2, 4.4, and 4.9 µM, respectively (Bessoff et al., 2013). Similarly, the effect of vorinostat, a histone deacetylase inhibitor that reduces parasite growth, has been demonstrated with an IC_{50} of 0.203 µM (Guo et al., 2018). Finally, vatalanib, a tyrosine kinase inhibitor, can reduce *Cryptosporidium* growth with an IC_{50} of 0.322 µM (Guo et al., 2018). On the other hand, obatoclax, an inhibitor of the Bcl-2 protein family that induces apoptosis in cancer cells, shows activity against trophozoites and *E. histolytica* cyst with an IC_{50} of 1 µM; in addition, obatoclax is also active against parasites resistant to metronidazole (Ehrenkaufer et al. 2020; Ehrenkaufer et al. 2018). Among the anti-neoplasics that target phosphatidylinositol 3-kinase, omipalisib inhibits the growth of *A. ceylanicum* with an IC_{50} of 0.17 µM and *T. muris* with an IC_{50} of 0.5 µM, while serabelisib inhibits the growth of *T. muris* with an IC_{50} 2.8 µM (Tyagi et al., 2019).

10.3.9 Anti-inflammatories

Recently, the anti-*Giardia* effect of acetylsalicylic acid, a well-known cyclooxygenase II inhibitor, with an IC_{50} of 0.29 mM, has also been shown to affect the expression of HSP70, damaging parasite membrane and morphology (Ochoa-Maganda et al., 2020). On the other hand, Daflon, an anti-inflammatory containing diosmin and hesperidin, has been shown to reduce the number of trophozoites and cysts in an *in vivo* mouse model (Fahmy et al., 2021). In the case of *Cryptosporidium*, indomethacin has shown that it can inhibit its growth with an IC_{50} of 5.4 µM; on the other hand, tegaserod reduces the growth of *Cryptosporidium* with an IC_{50} of 4.8 µM (Bessoff et al., 2013).

10.3.10 Antidepressants

Within this group, two antidepressants belonging to the group of selective serotonin reuptake inhibitors have shown an effect on *T. trichiura* and *A. caninum*, sertraline with an IC_{50} of 7.2 and 4.3

μM, respectively, and paroxetine with an IC_{50} of 5.4 and 6.4 μM respectively (Weeks et al., 2018). On the other hand, neuroleptic chlorpromazine used to treat psychotic disorders, such as schizophrenia or manic depression in adults, can reduce the motility of *T. muris* in adults and *T. trichiura*, it has an IC_{50} of 0.7 μM, while in *A. caninum* it prevents hatching and development and has an IC_{50} of 6.1 μM (Weeks et al., 2018).

10.3.11 Phosphodiesterase Inhibitors

Within the group of phosphodiesterase inhibitors, those with activity against *A. ceylanicum* and *T. muris* stand out, such as rolipram (IC_{50} of 2.08 and 2.83 μM, respectively), zaprinast (IC_{50} of 2.08 and 3.00 μM, respectively), sildenafil (IC_{50} of 2.15 and 3.00 μM, respectively), vardenafil (IC_{50} of 2.67 and 2.80 μM, respectively), tadalafil (IC_{50} of 0.67 and 0.80 μM, respectively), avanafil (IC_{50} of 2.80 μM for *T. muris*), and roflumilast (IC_{50} of 3.00 μM for *T. muris*) (Tyagi et al., 2019).

10.3.12 Immunomodulators

Within the group of immunomodulators, the inhibitors of the dihydroorotate dehydrogenase enzyme stand out as inhibitors of *A. ceylanicum* growth, where brequinar and teriflunomide have an IC_{50} of 2.25 and 2.00 μM, respectively (Tyagi et al., 2019).

10.4 NOVEL STRATEGIES FOR DRUG TARGET DISCOVERY IN INTESTINAL PARASITES

10.4.1 Identification of New Molecular Targets by Bioinformatic Assays

The discovery of suitable drug targets to design selective and effective drugs against intestinal diseases is a major challenge. Computational and experimental approaches are important tools for the development of drugs without damage to host. The elucidation of a potential target, absent or different from the host homolog, and detection of molecules for the pathogen survival are essential criteria (Raj et al., 2020).

Molecular docking is an *in silico* structure-based method widely used in drug discovery. Docking enables identification of novel compounds of therapeutic interest, predicting ligand-target interactions at a molecular level, or delineating structure-activity relationships (SAR), without knowing *a priori* the chemical structure of other target modulators (Pinzi and Rastelli, 2019). Computational approaches now make possible the virtual screening of millions of compounds in a short amount of time. This process is generally accomplished by first predicting the molecular orientation of a ligand within a receptor, and then estimating their complementarity using a scoring function (Pinzi and Rastelli, 2019). General processes for *in silico* approaches are represented in Figure 10.1.

Conserved pathways are suitable molecular targets in intestinal parasites. Several molecules involved in conserved pathways (lipid synthesis, glycolysis, polyamine, redox metabolism, among others) are potential targets. The pentose phosphate pathway participates in the generation of NADPH and several intermediate metabolites for nucleic acid synthesis. The fused *G. lamblia* glucose-6-phosphate dehydrogenase (G6PD) and 6-phosphogluconolactonase (6PGL) were similar to the 3D structure of *H. sapiens*. By biochemical analysis and the computational approach, the possible structural NADP+ binding site of G6PD and 6PGL was determined. They conclude that the structural NADP+ binding site of *G. lamblia* is partially conserved between humans and other species (Morales-Luna et al., 2020). Virtual screening of the homology model of *G. intestinalis* arginine deiminase (GiADI) and druggability analysis were performed obtaining a variety of selective inhibitors, and this enzyme is an attractive target for drug design against *G. intestinalis* because of its metabolic function involved in the energy production and defense of this protozoan parasite and the

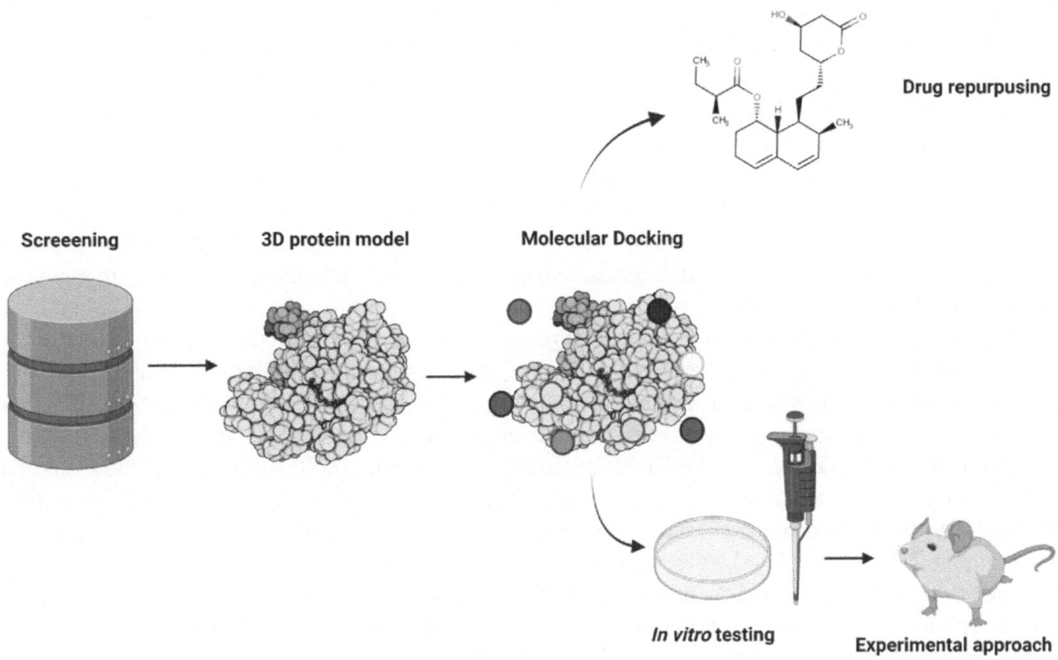

FIGURE 10.1 Computational approaches in pharmaceutical research. The integration of *in silico* and experimental techniques are essential in the identification of molecular targets and drug discovery.

lack of the enzyme in the human host (Trejo-Soto et al., 2016). Similarly, *in silico* approaches identified 11 potential inhibitors to another purine salvage enzyme. Guanine phosphoribosyltransferase (GPRT) of *G. lamblia* has a critical role on the synthesis of DNA and RNA. These compounds suggested a better selectivity for the parasite enzyme by avoiding steric bumps with a flexible loop in the human enzyme binding site (das Neves Gustavo et al., 2017).

The TATA-binding protein (TBP) is a central transcription factor in eukaryotes; thus, affecting these interactions will be lethal for any living being. Structural and dynamic computer analysis of surface properties on TBPs of parasites like *E. histolytica*, *T. solium*, and *N. americanus* showed their difference compared to their human homolog. The TFIIA- and NC2-binding sites are important to regulate protein function. Thus, inhibiting interaction at these sites could be lethal for the parasites as shown in druggability assays (Santiago et al., 2020). Histone Deacetylases (HDACs) are enzymes that regulate many essential biological processes and represent targets in cancer therapy. In *E. histolytica*, the homology model and *in silico* analysis suggest that EhHDAC could regulate many important cellular functions like those that have been explored in malaria, leishmaniasis, and schistosomiasis (Montaño et al., 2020). *E. histolytica* adenosine 5'-phosphosulfate kinase (EhAPSK) is an essential enzyme in *Entamoeba* sulfolipid metabolism, and it contributes to fatty alcohol disulfates and cholesteryl sulfate, which play important roles in trophozoite proliferation and cyst formation. Through molecular docking analysis, 15 EhAPSK inhibitors (from 400 compounds tested) were identified, and three of them showed high affinity, dose-dependent effects, and halted not only *Entamoeba* trophozoite proliferation but also cyst formation (Mi-Ichi et al., 2019). The use of structural bioinformatics on *G. lamblia* identified a hypothetical protein GiK as a potassium channel, these proteins have roles on life cycle, growth, and stress response in parasites. By the 3D-GiK structure and virtual screening of 290 compounds, 110 different potassium inhibitors as potential antigiardiasic agents were identified. Additionally, its structure and low homology of GiK with proteins in the human genome make it a specific target for drug design (Palomo-Ligas et al., 2019).

Calcium-dependent protein kinases (CDPK) are found in the apicomplexan, algae, and plants; they are not described in vertebrates as being selective targets. CDPK family-proteins could be important in the activation of motility, protein secretion, cell egress, and cell invasion in parasites. Protein model and virtual screening of inhibitors confirmed that CpCDPK6 protein of *C. parvum* was used as a suitable target for anti-Cryptosporidium drug design (Dhal et al., 2020).

Lactate Dehydrogenase plays a vital role on regulation of glycolysis; *C. parvum* depends mainly on this pathway for energy production. In other parasites, the blockage of glycolysis is a promising tool for treatment. By structure-based, ligand-based, and *de novo* drug design they find an effective and selective inhibitor from 40 compounds. Experimental confirmations are necessary to evaluate this activity (Dhal et al., 2018).

Molecular docking simulation is an appropriate method to determine the stability of a ligand-protein complex. A multi-subunit vaccine candidate for *C. parvum* treatment was constructed with 250 amino acid residues with a probability of antigenicity and allergenicity, and with different epitopes like B-cell, Helper T-lymphocytes, and cytotoxic T- lymphocytes epitopes that confers stability and binding affinity to the TLR2 receptor. The overall results showed that the designed complex could be an efficient vaccine candidate against *C. parvum* (Dhal et al., 2019). A similar strategy was performed in *E. tenella*, where the micronemal protein of apicomplexan parasites, apical membrane antigen 1 (AMA1), appears to be essential during invasion of host cells. Recently, peptides that specifically bind *E. tenella* AMA1 (EtAMA1) were screened from a phage display peptide library. Via docking tests, two phages strongly interacted with EctoAMA1 and sporozoites protein showing inhibition of the parasite (Ma et al., 2019).

As mentioned before, searching for differences in the potential target is important for the development of selective toxicity. Many different types of acetylcholine receptors (AChRs) are expressed by nematodes. At least three subtypes of nAChR receptors are found on the musculature of adult nematodes such as *A. suum* which are of interest for potential therapeutic exploitation. By a ligand-based 3D pharmacophore model and virtual screening, 23,344 compounds with potential anthelmintic activity were studied. Three new compounds exhibit a modulatory effect on nematode contractions, and this was experimentally tested. These efforts support the idea of using new tools for rational drug design (Stevanovic et al., 2021).

10.4.2 Gene Editing Approaches

Genetic manipulation of parasites through several techniques that modify gene expression at DNA, RNA, or protein level has become an emergent and changing strategy to elucidate the biology, gene function, and identification of potential therapeutic options (Briquet et al., 2021). CRISPR-Cas 9 technologies allow us to introduce site-specific modifications in the cells' genomes. A key property of Cas9 is its ability to bind to DNA at sites defined by the guide RNA sequence and the PAM, allowing applications beyond permanent modification of DNA (Doudna and Charpentier, 2014). The CRISPR approach has been used to induce a variety of modifications including directed insertion/deletion (InDel) of bases, gene *knock-in*, introduction of mutations in both alleles of a target gene, and deletion of small DNA fragments. Thus, CRISPR technology offers a precise molecular tool for directed genome modification with a range of potential applications; further, its high mutation efficiency, simple process, and low cost provide additional advantages over prior editing techniques (Cui and Yu 2016; Sarma et al. 2021).

Differentiation of *G. lamblia* is essential for survival and pathogenesis. For cyst formation, CWPs synthesis is fundamental. Leukemia factor (MLF) proteins are involved in cell differentiation in mammals. In *Giardia*, MLF-like protein has been identified. The *knockdown* of *MLF* gene by the CRISPR/Cas9 system resulted in a significant decrease of *CWP* gene expression and cyst formation, suggesting a positive role of MLF in encystation (Lin et al., 2019). CRISPR interference (CRISPRi) is a modification of the CRISPR/Cas9 system that directs catalytically inactive Cas9 (dCas9) to target loci for stable transcriptional repression. In contrast to reverse genetic approaches, the use of untargeted, genome-wide CRISPRi was proposed in *G. lamblia* as a tool to identify essential genes critical for growth, differentiation, and pathogenesis. Recently, an efficient and stable CRISPRi-mediated transcriptional repression of exogenous and endogenous genes in *Giardia* was done. CRISPRi *knockdown* of *kinesin-2a* and *kinesin-13* causes severe flagellar-length defects that mirror defects. *Knockdown* of the ventral disk MBP protein also causes severe structural alterations that are highly prevalent and persist in the population more than 5 days longer than modifications associated with a transient morpholino-based *knockdown*. This methodology in polyploid *Giardia* is proposed as a rapid evaluation of *knockdown* phenotypes (McInally et al., 2019). The successful implementation of CRISPR-Cas 9 was also evaluated recently in other parasites. In *E. histolytica* CRISPR/Cas9 mediated recombination by restoration of luciferase activity only in presence of all necessary components (Cas9, luciferase guide RNA, and donor luciferase DNA) (Kangussu-Marcolino et al., 2021). On another side, this strategy was implemented to tag the putative *E. tenella* dense granule protein 9 (EtGRA9) and monitored its expression throughout parasite life cycle. Additionally, a large-scale approach was implemented with disruption of 33 members of the AP2 transcription factors family. The manipulation was effective in individual and multiple genes. They demonstrated that EtGRA9 is a secreted protein present in all stages of life and disruption of AP2 revealed that 23 of 33 factors are involved on parasite survival (Hu et al., 2020). Calcium-dependent protein kinase-1 (CDPK1) is one of the leading drug targets in *C. parvum*, by CRISPR/Cas9 system the *cdpk1* gene were deleted confirming its importance for parasite survival. In addition, the effect of CDPK1 *knockdown* on a mouse infection model revealed the essential role of this kinase in parasite proliferation (Choudhary et al., 2020). These examples show that CRISPR is a promising genomic editing strategy for validation of new targets in intestinal diseases caused by parasites.

Nek kinases have a roll regulating microtubule organization, *G. lamblia* possess 198 Nek kinases whereas humans have only 11, due their complex cytoskeleton organization.

Cytokinesis defects in *Nek8445*-depleted *Giardia* cells were performed by *knockdown* of the *Nek8445*gene, founding that depletion results in short flagella, aberrant ventral disk organization, loss of the funis, defective axoneme exit, and altered cell shape. In addition, presence of a small gatekeeper residue in the kinase makes it a potential target for design of specific inhibitors with relatively low activity against human kinases (Hennessey et al., 2020).

Phagocytosis is intimately associated with the biology of *E. histolytica* and is required for the amebic virulence. Ca^{+2} plays a vital role in the process interfering with contractile proteins (Choudhary et al., 2020).The blockage of calmodulin-like calcium binding protein (EhCaBP1) by inducible expression of the antisense RNA resulted in major changes of actin organization (Sahoo et al., 2004). Formation of phagosome complex also involves participation of EhCaBP3, and EhCaBP5. The overexpression of Ca^{+2}-binding-defective mutants of all these proteins helped to delineate the participation of Ca^{+2} in different steps (Babuta, Bhattacharya, and Bhattacharya 2020).

Monoclonal antibody-screening was conducted to identify *E. histolytica* surface proteins that participate in phagocytosis. The screen identified one antibody that dramatically blocks phagocytosis by inhibiting *E. histolytica*'s ability to adhere to apoptotic cells. The antigen to which it binds is the serine-rich *E. histolytica* protein (SREHP), member of a family of immunogenic amebic surface proteins (Teixeira and Huston, 2008). Also, recently a classification of *E. histolytica* strains according to SREHP genes confirm their heterogeneity and influence on the infection, participating on the genetic profile of infecting strains (Ngobeni, 2021).

Otherwise, novelty strategies with the aim of developing new treatments for parasite infections are necessary. Microneme proteins play a crucial role on invasion of host cells by *Eimeria* parasites: between them, EtMIC3 contains seven tandem microneme adhesive repeat regions (MARRa-b). MARb and one of the four repeated MARc domains of EtMIC3 protein were selected as ligand. Results showed that three phages bind to EtMIC3-bc1 protein. The phages were orally inoculated in chickens and revealed that they offered protection against *E. tenella* challenge and effectively bound to EtMIC3 protein secreted by *E. tenella* sporozoites in the intestinal tract (Chen et al., 2021).

Gene expression regulation is an important event for eukaryotic cell biology and survival, the polyadenylation factor EhCFIm25 in *E. histolytica* was silenced by double-stranded RNA (dsRNA), producing a significant acceleration on parasite proliferation and cell death; moreover, cells were larger and multinucleated, also, their ability to move and phagocyte erythrocytes was reduced, therefore, indicating loss of virulence. The absence of EhCFIm25 had an impact on ~20 genes and proteome expression (Salgado-Martínez et al., 2021).

10.5 NANOTECHNOLOGY ON NEW ALTERNATIVES OF TREATMENT AGAINST INTESTINAL PARASITE INFECTION

The use of nanotechnology for medical purposes is known as nanomedicine and is described as the application of nanomaterials for diagnosis, control, monitoring, prevention, and treatment of diseases (Prasad et al. 2016). Nanomaterials can be defined as materials with a size between 1–100 nm and can have an inorganic (such as metal and metal oxide) or organic structure (peptide, lipid, and polymer-based nanoparticles) (Almohammed et al., 2021; Patra et al., 2018; Prasad et al. 2017). Characteristics of both types are represented in Figure 10.2.

Nanomaterials have shown to be promising carriers of drugs and bioactive molecules for the treatment of parasitic infections because they improve pharmacokinetic parameters such as poor cell permeability, non-specific distribution, poor bioavailability, and rapid elimination of the antiparasitic agent from the human body (Sun et al., 2019). Likewise, the use of nanoparticles as antimicrobial agents has been reported due to the cytotoxic effect that they can exert mainly in protozoa such as *G. lamblia*, *E. histolytica*, *Cryptosporidium spp*, *C.cayetanensis*, and *B. hominis* (Abaza, 2016;

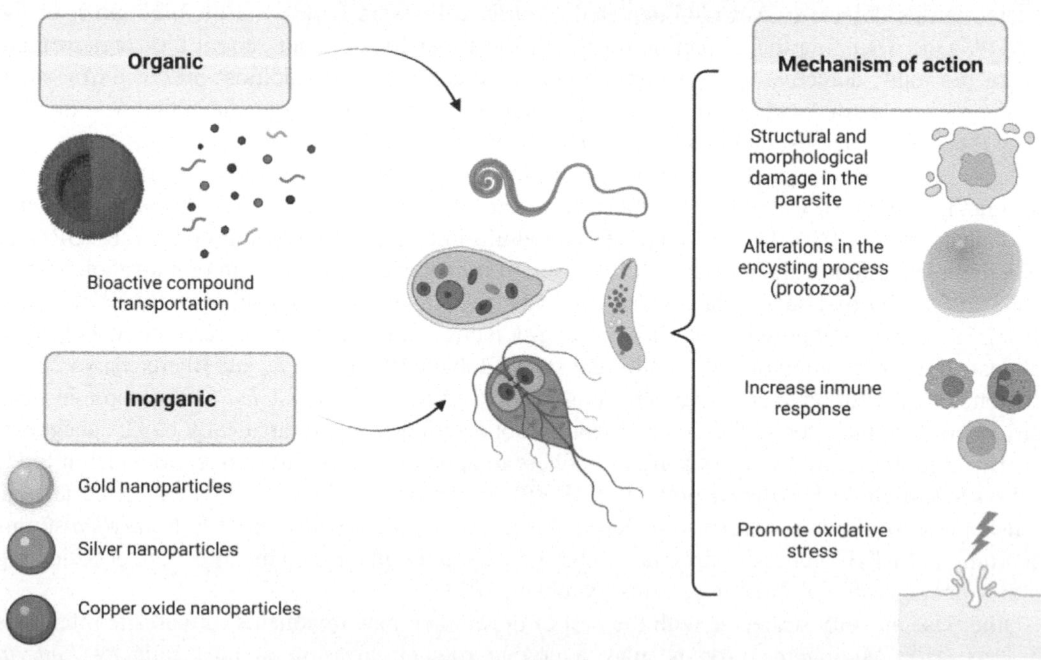

FIGURE 10.2 Damage to intestinal parasites caused by nanomaterials. *Left.* A general classification of nanomaterials (organic and inorganic compounds) is summarized. *Right.* Proposed mechanisms of action caused by nanomaterials are: 1) Structural and morphological changes, 2) alterations in differentiation process, 3) increased immune response, and 4) promote oxidative stress.

Gaafar et al., 2019; Norouzi, 2017). The mechanisms of action proposed for the inhibition of parasite growth include: 1) increased immune response, 2) structural and morphological changes in the parasite, 3) promote oxidative stress, and 4) alterations in the encysting process (Brakat and Sharaf-El-Deen, 2020; Malekifard et al., 2020; Younis et al., 2020).

10.5.1 Inorganic Nanoparticles

Inorganic nanoparticles are generally non-toxic, hydrophilic, biocompatible, and highly stable compared to organic materials. This type of nanoparticles includes gold, silver, iron oxide, copper oxide, silica, and hydroxides nanoparticles.

Gold nanoparticles (AuNPs) have been widely used as important tools in therapy, drug delivery, targeting, and imaging. Regarding the antiparasitic effect, Bavand et al. (2014) described the *in vitro* effect of AuNPs on *G. lamblia* cysts at 0.05, 0.1, and 0.3 mg/mL for 5, 15, 30, 60, and 180 min. The effect of AuNPs was concentration-time dependent. The concentration of 0.3 mg/mL at 180 min presented a lethality of 96% on *Giardia* cysts. In another study, Al-Ardi, (2020) analyzed the effect of AuNPs and *Citrullus colocynthis* aqueous extract nanoparticles against *G. lamblia* in a Swiss Albino mice infected model. Nanoparticles were administered orally at a dose of 20 µg daily for 8 days and both nanoparticles were administered separately and in combination. The combination of both nanoparticles showed greater efficacy in the treatment of giardiasis (93.2% reduction of trophozoites) compared to the use of AuNPs alone (61.2%).

Silver nanoparticles (AgNPs) are among the most studied and applied nanomaterials in the biomedical field. AgNPs exhibited antimicrobial and cytotoxicity activities which depend on factors including surface chemistry, size, size distribution, shape, particle morphology, particle composition, agglomeration, and dissolution rate (Zhang et al., 2016; Aziz et al. 2016, 2019). Several studies show the antiparasitic effect of AgNPs, and Said et al. (2012) reported that AgNPs at a dose of 50 µg for 8 days reduced 72.7% the parasite count in stool samples of rats with giardiasis. Another study *in vivo* conducted by Idan and Ardalan, (2020) described the effect of AgNPs as an anti-*Giardia* agent in infected mice. They demonstrated that a dose of 100 µg/g for 3 days reduced 66.6 and 83.3% of *Giardia* cysts after 24 and 48 h, respectively. In another study the effect of AgNPs on *E. histolytica* cysts and *C. parvum* oocysts *in vitro* was analyzed. The LC_{50}-180 min was 0.34 mg/L and 0.54 mg/L for *E. histolytica y C. parvum*, respectively (Saad et al., 2015). The AgNPs were shown to exert direct damage on the oocyst wall of *C. parvum* by showing alterations in its morphology and this decreased the protozoa viability. Cameron et al. (2016) also described the antiparasitic effect *in vitro* of AgNPs against *C. parvum*. They found an excystation of 33.3% of the oocysts when used at a concentration of 500 µg/mL. They propose that AgNPs interact with the cell wall and, at high concentrations, can fully break the oocyst wall, and therefore decrease viability of sporozoites. In another study, Valenzuela-Salas et al. (2021) described the effect exerted by protein-coated AgNPs on trophozoites of *E. histolytica* HM1-IMSS strain. The AgNPs formulation presented moderate antiproliferative activity against *E. histolytica* trophozoites cultures, with an IC_{50}=69.2 µM (7.4 µg/mL). The possible mechanism of action is due to an overproduction of ROS. The effects of AgNPs on *Blastocystis hominis* cysts have also been studied. Abdel-Fattah Ahmed et al. (2015) found that a concentration of 150 µg/mL of AgNPs inhibited growth of 83.5% after 72 hours of exposure. Also, Younis et al. (2020) reported the effect of AgNPs against *B. hominis* with a concentration of 150 µg/mL but with 3 hours of exposure. They report inhibition of 71.7% on the growth of the parasite, finding a greater inhibitory effect compared to metronidazole. In addition, Gaafar et al. (2019) reported the effect of AgNPs in mice infected with *C. cayetanensis*. AgNPs were administered daily (10 µg i.p.) for 7 days starting from the 6th day after infection. The nanoparticles reduced the oocyst count in the feces of immunocompetent and immunosuppressed mice by 96.9% and 92.4% at day 14 after infection, respectively. The proposed mechanism could be attributed to the immunomodulatory efficacy of the AgNPs that recruit and activate leukocytes, especially macrophages, as well as the increase in IFN-γ levels and thus promote parasite elimination. Metal oxide nanoparticles are key constituents in catalysis, diagnosis, drug delivery, semiconductors, sensing, and solid oxide fuel cells. Its antimicrobial effect has also been reported. The principal mode of antimicrobial activity is the generation of ROS. Besides ROS production, there is damage to cell membranes by electrostatic interaction, disturbance in metal/metal ion homeostasis, protein and enzyme dysfunction (Raghunath and Perumal, 2017). Brakat and Sharaf-El-Deen (2020) reported that zinc oxide (ZnO) nanoparticles at a dose of 10 mg/kg administered daily for 7 days reduced the presence of *G. lamblia* cysts in mice by up to 93.7%, and if combined with metronidazole (500 mg/kg) there is a lethality of 100%. They suggested that the mechanism of action is possibly by potentiation of the intestinal immune response by increasing the levels of IgA and the cellular immune response. Another study by Malekifard et al. (2020) evaluated the effect of selenium and copper nanoparticles on *Giardia lamblia* cysts *in vitro*. The selenium and copper oxide (CuO) nanoparticles had cytotoxic effects against *Giardia* cysts at 0.3 mg/mL and 0.6 mg/mL after 180 min of exposure, respectively, and this effect was like that exerted by metronidazole at a concentration of 5 mg/mL. Another study by Saad et al. (2015) reported the use of CuO nanoparticles against *E. histolytica* and *C. parvum*. The nanoparticles presented a cytotoxic effect with a LC_{50}-3 h of 0.13 mg/mL on *E. histolytica* cysts and 0.72 mg/mL on *C. parvum* oocyts. Another type of metal oxide nanoparticles reported with antimicrobial activity are magnesium oxide (MgO) nanoparticles. MgO has some advantages over other metals, such as non-mutagenic effect, good stability, and it is a normal component of the human body. Hussein et al. (2018) described the *in vitro* antiprotozoal activity of MgO nanoparticles against *C. cayetanensis* oocysts. The lethal concentration on the oocytes was 15 mg/mL after 24 hours of incubation, in addition to causing a failure in the sporulation process (≥90%). The possible mechanism of action may be referred to as the formation of superoxide anions O^{-2} in the MgO surface. MgO is hydrated, forming $Mg(OH)_2$ which covers and increases the surfaces of MgO and OH^- concentration on the oocysts' surface. This, in turn, increases O^{-2} concentration that leads to more effective destruction on the oocysts wall. Therefore, loss of oocysts' wall integrity means the destruction of the oocysts and a reduction in the viability of the sporozoites. Thus, the use of metal oxide nanoparticles can be an alternative for water treatment to remove the infective phase of various intestinal parasites.

10.5.2 Organic Nanoparticles

There is a growing interest on development of organic nanomaterials for biomedical applications. Organic nanoparticles are an interesting alternative for transporting bioactive molecules due to their non-toxicity, the ease of their preparation, and their high stability in biological fluids and during storage. Some examples of organic nanoparticles are chitosan, silk fibroin, or other biodegradable polymers, including poly (lactic-co-glycolic acid) (PLGA) (Virlan et al., 2016). Chitosan is a non-toxic polysaccharide derived from shrimp and crabs with natural chitin alkaline deacetylation. Chitosan and its derivatives have some exceptional properties such as minimum toxicity, biocompatibility, and biodegradability, and have been broadly used as an immunomodulatory, anti-cancer, anti-nociceptive, antioxidant, anti-inflammatory, and antimicrobial agents (Almohammed et al., 2021). Chitosan nanoparticles (CsNPs) can be used for drug delivery, controllable drug release, and recently, CsNPs have been proven to be an effective anti-fungal, anti-bacterial, and anti-protozoal agent. Chabra et al. (2019) described the *in vitro* effect of CsNPs on *G. lamblia* trophozoites, as well as, their effect on cyst excretion in infected mice. The concentrations of 100, 200, and 400 µg/mL significantly reduced *G. lamblia* trophozoites, ranging from 89 to 100%, after 3 h of exposure. Furthermore, cyst excretion was reduced after oral administration of a concentration of 50 µg/kg of CsNPs daily in mice with giardiasis. Also, Elmi et al. (2020) reported that CsNPs at 50 µg/mL were able *in vitro* to eliminate 31.3% of *G. lamblia* cysts after 180 min and the IC_{50} was calculated at 83 µg/mL. In addition, Yarahmadi et al. (2016) analyzed the effect of CsNPs on *G. lamblia* cysts viability. They used concentrations of 50, 100, 200, and 400 µg/mL with an exposure time of 180 min. The concentration of 400 mg/mL exhibited a 100% reduction of cyst viability. Also, Said et al. (2012) found that CsNPs at a dose of 50 µg for 8 days presented a reduction of 68.2% in the parasite count in stool samples of mice with giardiasis. In addition, they evaluated the synergy effect of CsNPs, AgNPs, and curcumin nanoparticles. The CsNPs + AgNPs showed a reduction of 90.2% and the use of three types of nanoparticles caused a total eradication from stool and intestine. The cytotoxic effect of CsNPs on *C. parvum* has also been reported. Ahmed et al. (2019) reported *C. parvum* oocyst *in vitro* destruction (97.3%) after 72 h at a concentration of 5 mg/mL. The oocysts were cracked, and their structure had shrunk, their wall was shredded, and their content expulsion was either seen attached to the wall or distributed near the oocysts that were damaged. They suggest that the mechanism might be due to the small size and excellent film-forming ability of CsNPs. The destructive effect could therefore potentially increase electrostatic interaction between Cs NPs and oocysts. In addition, they evaluated the effect on mice infected with oocysts of *C. parvum*. The animals received an inoculum of oocytes with the CsNPs at different doses (0.5, 1.5, 3, 5, and 7 mg/mL), finding that after 10 days post-infection, doses of 3 mg/mL or more prevented the infection by the parasite. Likewise, there are studies where CsNPs are loaded with antiparasitic drugs such as metronidazole and nitazoxanide. El-Gendy et al. (2021) described the effect of CsNPs and CsNPs loaded with metronidazole in hamsters infected with *G. lamblia*. The CsNPs dose was 50 µg/day for 7 consecutive days. The metronidazole-CsNPs were administered at half the dose (25 µg/day). The highest percentage of reduction in the *Giardia* cyst and trophozoite counts were in the group that received metronidazole-CsNPs (94.69%, 94.29%). Lower percentages of reduction were recorded for metronidazole treated group (90.15%, 89.52%) and CsNPs treated group (63.64%, 75.24%). In another study, Moawad et al. (2021) improved the efficacy of nitazoxanide by loading it to CsNPs against *C. parvum*. The dose used was 200 mg/kg/day given to mice seven days post-infection and continued for 3 days in immunocompetent and 6 days in immunosuppressed mice. Treatment of *Cryptosporidium* infected mice with nitazoxanide loaded on CsNPs resulted in the highest significant reduction in oocysts shedding in both immunocompetent (75.7%) and immunosuppressed (67.3%) groups followed by treatment with nitazoxanide (57.1 and 41.0%). Thus, the loading of drugs on CsNPs increased the loaded drug efficacy against the target organism and improved the healing of intestinal pathological changes induced by protozoa infection.

Curcumin (CUR) is a natural polyphenol compound derived from turmeric rhizome. CUR has excellent pharmacological properties by interacting with various molecular targets. However, due to its poor pharmacokinetic properties and because it is an unstable molecule, which undergoes degradation by light and temperature, its use is limited. To overcome these problems, novel nano systems have been proposed (Zielińska et al., 2020). Layered double hydroxides (LDH) are synthetic minerals with positively charged brucite-type layers containing divalent and trivalent cations ($M2^+$ and $M3^+$). They are considered promising inorganic nanocarriers that have several attractive features for their uses in drug delivery. Gutiérrez-Gutiérrez et al. (2020) evaluated the effect of LDH nanoparticles prepared with zinc and magnesium cations as a vehicle for CUR against *G. lamblia* and *E. histolytica*. LDH potentiates the effect of CUR by reducing the growth of *G. lamblia* trophozoites, with an IC_{50} of 13.15 µM for CUR, 10.96 µM for Mg-LDH-CUR, and 12.21 µM for Zn-LDH-CUR. In the case of *E. histolytica*, Mg-LDH-CUR drastically improved the effect of CUR on the growth of *E. histolytica* trophozoites, with an IC_{50} value of 205.71 µm for CUR, 147.87 µM for Mg-LDH-CUR, and 190.53 µM for Zn-LDH-CUR. Therefore, LDH improved stability, resistance to degradation, solubility, and cytotoxic activity of CUR. Poly (lactic-co-glycolic acid) (PLGA) is one of the most successfully used biodegradable polymers because its hydrolysis leads to metabolite monomers, lactic acid, and glycolic acid. As these two monomers are endogenous and simply metabolized by the body via the Krebs cycle, negligible systemic toxicity is associated with the use of PLGA for drug delivery or biomaterial applications (Fatemeh Sadat Tabatabaei Mirakabad, 2014). Said et al. (2012) reported the incorporation of CUR into PLGA and its greater effectiveness against *G. lamblia*. CUR (50 µg/

daily for 8 days) reduced the number of *Giardia* cysts count in a stool sample (13.1%) and when CUR was coupled to the PLGA NPs the reduction was 54.6%. Also, they evaluated the synergistic effect of CsNPs, AgNPs, and CUR-PLGA NPs. The greatest synergistic effect was observed when combining CsNPs with CUR-PLGA NPs, with a reduction in the presence of cysts of 81.3% in mice with giardiasis. Furthermore, when the three nanoparticles were combined, complete eradication of the protozoa was obtained. In another study, Madbouly Taha et al. (2017) evaluated encapsulation of metronidazole in PLGA NPs in mice infected with *G. lamblia*. Metronidazole was administrated at 60 and 120 mg/kg/daily for five days post-infection and metronidazol was loaded on PLGA NPs at the same doses. Metronidazole reduced the amount of *G. lamblia* trophozoites in the small intestine by 87.2 and 93.4%, and regarding metronidazole-PLGA-NPs, the reduction was 100 and 87.5% at the 60 and 120 mg/kg doses, respectively. Thus, the maximal anti-Giardia effect and complete cure were recorded with the lowest dose of metronidazole-PLGA-NPs. Therefore, these studies show the usefulness of nanoparticles on the treatment of diseases caused by intestinal protozoa, either through a direct effect of the nanoparticle or by loading another bioactive molecule. A compendium of strategies mentioned previously is showed in Table 10.1.

10.6 ANTIMICROBIAL PEPTIDES AS ANTIPARASITIC AGENTS

In general terms, antimicrobial peptides (AMPs) are small molecules of amino acids, generally between 10 and 60 amino acid residues, which participate on the response to pathogens infection. They have shown a wide range of mechanisms of antimicrobial activity against bacteria, fungi, parasites, and viruses with additional immunomodulatory effects (Verdugo-González et al., 2019). AMPs exist in various organisms, like bacteria, fungi, animals, and plants as showed in Figure 10.3. Most of them are cationic AMPs that play the key on antimicrobial roles (Lei et al., 2019). The principal advantage of AMPs is that the antimicrobial mechanisms that cause anti-parasitic properties are different from the drug-related mechanisms, because of that, AMPs can be applied to treat multiple parasites and even the drug-resistant ones. Despite the considerable potential of AMPs therapeutic agents, some limitations must be considered for their development and commercial application. The main obstacles so far identified for peptide drug development are the high production costs, as compared with small drugs. Beside these general issues, development of peptide-based antiprotozoal drugs presents some specific challenges related to the complex life cycles of protozoa, which often involve multiple stages with significant differences in metabolism, protein expression, and membrane composition (Giovati et al., 2018). Several antimicrobial peptides have been tested, for example, the evaluation of *in vitro* and *in vivo* anti-apicomplexan properties of a synthetic peptide (cNK-2). The peptide exhibited dose- and time-dependent cytotoxic activity against *Eimeria acervuline* and *Eimeria tenella* sporozoites through disruption of membrane integrity and release of intracellular content. In addition, *in vitro* peptide administration increased protection against *Eimeria* infection. The potential utility of the cNK-2 peptide is a complementary strategy for treatment of economically costly infectious diseases of poultry and other food animals (Lee et al., 2013). The recombinant molecules combining monoclonal antibodies targeting *C. parvum* fused to bioactive peptides proved *in vitro* to efficiently kill *C. parvum* sporozoites and inhibit infection when administered orally in a prophylactic murine model of cryptosporidiosis (Imboden et al., 2010). In other study, the cytotoxic properties of CM11 (a chimeric peptide that is derived from bee venom and butterfly compounds) against *E. histolytica* and co-culture with human colonic carcinoma (Caco-2) were tested. The IC_{50} values were 4.904 μg/mL on CM11 treated cells and 10.123 μg/mL on metronidazole, the results revealed that de CM11 peptide require less concentration to make its anti-amoebic effect compared with metronidazole. Therefore, the study of CM11 for antiparasitic effects can make a difference in the amebiasis treatment worldwide (Mahdavi Abhari et al., 2019). on the other hand, a novel antimicrobial peptide isolated from American bullfrog, *Rana catesbeiana*, skin, was called renalexin. The antiparasitic effects of this peptide were tested against

TABLE 10.1
Summary of nanoparticles against intestinal parasite infection

Nanoparticle	Parasite	Condition	Dose	Time	Effect	Ref
AuNPs	*Giardia lamblia* (cysts)	*In vitro*	0.05, 0.1, and 0.3 mg/mL	5, 15, 30, 60, 180 min	96% of lethality on cysts with a concentration of 0.3 mg/mL at 180 min	(Bavand et al., 2014)
AuNPs+*Citrullus colocynthis* NPs	*Giardia lamblia*	*In vivo* (mice)	20 µg/day	8 days	AuNPs alone eliminated 61.2%. Combination of both nanoparticles showed greater reduction (93.2%)	(Al-Ardi, 2020)
AgNPs	*Giardia lamblia*	*In vivo* (rats)	50 µg/day	8 days	Reduced 72.7% cysts and trophozoites count	(Said et al., 2012)
AgNPs	*Giardia lamblia*	*In vivo* (mice)	100 µg/g/day	3 days	Reduced 83.3% of *Giardia* cysts after 48 h	(Idan and Ardalan, 2020)
AgNPs	*Entamoeba histolytica* (cysts) *Cryptosporidium parvum* (oocysts)	*In vitro*	0.5, 1.0, 2.0 and 4.0 mg/L	180 min	The LC_{50}-180 min was 0.34 mg/L and 0.54 mg/L for *E. histolytica* and *C. parvum* respectively	(Saad et al., 2015)
AgNPs	*Cryptosporidium parvum* (oocysts)	*In vitro*	0.5, 5, 50 and 500 µg/mL	0, 1, 3 and 7 days	Reduction of 33.3% of excystation when using a concentration of 500 µg/mL	(Cameron et al., 2016)
Protein-coated AgNPs	*Entamoeba histolytica* (trophozoites)	*In vitro*	27.8, 55.6, and 111.2 µM	72 h	Antiproliferative effect. IC_{50}=69.2 µM (7.4 µg/mL)	(Valenzuela-Salas et al., 2021)
AgNPs	*Blastocystis hominis* (cysts)	*In vitro*	50, 100, 150 µg/mL	24, 48 and 72 h	AgNPs showed growth inhibition at 150 µg/mL after 72 h	(Ahmed et al., 2015)
AgNPs	*Blastocystis hominis* (cysts)	*In vitro*	150 µg/mL	180 min	Inhibition of 71.7% on the growth of the parasite	(Younis et al., 2020)
AgNPs	*Cyclospora cayetanensis*	*In vivo* (mice)	10 µg/day	7 days	Reduced the oocyst count in the feces of immunocompetent and immunosuppressed mice by 96.9% and 92.4% at day 14 after infection, respectively	(Gaafar et al., 2019)
ZnO NPs	*Giardia lamblia* (cysts)	*In vivo* (mice)	10 mg/kg/day	7 days	Reduced the presence of *Giardia lamblia* cysts in mice by up to 93.7%, and if combined with metronidazole (500 mg/kg) there is a lethality of 100%	(Brakat et al, 2020)
CuO NPs and Se NPs	*Giardia lamblia* (cysts)	*In vitro*	0.15, 0.3 and 0.6 mg/mL	10, 15, 30, 60 and 180 min	CuO NPs exhibited a mortality rate of 97% at 180 min with 0.6 mg/mL; and Se NPs presented a mortality rate of 100% at 0.3 mg/mL and 180 min	(Malekifard et al., 2020)
CuO NPs	*Entamoeba histolytica* (cysts) *Cryptosporidium parvum* (oocysts)	*In vitro*	0.5, 1.0, 2.0 and 4.0 mg/L	180 min	Cytotoxic effect with a LC_{50}-3h of 0.13 mg/mL on *E. histolytica* cysts and 0.72 mg/mL on *C. parvum* oocyts	(Saad et al., 2015)
MgO NPs	*Cyclospora cayetanensis* (oocysts)	*In vitro*	1.25, 2.5, 5, 10, 12.5, and 15 mg/mL	24, 48 and 72 h	Lethal concentration on oocytes was 15 mg/mL after 24 hours of incubation	(Hussein et al., 2018)

(continued)

TABLE 10.1 (Continued)
Summary of nanoparticles against intestinal parasite infection

Nanoparticle	Parasite	Condition	Dose	Time	Effect	Ref
CsNPs	*Giardia lamblia* (trophozoites and cysts)	*In vitro* and *in vivo* (mice)	100, 200, and 400 µg/mL and 10, 50 and 100 µg/mL/day	30, 60 and 180 min; 24, 48, and 72 h	Reduced *G. lamblia* trophozoite count, ranging from 89 to 100% respect the control. Cyst excretion was reduced after oral administration of a concentration of 50 µg/kg	(Chabra et al., 2019)
CsNPs	*Giardia lamblia* (trophozoites and cysts)	*In vitro*	1, 5, 10, 20, 40, and 50 µg/mL	180 min	The IC_{50} was calculated at 83 µg/mL	(Elmi et al., 2020)
CsNPs	*Giardia lamblia* (cysts)	*In vitro*	50, 100, 200, and 400 µg/mL	180 min	The concentration of 400 mg/mL exhibited a 100% reduction in the viability of the cysts	(Yarahmadi et al., 2016)
CsNPs CsNPs+AgNPs CsNPs+AgNPs+CURNPs	*Giardia lamblia*	*In vivo* (mice)	50 µg/day	8 days	Reduced 68.2% the parasite count in stool samples CsNPs + AgNPs showed a reduction of 90.2%, and the three NPs completely eradicated *Giardia*	(Said et al., 2012)
CsNPs	*Cryptosporidium parvum* (oocysts)	*In vitro* and *in vivo* (mice)	0.5, 1.5, 3, 5, and 7 mg/mL	2, 24 and 72 h One dose (500 µL)	Oocyst destruction (97.3%) after 72 h at a concentration of 5 mg/mL. Doses of ≥3 mg/mL prevented the infection by the parasite	(Ahmed et al., 2019)
CsNPs CsNPs+Metronidazole	*Giardia lamblia*	*In vivo* (hamsters)	50 µg/day	7 days	CsNPs+Metronidazole reduced the *Giardia* cyst and trophozoite count (94.6-94.3%)	(El-Gendy et al., 2021)
CsNPs+Nitazoxanide	*Cryptosporidium parvum*	*In vivo* (mice)	200 mg/kg/day	10-13 days	Nitazoxanide loaded on CsNPs reduced oocysts shedding in both immunocompetent (75.7%) and immunosuppressed (67.3%) mice	(Moawad et al., 2021)
Mg-LDH-CUR Zn-LDH-CUR	*Giardia lamblia* (trophozoites) *Entamoeba histolytica* (trophozoites)	*In vitro*	7.5 and 15 µM	12, 24, 48 and 72 h	*G. lamblia*: IC_{50} 10.96 µM for Mg-LDH-CUR and 12.21 µM for Zn-LDH-CUR *E. histolytica*: IC_{50} 147.87 µM for Mg-LDH-CUR and 190.53 µM for Zn-LDH-CUR	(Gutiérrez-Gutiérrez et al., 2020)
CUR-PLGA NPs	*Giardia lamblia*	*In vivo* (rats)	50 µg/day	8 days	Reduced the number of *Giardia* cysts count in stool sample (54.6%)	(Said et al., 2012)
Metronidazole-PLGA-NPs	*Giardia lamblia*	*In vivo* (mice)	60 and 120 mg/kg/day	5 days	Metronidazole-PLGA NPs, reduced the amount of *G. lamblia* in the small intestine, 100 and 87.5% at the 60 and 120 mg/kg doses respectively	(Madbouly et al., 2020)

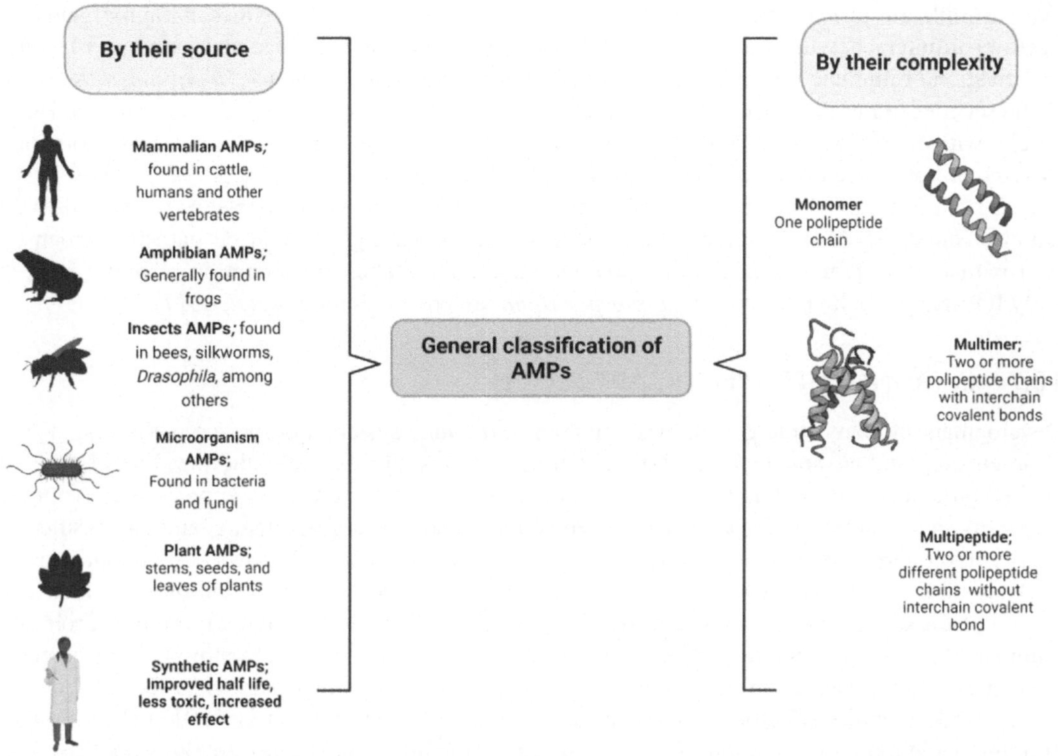

FIGURE 10.3 Antimicrobial peptides general classification. These molecules are widely extended in nature and could be classified by their source or structure.

C. parvum by inoculation of the isolate on infected cell monolayers. Renalexin was used alone and in combination with other peptides like azithromycin, clarithromycin, rifampin, indolicidin, and rifabutin. Renalexin alone presented the most relevant effect on the infected monolayer, suppressing cell growth 40% approximately, however the effect was enhanced with combination of renalexin and other molecules, obtaining 94.1% of parasite-cell growth inhibition with the renalexin-indolicidin combination. In addition, the fractionated proteins obtained from natural honey, which contain an array of glycoproteins, proteoglycans, and glycopeptides, exhibited *in vitro* activities against *G. lamblia* with IC_{50} values ≤ 25 µg/mL. Therefore, indicating the potential of honey proteins and peptides as novel antigiardial agent (Mohammed et al., 2015).

On the other hand, the prototypic cyclotide kalata B1 was investigated due it's toxic effect against two economically important gastrointestinal nematode parasites of sheep, *Haemonchus contortusand* and *Trichotrongylus colubriformi*. In previous studies, the removal of charge through acetylation of lysine residues resulted in reduced antiparasitic activity, highlighting the importance of positively charged residues. The anthelmintic properties of lysine mutants were evaluated by their ability to interfere with development of eggs through to L3 larvae stage. Results showed that substitution of Gly-18, Thr-20, Ser-22, Thr-27, Asn-29, or Gly-1 sites with lysin, enhance the anthelmintic activity of cyclotide as shown by the IC_{50} values of the mutants; 1.1, 0.9, 2.3, 0.4 and 0.5 µM compared with 2.7 µM in the kB1 *wild type* against *H. contortus*. Also, the 1.8, 1.7, 2.7, 3.4, 1.2 and 1.5 µM concentrations compared with 4.5 µM in the kB1 *wild type* against *T. colubriformis*. Gastrointestinal nematodes cause major losses to livestock industries worldwide, therefore development of new treatments against these parasites can help with major economic issues preventing losses and increasing product quality (Huang et al., 2010). Lactoferrin is a multifunctional glycoprotein found in breastmilk, this protein chelates iron which is necessary for essential enzymes in microorganisms. Lactoferrin interacts with molecular and cellular components of both host and pathogens leading to activation of immune responses, therefore its antiparasitic activity against *C. parvum* was tested. Different concentrations of lactoferrin: 0, 1, 2.5, 5, and 10 mg/mL were applied to co-cultured HCT-8 cells with *C. parvum* sporozoites to measure viability of cells. Results showed that sporozoite exposed to low concentrations of lactoferrin were able to infect HCT-8 cells. However, sporozoites treated with 5 mg/mL lactoferrin had significantly reduced numbers of parasites present after 24 hours of infection. Also, lactoferrin at a concentration of 10 mg/mL reduced parasite burden by approximately 60% in comparison to untreated sporozoites. Thus, consumption of breast milk at early life stages can help to reduce *Cryptosporidium* infections (Paredes et al., 2017).

10.7 IMMUNOTHERAPEUTIC APPROACH

Development of many strategies to evade the host's immune system and survive within the host is an interesting and advantageous evolution step in parasites. For example, the first line of defense against protozoan parasites is the complement system. The complement is a major part of innate immunity that is activated by a robust and efficient proteolytic cascade that eventually results on opsonization and lysis of many invading pathogens (Holers, 2014). However, immunoproteomic analysis of *T. spiralis* adult worms showed the presence of serine protein inhibitors, which can invalidate the complement system activation (Song et al., 2018). In addition, arginine is a relevant amino acid in host-immune response, lower levels of arginine have shown to inhibit T-cell function (Zea et al., 2004). Depletion of arginine is a recognized strategy that pathogens like *G. lamblia* use to evade immune effector mechanisms as demonstrated by Banik et al. (2013) showing that arginine depletion by *G. lamblia* arginine deiminases modulate the surface markers and cytokine response of *in vitro*-activated human dendritic cells. Immune response evasion mechanisms vary among parasites. Several mechanisms had been described as shown in Figure 10.4.

An example of these mechanism is secretion of proteases of *E. histolytica* which degrade the intestine mucosal layer, or removal of infiltrating immune cells by phagocytosis. Another example is secretion of immune modulators by helminths, causing blockage of alarmins, which are released in response to tissue injury for a consequent activation of immune response (Coakley et al., 2016). Nevertheless, the *in vitro* study of parasite and host gene expression during the early stages of *E. tenella* infection of the macrophage cell line HD11 showed that pathogen recognition receptors, MRC2, TLR15 and NLRC5 of *E. tenella*, present distinct chemokine and cytokine induction patterns. This could be because of self-immune regulation or the effect of *Eimieria´s* mechanism of evasion (Sandholt et al., 2021). Also, antigenic variation on protozoa, which is a result of emergence of new sequence varieties within a species. Deitsch et al. (2009) give as a result a different phenotypic pattern, which cannot be recognized for the specific secondary immune response. Immunotherapy is a therapeutic approach that manipulates the adaptive and innate immune system for prophylactic and/or therapeutic purposes through passive and active mechanisms. Passive immunotherapies harness administration of antibodies and cells, while active immunotherapies stimulate immune response (Naran et al., 2018; Yousofi Darani et al., 2016). Although most of the immunotherapies are focused on cancer treatment, these can be used against infectious diseases such as parasitic infections (Naran et al., 2018).

Therapeutic vaccines are treatments that use the patient's immune system to attack an existing disease, unlike prophylactic vaccines that prevent disease (Shimasaki 2020). However, results obtained using inactivated parasite therapeutic vaccines against intestinal parasitic diseases have not been entirely successful. For example, to treat giardiasis, Olson et al. (2001) vaccinated dogs that had not been able to be cured by chemotherapy with a commercial vaccine. The clinical signs resolved between 16 and 42 days after vaccination and stopped shedding fecal cysts between 21- and 70-days post vaccination. Nonetheless, when Anderson et al. (2004) replicated the experiment

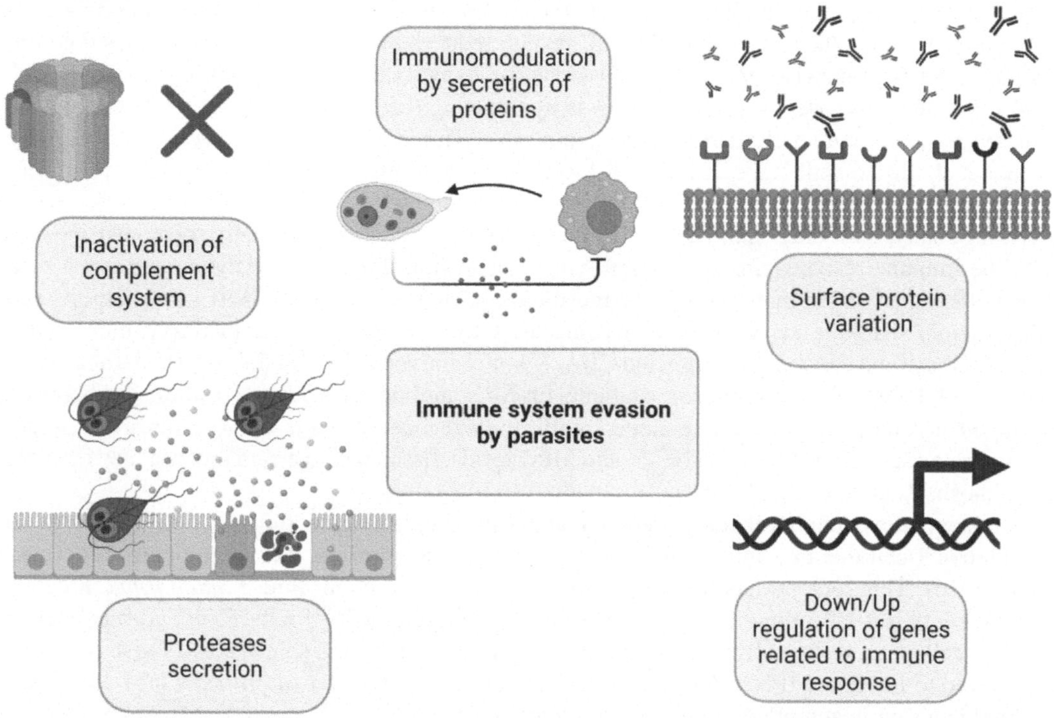

FIGURE 10.4 Mechanisms of immune invasion caused by parasites. These pathogens developed several strategies to evade the host´s immune system. 1) Inactivation of the complement system, 2) impairment of biomolecules by proteases secretion, 3) genetic modulation, 4) surface proteins modification to cause antigenic variation, and 5) ability to block immune response by protein secretion.

using GiardiaVax ® the treatment didn't result effective. The same lack of effect was seen in cats infected with *Giardia spp.* when treated with a commercial inactivated parasite vaccine (Stein et al., 2003). It should be noted that, since *Giardia spp.* suffers antigenic variation, the use of variable surface proteins (VSP) seems feasible to use antigens in therapeutic vaccines against this infection. Serradell et al. (2016) tried a vaccine with the complete repertoire of VSP immunopurified or in isolated plasmatic membranes of transgenic trophozoites. This vaccine reduced or cured giardiasis in dogs. However, recently, it has been proposed that metabolic pathways proteins, like enolase and ornithine carbamoyl transferase which are involved in glycolysis and arginine metabolism are promising antigens due to their conserved expression, lack of closely with human proteins, and immunogenicity capacity. In a giardiasis model in mice, an induction of IgG1 and IgG2b strong systemic response was observed with both antigens, and with enolase immunization a cyst excretion reduction was also detected. These results increase the panorama of therapeutic options for the control of giardiasis (Ihara et al., 2022).

Therapeutic helminth vaccines have been developed with mild satisfactory results. Al-Attar et al. (2020) infected mice with *T. spiralis* and treated with excretory/secretory antigens from adult worms or larvae administered intraperitoneally. Vaccines with antigens from adult worms decreased excretion of worms by 80.1% and larvae by 63.5%. Vaccines with larval antigens reduced excretion of adult worms by 86.9% and larvae by 87.96%, indicating that larval secretory/excretory antigens have a better therapeutic effect than adult worm antigens.

Antibody therapy is a passive immunotherapy where antibodies are used as a treatment to help fight a disease. A similar effect was observed when monoclonal antibodies were administered to

IFN-γ gene *knockout* adult SCID mice infected with *C. parvum*. The most effective antibody was the one directed against the surface protein CSL, which reduces intestinal infection but does not eliminate the parasite. Likewise, anti-GP25-200 and anti-P23 antibodies were also effective (Riggs et al., 2002). An alternative way to use antibodies is by targeting them against cell receptors to modulate immune response and ameliorate negative symptoms of infections. This effect was observed when Arendt et al. (2019) antibodies anti-IL10 R2 were orally administered to *Eimeria* infected chickens, which overcame the infection and had similar weight to uninfected chicks.

Additionally, antibodies can be used to block virulence factors, especially those that interfere with the immune response in conjunction with antiparasitic drugs. *E. histolytica* secretes a cytokine homolog of inflammatory macrophage migration inhibitory factor (MIF) that inhibits pro-inflammatory cytokine secretion which contributes to host damage and therefore more severe disease. Ghosh and collaborators infected CBA- /- J mice intracecally with *E. histolytica* trophozoites and treated them with metronidazole and anti-Eh-MIF antibodies. Addition of these antibodies to treatment instead of metronidazole alone significantly reduced parasite-induced intestinal inflammation, as measured by CXCL1, TNF-α, and MPO levels, tissue damage, and mucosal barrier integrity (Ghosh, Padalia, et al., 2020).

Another passive immunotherapy used to treat intestinal parasite infection is adoptive cell transfer. In adoptive T-cell therapy, antigen-specific T cells are transferred for therapeutic purposes (Busch et al., 2016). This therapy has been shown to be effective in controlling *E.falciformis* infection. Pogonka et al. (2010) immunized mice deficient in transgenic IFN-γ with *E. falciformis* oocysts. CD8+ T cells were purified from mesenterial lymph node (MLN) and spleen cells. These cells were transferred to irradiated IFN-γ deficient transgenic mice. The transfer of primed CD8+ MLN cells resulted in a significant reduction in oocyst production.

Transfer of activated cells prior to *C. parvum* infection appears to be an effective prophylactic treatment, as demonstrated by Tessema et al. (2009) when transferring intraepithelial lymphocytes (IEL) and CD4+ T cells from the spleen and mesenteric lymph nodes of interferon and interleukin-12p40-deficient C57BL/6 mice infected with *C. parvum* to naive mice. This treatment reduced the parasite load and excretion by up to 70%. On the other hand, the opposite effect was shown when the adoptive transfer of CD4 helper T-lymphocytes from the mucosa prepared with *Gymnophalloides seoi* to mice prior to infection caused strong goblet cell hyperplasia and markedly accelerated worm shedding (Guk et al., 2009). This exemplifies the importance of knowing the pathogenesis bases of the parasite and host-immune response to establish an adequate immunotherapy. As is the case of Wang et al. (2018), who proposed affecting Foxp3 + Tregs, given that during *E. multilocularis* infection, they favor survival of metacestodes by affecting antigen presentation and suppressing Th1-type immune responses.

Generating successful immunotherapies for intestinal parasitic infections has been complicated. This is largely due to the complexity of these pathogens, as well as that of the host-parasite relationship. The parasite can modulate the host's immune responses toward "ineffective" responses, young animals do not respond adequately (which is the stage of life when vaccines must be effective in blocking the early production associated with over-losses) and, if recombinant proteins are used, persist the challenge of maintaining the correct conformation of these molecules (Matthews et al., 2016).

10.8 CONCLUSION

Intestinal diseases constitute a global public health problem, with parasites being the main causative agent. Distribution and prevalence of these organisms is variable, but the main affected population includes children and immunocompromised individuals. Other important factors are socioeconomic and geographic status, climate, inadequate water supplies, and unsuitable sanitation. Due to the impact of these diseases on health, studies that contribute to cellular biology knowledge on parasites are necessary to develop new molecules with an efficient and safe profile. In this sense, the main approaches are the use of computational tools to identify essential molecular targets for parasite survival and rational drug design, like molecules that are involved in metabolic pathways. Also, use of novel technologies in genetic edition, like CRISPR, have a potential field of application because it allows to determine essential genes critical for growth, differentiation, and pathogenesis processes. In addition, other edition techniques can induce modifications for gene or protein disturbance that could be lethal for intestinal parasite pathogens. Another side is that repurposing of existing drugs is another method that could be helpful in time and cost approval process for a second use drug. These characteristics make it an ideal strategy to consider its resurgence using drugs that are already approved or used commercially.

Besides, in the search of innocuous treatment alternatives, several bioactive compounds have been shown their antiparasitic potential. Between them, plant extracts, isolated and purified compounds or semisynthetic derivates have been tested. Some of these molecules have shown greater effectiveness than reference drugs, others are used as adjuvants, and some have even shown prophylactic effects. The heterogeneity of natural compounds with antiparasitic activities include the use of peptides, antibodies, and other metabolites. For example, antimicrobial peptides have been found to be effective in drug-resistant strains and to be effective in stages of differentiation in parasites, a process closely related to its infectivity and survival. On the other hand, peptides, nanotechnology, and immunotherapy promote an integral therapeutic approach in which parameters like technological drug development or drug response are considered. The functionalization of compounds allows improving the physicochemical parameters of the molecules, such as their stability, in addition to their pharmacokinetic profile. This allows improving the effectiveness of these potential treatments. In the case of immunomodulatory therapies, there has recently been a resurgence in its use as antiparasitic agents. New antigens that are less variable, have less homology to humans, and induce an immune response are being explored. Passive therapies using antibodies are also promising and ameliorating negative symptoms of infections. In general, these therapies can be combined to improve their effectiveness. The continuous scientific advances will allow us to make efforts to contribute to the cellular and molecular understanding of intestinal parasites and to improve the antiparasitic treatment available.

REFERENCES

Abaza, S. (2016). Applications of nanomedicine in parasitic diseases. *J. Egypt. Soc. Parasitol. 9*(1), 1. https://doi.org/10.4103/1687-7942.192997

Ahmed, M., Ismail, K., Ahmed, S., Ibrahim, A., and Gohar, Y. (2015). In vitro activity of curcumin and silver nanoparticles against blastocystis hominis. *Infect Dis Clin Pract. 23*, 135–140. https://doi.org/10.1097/IPC.0000000000000242

Ahmed, S. A., El-Mahallawy, H. S., and Karanis, P. (2019). Inhibitory activity of chitosan nanoparticles against Cryptosporidium parvum oocysts. *Parasitol. Res. 118*(7), 2053–2063. https://doi.org/10.1007/s00436-019-06364-0

Al-Ardi, M. H. (2020). The uses of gold nanoparticles and Citrullus colocynthis L. nanoparticles against Giardia lamblia in vivo. *Clin. Epidemiol. Glob. Health. 8*(4), 1282–1286. https://doi.org/10.1016/j.cegh.2020.04.028

Al-Attar, T., El-Kersh, W., Sadek, G., Harba, N., Osheiba, S., and Brakat, R. (2020). A study of immunotherapeutic efficacy of trichinella spiralis excretory-secretory proteins in murine trichinellosis. *J. Egypt. Soc. Parasitol. 50*(2), 281–292. https://doi.org/10.21608/jesp.2020.113049

Aldaba-Muruato, L. R., Muñoz-Ortega, M. H., Macías-Pérez, J. R., Pulido-Ortega, J., Martínez-Hernández, S. L., and Ventura-Juárez, J. (2017). Adrenergic regulation during acute hepatic infection with Entamoeba histolytica in the hamster: involvement of oxidative stress, Nrf2 and NF-KappaB. *Parasite (Paris, France). 24*, 46. https://doi.org/10.1051/parasite/2017048

Alkorta, M., Manzanal, A., and Zeberio, I. (2022). A case of severe diarrhoea caused by Cyclospora cayetanensis in an immunocompromised patient in northern Spain. *Access Microbiol. 4*(2). https://doi.org/10.1099/acmi.0.000313

Al-Megrin, W. A. (2017). In vivo study of pomegranate (Punica granatum) peel extract efficacy against Giardia lamblia in infected experimental mice. *Asian Pac. J. Trop. Biomed.* *7*(1), 59–63. https://doi.org/https://doi.org/10.1016/j.apjtb.2016.08.018

Almohammed, H. I., Albalawi, A. E., Sadoun, H. al, Bakhtiari, N., Amraei, M., Moghaddam, A., and Al-Awsi, G. R. L. (2021). Applying nanoparticles for treating giardia infection: A systematic review. *Int J App Pharm.* *13*(5), 15–19. https://doi.org/10.22159/ijap.2021v13i5.42466

Almoradie, A. M., Angeles, R. J., Beltran, E. V., Ugali, M., Valles, N. S., Los Banos, Z. D., Mahboob, T., Barusrux, S., and Nissapatorn, V. (2018). Cryptosporicidal activity of plant extracts against cryptosporidium parvum and cryptosporidium hominis. *Asian J. Pharmacogn.* *2*(3), 22–31.

Alnomasy, S., Al-Awsi, G. R. L., Raziani, Y., Albalawi, A. E., Alanazi, A. D., Niazi, M., and Mahmoudvand, H. (2021). Systematic review on medicinal plants used for the treatment of Giardia infection. *Saudi J Biol Sci.* *28*(9), 5391–5402. https://doi.org/10.1016/j.sjbs.2021.05.069

Anderson, K. A., Brooks, A. S., Morrison, A. L., Reid-Smith, R. J., Martin, S. W., Benn, D. M., and Peregrine, A. S. (2004). Impact of Giardia vaccination on asymptomatic Giardia infections in dogs at a research facility. *Can. Vet. J.* *45*(11), 924–930.

Andrade, R. M., and Reed, S. L. (2015). New drug target in protozoan parasites: The role of thioredoxin reductase. *Front. Microbiol.* *6*, 975. www.frontiersin.org/article/10.3389/fmicb.2015.00975

Arendt, M. K., Knoll, L. J., and Cook, M. E. (2019). Oral antibody to interleukin-10 receptor 2, but not interleukin-10 receptor 1, as an effective Eimeria species immunotherapy in broiler chickens. *Poult. Sci.* *98*(9), 3471–3480. https://doi.org/10.3382/ps/pez064

Aschale, Y., Minwuyelet, A., Akalu, T. Y., and Talie, A. (2022). Prevalence of intestinal parasite infections and associated factors among pregnant women in Northwest Ethiopia. *J. Parasitol Res.* *2022*, 1–7. https://doi.org/10.1155/2022/9065425

Ashburn, T.T. and Thor, K.B. (2004) Drug Repositioning: Identifying and Developing New Uses for Existing Drugs. *Nature Reviews Drug Discovery*, 3, 673–683. http://dx.doi.org/10.12688/f1000research.6653.1

Aziz, N., Faraz, M., Sherwani, M. A., Fatma, T., Prasad, R. (2019). Illuminating the anticancerous efficacy of a new fungal chassis for silver nanoparticle synthesis. Front Chem. 7, 65. DOI: 10.3389/fchem.2019.00065

Aziz, N., Pandey, R., Barman, I., Prasad, R. (2016). Leveraging the attributes of Mucor hiemalis-derived silver nanoparticles for a synergistic broad-spectrum antimicrobial platform. Front Microbiol. 7, 1984. DOI: 10.3389/fmicb.2016.01984

Babuta, M., Bhattacharya, S., and Bhattacharya, A. (2020). Entamoeba histolytica and pathogenesis: A calcium connection. In *PLoS Pathog.* (Vol. 16, Issue 5). Public Library of Science. https://doi.org/10.1371/journal.ppat.1008214

Balta, I., Marcu, A., Linton, M. et al. The in vitro and in vivo anti-virulent effect of organic acid mixtures against Eimeria tenella and Eimeria bovis. *Sci Rep.* *11*, 16202 (2021). https://doi.org/10.1038/s41598-021-95459-9

Banerjee, T., Singh, A., Kumar, S., Dhanani, T., Gajbhiye, N. A., Koley, T. K., Maurya, A., and Filgona, J. (2019). Ovicidal and larvicidal effects of extracts from leaves of Andrographis paniculata (Burm. f.) Wall.ex Nees against field isolates of human hookworm (Ancylostoma duodenale). *J. Ethnopharmacol.* *235*, 489–500. https://doi.org/https://doi.org/10.1016/j.jep.2019.02.021

Banik, S., Viveros, P. R., Seeber, F., Klotz, C., Ignatius, R., and Aebischer, T. (2013). Giardia duodenalis arginine deiminase modulates the phenotype and cytokine secretion of human dendritic cells by depletion of arginine and formation of ammonia. *Infect. Immun.* *81*(7), 2309–2317. https://doi.org/10.1128/IAI.00004-13

Barbosa, C. V., Barreto, M. M., de Jesus Andrade, R., Sodré, F., D'Avila-Levy, C. M., Peralta, J. M., Igreja, R. P., de Macedo, H. W., and Santos, H. L. C. (2018). Intestinal parasite infections in a rural community of Rio de Janeiro (Brazil): Prevalence and genetic diversity of Blastocystis subtypes. *PLoS One.* *13*(3). https://doi.org/10.1371/journal.pone.0193860

Bavand, Z., Gholami, S., Honary, S., Rahimi-Esboei, B., Torabi, N., and Barabadi, H. (2013). In vitro evaluation of the effect of gold nanoparticles on Giardia lamblia cyst. *J Arak Uni Med Sci.* *16*(79).

Bavand, Z., Gholami, S., Honari, S., Rahimi Esboei, B., Torabi, N., and Borabadi, H. (2014). Effect of gold nanoparticles on giardia lamblia cyst stage in in vitro. *J Arak Uni Med Sci.* *16*(10), 27–37. http://jams.arakmu.ac.ir/article-1-2461-en.html

Beshay, E. V. N. (2018). Therapeutic efficacy of Artemisia absinthium against Hymenolepis nana: In vitro and in vivo studies in comparison with the anthelmintic praziquantel. *J. Helminthol. 92*(3), 298–308. https://doi.org/DOI: 10.1017/S0022149X17000529

Bessoff, K., Sateriale, A., Lee, K. K., and Huston, C. D. (2013). Drug repurposing screen reveals FDA-approved inhibitors of human HMG-CoA reductase and isoprenoid synthesis that block Cryptosporidium parvum growth. *Antimicrob. Agents Chemother. 57*(4), 1804–1814. https://doi.org/10.1128/AAC.02460-12

Bethesda. (2020). Nitazoxanide in *LiverTox: Clinical and Research Information on Drug-Induced Liver Injury.* National Institute of Diabetes and Digestive and Kidney Diseases. www.ncbi.nlm.nih.gov/books/NBK548943/

Brakat, R., and Sharaf-El-Deen, S. (2020). Zinc oxide nanoparticles kill giardia and protect against intestinal damage. *Egypt. J. Med. Microbiol.* 28: 95–103. www.ejmm-eg.com

Briquet, S., Gissot, M., and Silvie, O. (2021). A toolbox for conditional control of gene expression in apicomplexan parasites. *Mol Microbiol. 117*(3), 618–631. https://doi.org/10.1111/mmi.14821

Busch, D. H., Fräßle, S. P., Sommermeyer, D., Buchholz, V. R., and Riddell, S. R. (2016). Role of memory T cell subsets for adoptive immunotherapy. *Semin Immunol.* 28(1): 28–34. doi:10.1016/j.smim.2016.02.001

Calzada, F., and Bautista, E. (2020). Plants used for the treatment of diarrhoea from Mexican flora with amoebicidal and giadicidal activity, and their phytochemical constituents. *J. Ethnopharmacol. 253*, 112676. https://doi.org/https://doi.org/10.1016/j.jep.2020.112676

Cameron, P., Gaiser, B. K., Bhandari, B., Bartley, P. M., Katzer, F., and Bridle, H. (2016). Silver nanoparticles decrease the viability of Cryptosporidium parvum oocysts. *Appl Environ Microbiol.* 82(2), 431–437. https://doi.org/10.1128/AEM.02806-15

Capparelli, E. v, Bricker-Ford, R., Rogers, M. J., McKerrow, J. H., and Reed, S. L. (2016). Phase I clinical trial results of auranofin, a novel antiparasitic agent. *Antimicrob. Agents Chemother. 61*(1), e01947–16. https://doi.org/10.1128/AAC.01947-16

Chabra, A., Rahimi-Esboei, B., Habibi, E., Monadi, T., Azadbakht, M., Elmi, T., Valian, H. K., Akhtari, J., Fakhar, M., and Naghshvar, F. (2019). Effects of some natural products from fungal and herbal sources on Giardia lamblia in vivo. *Parasitology*. 146(9), 1188–1198. https://doi.org/10.1017/S0031182019000325

Chai, J. Y., Jung, B. K., and Hong, S. J. (2021). Albendazole and mebendazole as anti-parasitic and anti-cancer agents: An update. *Korean J Parasitol.* 59(3), 189–225. https://doi.org/10.3347/kjp.2021.59.3.189

Chama, M. A., Dziwornu, G. A., Waibel, R., Osei-Safo, D., Addae-Mensah, I., Otchere, J., and Wilson, M. (2016). Isolation, characterization, and anthelminthic activities of a novel dichapetalin and other constituents of Dichapetalum filicaule. *Pharm. Biol. 54*(7), 1179–1188. https://doi.org/10.3109/13880209.2015.1059861

Choudhary, H. H., Nava, M. G., Gartlan, B. E., Rose, S., and Vinayak, S. (2020). A conditional protein degradation system to study essential gene function in cryptosporidium parvum. *mBio. 11*(4), e01231–20. https://doi.org/10.1128/mBio

Chen, W., Ma, C., Li, G., Jia, Z., Yang, X., Pan, X., and Ma, D. (2021). Specific EtMIC3-binding peptides inhibit Eimeria tenella sporozoites entry into host cells. *Vet Res. 52*(1). https://doi.org/10.1186/s13567-020-00873-y

Coakley, G., Buck, A. H., and Maizels, R. M. (2016). Host parasite communications-Messages from helminths for the immune system: Parasite communication and cell-cell interactions. *Mol. Biochem. Parasitol. 208*(1), 33–40. https://doi.org/10.1016/j.molbiopara.2016.06.003

Crouch, A. A., Seow, W. K., and Thong, Y. H. (1986). Effect of twenty-three chemotherapeutic agents on the adherence and growth of Giardia lambiia in vitro. *Trans. R. Soc. Trop. Med. Hyg. 80*(6), 893–896. https://doi.org/https://doi.org/10.1016/0035-9203(86)90249-X

Cui, Y., and Yu, L. (2016). Application of the CRISPR/Cas9 gene editing technique to research on functional genomes of parasites. *Parasitol. Int. 65*(6), 641–644. https://doi.org/10.1016/j.parint.2016.08.011

Daneshmand, A., Young, P., Campbell, B., Kheravii, S. K., Sharma, N. K., Afshari, R., Dias, D. A., Flavel, M., Kitchen, B., and Wu, S.-B. (2021). In vitro inhibitory activities of sugarcane extract on avian Eimeria sporozoites. *Int J Parasitol Drugs Drug Resist. 17*, 1–4. https://doi.org/https://doi.org/10.1016/j.ijpddr.2021.07.001

Davoodi, J., and Abbasi-Maleki, S. (2018). Effect of origanum vulgare hydroalcoholic extract on giardia lamblia cysts compared with metronidazole in vitro. *Iran. J. Parasitol. 13*(3), 486–492. https://pubmed.ncbi.nlm.nih.gov/30483342

Debnath, A., Parsonage, D., Andrade, R. M., He, C., Cobo, E. R., Hirata, K., Chen, S., García-Rivera, G., Orozco, E., Martínez, M. B., Gunatilleke, S. S., Barrios, A. M., Arkin, M. R., Poole, L. B., McKerrow, J. H., and Reed, S. L. (2012). A high-throughput drug screen for Entamoeba histolytica identifies a new lead and target. *Nat. Med. 18*(6), 956–960. https://doi.org/10.1038/nm.2758

Debnath, A., Ndao, M., and Reed, S. L. (2013). Reprofiled drug targets ancient protozoans: drug discovery for parasitic diarrheal diseases. *Gut Microbes. 4*(1), 66–71. https://doi.org/10.4161/gmic.22596

Deitsch, K. W., Lukehart, S. A., and Stringer, J. R. (2009). Common strategies for antigenic variation by bacterial, fungal and protozoan pathogens. *Nat. Rev. Microbiol.* 7(7), 493–503. https://doi.org/10.1038/nrmicro2145

Dhal, A. K., Pani, A., Mahapatra, R. K., and Yun, S. I. L. (2018). In-silico screening of small molecule inhibitors against Lactate Dehydrogenase (LDH) of Cryptosporidium parvum. *Comput. Biol. Chem. 77*, 44–51. https://doi.org/10.1016/j.compbiolchem.2018.09.002

Dhal, A. K., Pani, A., Mahapatra, R. K., and Yun Soon I.L. (2019). An immunoinformatics approach for design and validation of multi-subunit vaccine against Cryptosporidium parvum. *Immunobiology. 224*(6), 747–757. https://doi.org/10.1016/j.imbio.2019.09.001

Dhal, A. K., Pani, A., Yun, S. il, and Mahapatra, R. K. (2020). In-silico analysis of Calcium Dependent Protein Kinase 6 of Cryptosporidium parvum through molecular modeling, docking, and dynamics simulation study. *J. Biomol. Struct. Dyn. 39*(15), 5461. https://doi.org/10.1080/07391102.2020.1790036

Divakaruni, Mahabir, Orrett, Adidam, Venkata, Adidam, and Divakaruni. (2019). Compliance and treatment outcomes of various regimens for trichomoniasis in Trinidad and Tobago. *Med Sci (Basel). 7*(10), 97. https://doi.org/10.3390/medsci7100097

Doudna, J. A., and Charpentier, E. (2014). The new frontier of genome engineering with CRISPR-Cas9. *Science.* 346. doi: 10.1126/science.1258096

Drinić, M., Raninger, A., Zraunig, A., Astelbauer, F., Leitsch, D., Obwaller, A., Walochnik, J., Greger, H., and Duchene, M. (2019). Activity of methylgerambullin from Glycosmis species (Rutaceae) against Entamoeba histolytica and Giardia duodenalis in vitro. *Int. J. Parasitol: Drug. 10*, 109–117. https://doi.org/https://doi.org/10.1016/j.ijpddr.2019.08.001

Ehrenkaufer, G. M., Suresh, S., Solow-Cordero, D., and Singh, U. (2018). High-throughput screening of Entamoeba identifies compounds which target both life cycle stages and which are effective against metronidazole resistant parasites. *Front Cell Infect Microbiol.* 8: 276. doi:10.3389/fcimb.2018.00276

Ehrenkaufer, G., Li, P., Stebbins, E. E., Kangussu-Marcolino, M. M., Debnath, A., White, C. v., Moser, M. S., Derisi, J., Gisselberg, J., Yeh, E., Wang, S. C., Company, A. H., Monti, L., Caffrey, C. R., Huston, C. D., Wang, B., and Singh, U. (2020). Identification of anisomycin, prodigiosin and obatoclax as compounds with broad-spectrum anti-parasitic activity. *PLOS Negl. Trop. Dis. 14*(3). https://doi.org/10.1371/journal.pntd.0008150

El-Gendy, A. M. L., Mohammed, M. A. A., Ghallab, M. M. I., Abdel Aziz, M. O., and Ibrahim, S. M. (2021). Therapeutic effect of chitosan nanoparticles and metronidazole in treatment of experimentally giardiasis infected hamsters. *Iran J Parasitol. 16*(1), 32–42. https://doi.org/10.18502/ijpa.v16i1.5509

El-Kady, A. M., Abdel-Rahman, I. A. M., Fouad, S. S., Allemailem, K. S., Istivan, T., Ahmed, S. F. M., Hasan, A. S., Osman, H. A., and Elshabrawy, H. A (2021). Pomegranate peel extract is a potential alternative therapeutic for giardiasis. *Antibiotics (Basel). 10*(6):705. https://doi.org/10.3390/antibiotics10060705

Elmi, T., Rahimi Esboei, B., Sadeghi, F., Zamani, Z., Didehdar, M., Fakhar, M., Chabra, A., Hajialiani, F., Namazi, M. J., and Tabatabaie, F. (2021). In vitro antiprotozoal effects of nano-chitosan on plasmodium falciparum, giardia lamblia and trichomonas vaginalis. *Acta Parasitol.* 66(1), 39–52. https://doi.org/10.1007/s11686-020-00255-6

El-Wakil, E. S., Salem, A. E., and Al-Ghandour, A. M. F. (2021). Evaluation of possible prophylactic and therapeutic effect of mefloquine on experimental cryptosporidiosis in immunocompromised mice. *J Parasit Dis. 45*(2), 380–393. https://doi.org/10.1007/s12639-020-01315-4

Fahmy, A., Abuelenain, G. L., Rasheed, N., and Abdou, A. (2021). 'de Novo' repurposing of Daflon as anti-intestinal parasitic drug in experimental giardiasis. *Exp Parasitol.* 226–227, 108124. https://doi.org/https://doi.org/10.1016/j.exppara.2021.108124

Fatemeh Sadat Tabatabaei Mirakabad. (2014). PLGA-based nanoparticles as cancer drug delivery systems. *Asian Pac J Cancer Prev. 15*(2), 517–535. http://journal.waocp.org/article_28622.html

Favennec, L., Chochillon, C., Magne, D. et al. (1992). A new screening assay for antigiardial compounds: effects of various drugs on the adherence of Giardia duodenalis to Caco2 cells. *Parasitol Res.* 78, 80–81. https://doi.org/10.1007/BF00936187

Fernández-Lainez, C., de la Mora-de la Mora, I., García-Torres, I., Enríquez-Flores, S., Flores-López, L. A., Gutiérrez-Castrellón, P., Yépez-Mulia, L., Matadamas-Martínez, F., de Vos, P., and López-Velázquez, G. (2021). Multilevel approach for the treatment of giardiasis by targeting arginine deiminase. *Int J Mol Sci. 22*(17), 9491. https://doi.org/10.3390/ijms22179491

Gaafar, M. R., El-Zawawy, L. A., El-Temsahy, M. M., Shalaby, Th. I., and Hassan, A. Y. (2019). Silver nanoparticles as a therapeutic agent in experimental cyclosporiasis. *Exp. Parasitol. 207*, 107772. https://doi.org/https://doi.org/10.1016/j.exppara.2019.107772

Galkin, A., Kulakova, L., Lim, K., Chen, C. Z., Zheng, W., Turko, I. v, and Herzberg, O. (2014). Structural basis for inactivation of Giardia lamblia carbamate kinase by disulfiram. *J. Biol. Chem. 289*(15), 10502–10509. https://doi.org/10.1074/jbc.M114.553123

García-Torres, I., de la Mora-de la Mora, I., Marcial-Quino, J., Gómez-Manzo, S., Vanoye-Carlo, A., Navarrete-Vázquez, G., Colín-Lozano, B., Gutiérrez-Castrellón, P., Sierra-Palacios, E., López-Velázquez, G., and Enríquez-Flores, S. (2016). Proton pump inhibitors drastically modify triosephosphate isomerase from Giardia lamblia at functional and structural levels, providing molecular leads in the design of new antigiardiasic drugs. *Biochim Biophys Acta Gen Subj.* 1860(1, Part A), 97–107. https://doi.org/https://doi.org/10.1016/j.bbagen.2015.10.021

Ghadage, D. M., Kshirsagar, P. R., Pai, S. R., and Chavan, J. J. (2017). Extraction efficiency, phytochemical profiles and antioxidative properties of different parts of Saptarangi (Salacia chinensis L.) – An important underutilized plant. *Biochem. Biophys. Rep. 12*, 79–90. https://doi.org/10.1016/j.bbrep.2017.08.012

Ghodeif, A. O., and Jain, H. (2022) Hookworm. In: A. O. Ghodeif and H. Jain, eds, *StatPearls*. Treasure Island: FL: StatPearls Publishing.

Ghosh, S., Farr, L., Singh, A., Leaton, L. A., Padalia, J., Shirley, D. A., Sullivan, D., and Moonah, S. (2020). Cop9 signalosome is an essential and druggable parasite target that regulates protein degradation. *PLoS Pathog. 16*(9). https://doi.org/10.1371/journal.ppat.1008952

Ghosh, S., Padalia, J., Ngobeni, R., Abendroth, J., Farr, L., Shirley, D. A., Edwards, T., and Moonah, S. (2020). Targeting parasite-produced macrophage migration inhibitory factor as an antivirulence strategy with antibiotic-antibody combination to reduce tissue damage. *J. Infect. Dis. 221*(7), 1185–1193. https://doi.org/10.1093/infdis/jiz579

Giovati, L., Ciociola, T., Magliani, W., and Conti, S. (2018). Antimicrobial peptides with antiprotozoal activity: Current state and future perspectives. *Future Med Chem. 10*(22): 2569–2572. doi:10.4155/fmc-2018-0460

Guk, S. M., Lee, J. H., Kim, H. J., Kim, W. H., Shin, E. H., and Chai, J. Y. (2009). Cd4 T-cell-dependent goblet cell proliferation and expulsion of gymnophalloides seoi from the intestine of C57bl/6 Mice. *J. Parasitol. 95*(3), 581–590. https://doi.org/10.1645/GE-1807.1

Guo, F., Zhang, H., McNair, N. N., Mead, J. R., and Zhu, G. (2018). The existing drug vorinostat as a new lead against cryptosporidiosis by targeting the parasite histone deacetylases. *The J. Infect. Dis. 217*(7), 1110–1117. https://doi.org/10.1093/infdis/jix689

Gutiérrez-Gutiérrez, F., Palomo-Ligas, L., Hernández-Hernández, J. M., Pérez-Rangel, A., Aguayo-Ortiz, R., Hernández-Campos, A., Castillo, R., González-Pozos, S., Cortés-Zárate, R., Ramírez-Herrera, M. A., Mendoza-Magaña, M. L., and Castillo-Romero, A. (2017). Curcumin alters the cytoskeleton and microtubule organization on trophozoites of Giardia lamblia. *Acta Trop. 172*, 113–121. https://doi.org/10.1016/j.actatropica.2017.04.027

Gutiérrez-Gutiérrez, F., Puebla-Pérez, A. M., González-Pozos, S., Hernández-Hernández, J. M., Pérez-Rangel, A., Alvarez, L. P., Tapia-Pastrana, G., and Castillo-Romero, A. (2017). Antigiardial activity of podophyllotoxin-type lignans from bursera fagaroides var. fagaroides. *Molecules. 22*(5). https://doi.org/10.3390/molecules22050799

Gutiérrez-Gutiérrez, F., Romo-Mancillas, A., Puebla-Pérez, A. M., Hernández-Hernández, J. M., and Castillo-Romero, A. (2019). Identification and molecular characterization of the tubulin-podophyllotoxin-type lignans binding site on Giardia lamblia. *Chem. Biol. Drug Des. 94*(6), 2031–2040. https://doi.org/10.1111/cbdd.13605

Gutiérrez-Gutiérrez, F., Sánchez-Jiménez, C., Rangel-Castañeda, I. A., Carbajal-Arízaga, G. G., Macías-Lamas, A. M., Castillo-Romero, A., and Parra-Saavedra, K. J. (2020). Encapsulation of curcumin into layered double hydroxides improve their anticancer and antiparasitic activity. *J. Pharm. Pharmacol. 72*(7), 897–908. https://doi.org/https://doi.org/10.1111/jphp.13266

Guzowska, M., Wawrzyniak, R., and Wasiak, W. (2022). Seasonal variability and effect of sample storage on volatile components in Calypogeia azurea. *Molecules. 27*(8), 2426. https://doi.org/10.3390/molecules27082426

Habibi, H., Firouzi, S., Nili, H., Razavi, M., Asadi, S. L., and Daneshi, S. (2016). Anticoccidial effects of herbal extracts on Eimeria tenella infection in broiler chickens: in vitro and in vivo study. *J. Parasit. Dis. 40*(2), 401–407. https://doi.org/10.1007/s12639-014-0517-4

Hahn, J., Seeber, F., Kolodziej, H., Ignatius, R., Laue, M., Aebischer, T., and Klotz, C. (2013). High sensitivity of Giardia duodenalis to tetrahydrolipstatin (orlistat) in vitro. *PLoS One. 8*(8), e71597–e71597. https://doi.org/10.1371/journal.pone.0071597

Hemphill, A., Müller, N., and Müller, J. (2019). Comparative pathobiology of the intestinal protozoan parasites giardia lamblia, entamoeba histolytica, and cryptosporidium parvum. *Pathogens. 8* (3): 116. https://doi.org/10.3390/pathogens8030116

Hennessey, K. M., Alas, G. C. M., Rogiers, I., Li, R., Merritt, E. A., and Paredez, A. R. (2020). Nek8445, a protein kinase required for microtubule regulation and cytokinesis in Giardia lamblia. *Mol. Biol. Cell. 31*(15), 1611–1622. https://doi.org/10.1091/mbc.E19-07-0406

Hezarjaribi, H. Z., Elmi, T., Dayer, M. S., Gholami, S., Fakhar, M., Akbariqomi, M., and Ghaffarifar, F. (2015). A systematic review of the effects of Iranian pharmaceutical plant extracts on Giardia lamblia. *Asian Pac J Trop Dis. 5*(12), 925–929. https://doi.org/https://doi.org/10.1016/S2222-1808(15)60959-8

Hill, D. E., and Fetterer, R. H. (1997). The effect of disulfiram in egg shell formation in adult Trichuris muris. *The J. Parasitol. 83*(5), 938–942. https://doi.org/10.2307/3284293

Holers, V. M. (2014). Complement and its receptors: New insights into human disease. *Annu. Rev. Immunol.* 32, 433–445. https://doi.org/10.1146/annurev-immunol-032713-120154

Hu, D., Tang, X., ben Mamoun, C., Wang, C., Wang, S., Gu, X., Duan, C., Zhang, S., Suo, J., Deng, M., Yu, Y., Suo, X., and Liu, X. (2020). Efficient single-gene and gene family editing in the apicomplexan parasite eimeria tenella using CRISPR-Cas9. *Front. Bioeng. Biotechnol. 8*. https://doi.org/10.3389/fbioe.2020.00128

Huang, Y. H., Colgrave, M. L., Clark, R. J., Kotze, A. C., and Craik, D. J. (2010). Lysine-scanning mutagenesis reveals an amendable face of the cyclotide kalata B1 for the optimization of nematocidal activity. *J. Biol. Chem. 285*(14), 10797–10805. https://doi.org/10.1074/jbc.M109.089854

Hunt, S., Santos, Y., Seeram, N., Rowley, D., Ma, H., and Espinsosa, A. (2018). Effects of purified plant extracts and homologous commercial derivatives on entamoeba histolytica growth. *FASEB J. 32*(S1), 531.23–531.23. https://doi.org/https://doi.org/10.1096/fasebj.2018.32.1_supplement.531.23

Hussein, E. M., Ahmed, S. A., Mokhtar, A. B., Elzagawy, S. M., Yahi, S. H., Hussein, A. M., and El-Tantawey, F. (2018). Antiprotozoal activity of magnesium oxide (MgO) nanoparticles against Cyclospora cayetanensis oocysts. *Parasitol. Int. 67*(6), 666–674. https://doi.org/https://doi.org/10.1016/j.parint.2018.06.009

Idan, E. M., and Ardalan, N. M. (2020). A multifaceted review journal in the field of pharmacy introducing silver nanoparticles as anti-giardial in experimentally infected mice: therapy versus toxicity. *Sys Rev Pharm. 11*, 2020. https://doi.org/10.31838/srp.2020.4.103

Ihara, S., Miyamoto, Y., Le, C. H. Y., Tran, V. N., Hanson, E. M., Fischer, M., Hanevik, K., and Eckmann, L. (2022). Conserved metabolic enzymes as vaccine antigens for giardiasis. *PLoS Negl Trop Dis. 16*(4), e0010323.

Imboden, M., Riggs, M. W., Schaefer, D. A., Homan, E. J., and Bremel, R. D. (2010). Antibodies fused to innate immune molecules reduce initiation of Cryptosporidium parvum infection in mice. *Antimicrob. Agents Chemother. 54*(4), 1385–1392. https://doi.org/10.1128/AAC.00754-09

Jiménez, B. M., Kranz, P., Lee, C. S., Gero, A. M., and O'Sullivan, W. J. (1989). Inhibition of uridine phosphorylase from Giardia lamblia by pyrimidine analogs. *Biochem. Pharmacol. 38*(21), 3785–3789. https://doi.org/https://doi.org/10.1016/0006-2952(89)90586-8

Jourdan, P. M., Lamberton, P. H. L., Fenwick, A., and Addiss, D. G. (2018). Soil-transmitted helminth infections. *The Lancet. 391*(10117), 252–265. https://doi.org/10.1016/S0140-6736(17)31930-X

Juárez-Saldivar, A., Barbosa-Cabrera, E., Lara-Ramírez, E. E., Paz-González, A. D., Martínez-Vázquez, A. v, Bocanegra-García, V., Palos, I., Campillo, N. E., and Rivera, G. (2021). Virtual Screening of

FDA-Approved Drugs against Triose Phosphate Isomerase from Entamoeba histolytica and Giardia lamblia Identifies Inhibitors of Their Trophozoite Growth Phase. *Int. J. Mol. Sci. 22*(11), 5943. https://doi.org/10.3390/ijms22115943

Kangussu-Marcolino, M. M., Morgado, P., Manna, D., Yee, H., and Singh, U. (2021). Development of a CRISPR/Cas9 system in Entamoeba histolytica: Proof of concept. *Int. J. Parasitol. 51*(2–3), 193–200. https://doi.org/10.1016/j.ijpara.2020.09.005

Kannigadu, C., and D N'Da, D. (2020). Recent advances in the synthesis and development of nitroaromatics as anti-infective drugs. *Curr Pharm Des.* 26(36): 4658–4674. doi:10.2174/1381612826666200331091853

Karademir, U., Ural, K., Aysul, N., Ayan, A., Toplu, S., Ortlek, O., Balıkci, C., Kunyeli, A., and Erdogan, H. (2016). The efficacy of chloroquine treatment against naturally occuring Giardia duodenalis infection in lambs. *Revista MVZ Córdoba. 21*(2). https://doi.org/10.21897/rmvz.599

Keiser, J., and Utzinger, J. (2008). Efficacy of current drugs against soil-transmitted helminth infections: Systematic review and meta-analysis. *JAMA. 299*(16), 1937–1948. https://doi.org/10.1001/jama.299.16.1937

Keystone, J. S., and Murdoch, J. K. (1979). Drugs five years later. *Ann. Intern. Med. 91*(4), 582–586. https://doi.org/10.7326/0003-4819-91-4-582

Krakovka, S., Ribacke, U., Miyamoto, Y., Eckmann, L., and Svärd, S. (2022). Characterization of metronidazole-resistant Giardia intestinalis lines by comparative transcriptomics and proteomics. *Front Microbiol. 13*, 834008. https://doi.org/10.3389/fmicb.2022.834008

Lee, S. H., Lillehoj, H. S., Tuo, W., Murphy, C. A., Hong, Y. H., and Lillehoj, E. P. (2013). Parasiticidal activity of a novel synthetic peptide from the core α-helical region of NK-lysin. *Vet. Parasitol. 197*(1), 113–121. https://doi.org/https://doi.org/10.1016/j.vetpar.2013.04.020

Lei, J., Sun, L., Huang, S., Zhu, C., Li, P., He, J., Mackey, V., Coy, D. H., and He, Q. (2019). The antimicrobial peptides and their potential clinical applications. *Am J Transl Res. 11*(7): 3919–3931.

de Lima Corvino D. F., and Horrall Shawn. (2021). *Ascariasis.* StatPearls. www.ncbi.nlm.nih.gov/books/NBK430796/

Lim-Leroy, A., and Chua, T. H. (2020). Prevalence and risk factors of geohelminthiasis among the rural village children in Kota Marudu, Sabah, Malaysia. *PLoS One. 15*(9 September). https://doi.org/10.1371/journal.pone.0239680

Lin, B.-C., Su, L.-H., Weng, S.-C., Pan, Y.-J., Chan, N.-L., Li, T.-K., Wang, H.-C., and Sun, C.-H. (2013). DNA topoisomerase II is involved in regulation of cyst wall protein genes and differentiation in Giardia lamblia. *PLOS Negl. Trop. Dis. 7*(5), e2218–e2218. https://doi.org/10.1371/journal.pntd.0002218

Lin, Z. Q., Gan, S. W., Tung, S. Y., Ho, C. C., Su, L. H., and Sun, C. H. (2019). Development of CRISPR/Cas9-mediated gene disruption systems in Giardia lamblia. *PLoS One. 14*(3). https://doi.org/10.1371/journal.pone.0213594

Loderstädt, U., and Frickmann, H. (2021). Antimicrobial resistance of the enteric protozoon Giardia duodenalis - A narrative review. In *Eur. J. Immunol.* 11(2), 29–43. https://doi.org/10.1556/1886.2021.00009

López-Velázquez, G., Fernández-Lainez, C., de la Mora-de la Mora, J. I., Caudillo de la Portilla, D., Reynoso-Robles, R., González-Maciel, A., Ridaura, C., García-Torres, I., Gutiérrez-Castrellón, P., Olivos-García, A., Flores-López, L. A., and Enríquez-Flores, S. (2019). On the molecular and cellular effects of omeprazole to further support its effectiveness as an antigiardial drug. *Sci. Rep. 9*(1), 8922. https://doi.org/10.1038/s41598-019-45529-w

Ma, D., Huang, Y., Ma, C., Zhang, L., Wang, J., Wang, D., Li, J., and Dalloul, R. A. (2019). Eimeria tenella: Specific EtAMA1-binding peptides inhibit sporozoite entry into host cells. *Poult. Sci. 98*(10), 4480–4491. https://doi.org/10.3382/ps/pez298

Madbouly Taha, N., Salah A. Yousof, H.-A., El-Sayed, S. H., Younis, A. I., and Ismail Negm, M. S. (2017). Atorvastatin repurposing for the treatment of cryptosporidiosis in experimentally immunosuppressed mice. *Exp. Parasitol. 181*, 57–69. https://doi.org/https://doi.org/10.1016/j.exppara.2017.07.010

Madbouly, N. A., Nashee, H., Elgendy, A. A., Rabee, I., El Amir, A. (2020). Encapsulation of Low Metronidazole Dose in Poly (D,L-lactide-co-glycolide) (PLGA) Nanoparticles Improves Giardia intestinalis Treatment. *Infect Chemother, 52*(4), 550. doi: 10.3947/ic.2020.52.4.550

Mahdavi Abhari, F., Pirestani, M., and Dalimi, A. (2019). Anti-amoebic activity of a cecropin-melittin hybrid peptide (CM11) against trophozoites of Entamoeba histolytica. *Wien. Klin. Wochenschr. 131*(17), 427–434. https://doi.org/10.1007/s00508-019-01540-9

Malekifard, F., Tavassoli, M., and Vaziri, K. (2019). In vitro assessment antiparasitic effect of selenium and copper nanoparticles on giardia deodenalis cyst. *Iran J Parasitol. 15*(3): 411–417. doi:10.18502/ijpa.v15i3.4206

Malekifard, F., Tavassoli, M., and Vaziri, K. (2020). In vitro assessment antiparasitic effect of selenium and copper nanoparticles on giardia deodenalis cyst. *Iran. J. Parasitol. 15*(3), 411–417. https://doi.org/10.18502/ijpa.v15i3.4206

Martínez-González, J. J., Guevara-Flores, A., Álvarez, G., Rendón-Gómez, J. L., and del Arenal, I. P. (2010). In vitro killing action of auranofin on Taenia crassiceps metacestode (cysticerci) and inactivation of thioredoxin–glutathione reductase (TGR). *Parasitol. Res. 107*(1), 227–231. https://doi.org/10.1007/s00436-010-1867-1

Matthews, J. B., Geldhof, P., Tzelos, T., and Claerebout, E. (2016). Progress in the development of subunit vaccines for gastrointestinal nematodes of ruminants. *Parasite Immunol. 38*(12), 744–753. https://doi.org/10.1111/pim.12391

McInally, S. G., Hagen, K. D., Nosala, C., Williams, J., Nguyen, K., Booker, J., Jones, K., and Dawson, S. C. (2019). Robust and stable transcriptional repression in Giardia using CRISPRi. *Mol. Biol. Cell. 30*(1), 119–130. https://doi.org/10.1091/mbc.E18-09-0605

Mehndiratta, M. M., Wadhai, S. A., Tyagi, B. K., Gulati, N. S., and Sinha, M. (2016). Drug repositioning. *Int. J. Epilepsy.* 3(2), 91–94. https://doi.org/10.1016/j.ijep.2016.09.002

Mi-Ichi, F., Ishikawa, T., Tam, V. K., Deloer, S., Hamano, S., Hamada, T., and Yoshida, H. (2019). Characterization of Entamoeba histolytica adenosine 5'-phosphosulfate (APS) kinase: Validation as a target and provision of leads for the development of new drugs against amoebiasis. *PLOS Negl. Trop. Dis. 13*(8). https://doi.org/10.1371/journal.pntd.0007633

Miró, M. v, Luque, S., Cardozo, P., Lloberas, M., Sousa, D. M., Soares, A. M. S., Costa-Junior, L. M., Virkel, G. L., and Lifschitz, A. L. (2020). Plant-derived compounds as a tool for the control of gastrointestinal nematodes: Modulation of abamectin pharmacological action by carvone. *Front. Vet. Sci. 7*, 1028. www.frontiersin.org/article/10.3389/fvets.2020.601750

Moawad, H. S. F., Hegab, M. H. A. E. H., Badawey, M. S. R., Ashoush, S. E., Ibrahim, S. M., and Ali, A. A. E. L. S. (2021). Assessment of chitosan nanoparticles in improving the efficacy of nitazoxanide on cryptosporidiosis in immunosuppressed and immunocompetent murine models. *J. Parasit. Dis.* 45(3), 606–619. https://doi.org/10.1007/s12639-020-01337-y

Moawad, H. S. F., Hegab, M. H. A. E. H., Badawey, M. S. R., Ashoush, S. E., Ibrahim, S. M., and Ali, A. A. E. L. S. (2021). Assessment of chitosan nanoparticles in improving the efficacy of nitazoxanide on cryptosporidiosis in immunosuppressed and immunocompetent murine models. *J. Parasit. Dis., 45*(3), 606–619. https://doi.org/10.1007/s12639-020-01337-y

Mohammed, S. E. A., Kabashi, A. S., Koko, W. S., and Azim, M. K. (2015). Antigiardial activity of glycoproteins and glycopeptides from Ziziphus honey. *Nat. Prod. Res. 29*(22), 2100–2102. https://doi.org/10.1080/14786419.2014.986659

Mohammed Labib El-Ghendy, A., Abdelmoneim Abdelhamid Mo-Hammed, M., Mohammed Ibrahim Ghalabb, M., Omar Abdel Aziz, M., and Mahmoud Ibrahim, S. (2021). Therapeutic effect of chitosan nanoparticles and metronidazole in treatment of experimentally giardiasis infected hamsters. *Iran J Parasitol. 16*(1), 32–42. https://doi.org/10.18502/ijpa.v16i1.5509

Montaño, S., Constantino-Jonapa, L. A., Sixto-López, Y., Hernández-Ramírez, V. I., Hernández-Ceruelos, A., Romero-Quezada, L. C., Ruvalcaba-Ledezma, J. C., Talamás-Rohana, P., and López-Contreras, L. (2020). Vorinostat, a possible alternative to metronidazole for the treatment of amebiasis caused by Entamoeba histolytica. *J. Biomol. Struct. Dyn. 38*(2), 597–603. https://doi.org/10.1080/07391102.2019.1578693

Morales-Luna, L., González-Valdez, A., Sixto-López, Y., Correa-Basurto, J., Hernández-Ochoa, B., Cárdenas-Rodríguez, N., Castillo-Rodríguez, R. A., Ortega-Cuellar, D., Arreguin-Espinosa, R., de la Cruz, V. P., Serrano-Posada, H., Centeno-Leija, S., Rocha-Ramírez, L. M., Sierra-Palacios, E., Montiel-González, A. M., Rufino-González, Y., Marcial-Quino, J., and Gómez-Manzo, S. (2020). Identification of the NADP+ structural binding site and coenzyme effect on the fused G6PD::6PGL protein from giardia lamblia. *Biomolecules. 10*(1). https://doi.org/10.3390/biom10010046

Movahedi, F., Li, L., Gu, W., and Xu, Z. P. (2017). Nanoformulations of albendazole as effective anticancer and antiparasite agents. *Nanomedicine. 12*(20), 2555–2575. https://doi.org/10.2217/nnm-2017-0102

Muñoz, M. de L., Moreno, M. A., Pérez-Garcia, J. N., Tovar, G. R., and Hernandez, V. I. (1991). Possible role of calmodulin in the secretion of Entamoeba histolytica electron-dense granules containing collagenase. *Mol. Microbiol. 5*(7), 1707–1714. https://doi.org/https://doi.org/10.1111/j.1365-2958.1991.tb01919.x

Nagaraja, S., and Ankri, S. (2019). Target identification and intervention strategies against amebiasis. *Drug Resist. Updat. 44*, 1–14. https://doi.org/https://doi.org/10.1016/j.drup.2019.04.003

Naran, K., Nundalall, T., Chetty, S., and Barth, S. (2018). Principles of immunotherapy: Implications for treatment strategies in cancer and infectious diseases. *Front. Microbiol. 9*(December), 1–23. https://doi.org/10.3389/fmicb.2018.03158

Nash, T., and Rice, W. G. (1998). Efficacies of zinc-finger-active drugs against giardia lamblia. *Antimicrob Agents Chemother. 42*(6): 1488–1492. doi:10.1128/AAC.42.6.1488

Neumayr, A., Schunk, M., Theunissen, C., van Esbroeck, M., Mechain, M., Hatz, C., Mørch, K., Soriano Pérez, M. J., Sydow, V., Sothmann, P., Kuenzli, E., Rothe, C., Bottieau, E., and TropNet. (2021). Efficacy and tolerability of quinacrine monotherapy and albendazole plus chloroquine combination therapy in nitroimidazole-refractory giardiasis: A TropNet Study. *Clin. Infect. Dis. 73*(8), 1517–1523. https://doi.org/10.1093/cid/ciab513

das Neves Gustavo M, P. Kagami Luciano, Rodriguez Ricardo P, da Silva Vinícius B., Eifler-Lima Vera L., and Kawano Daniel F. (2017). In silico design of new inhibitors of guanine phosphoribosyltransferase (GPRT) from Giardia lamblia as antiparasitic drug candidates. *Curr. Drug Discov. Technol. 14*(1), 8–24. https://doi.org/10.2174/1570163813666161213125622

Ngobeni, R. (2021). Molecular characterization of e. histolytica strains based on the serine rich entamoeba histolytica protein (srehp) and chitinase genes. https://doi.org/10.21203/rs.3.rs-665084/v1

Ni, Z. X., Cui, J. M., Zhang, N. Z., and Fu, B. Q. (2017). Structural and evolutionary divergence of aquaporins in parasites (review). *Mol Med Rep*. 15(6): 3943–3948. doi:10.3892/mmr.2017.6505

Njoku, M. O., Iloh, K. K., Okike, C. O., Njoku, G. C., and Ojinnaka, N. C. (2022). The prevalence and intensity of intestinal helminths among institutionalized children in three states of South-East Nigeria. *Niger J Clin Pract. 25*(5), 718–724. https://doi.org/10.4103/njcp.njcp_1880_21

Norouzi, R. (2017). A review on most nanoparticles applied against parasitic infections. *J. Biol. Today's World. 6*(10), 196–203. https://doi.org/10.15412/J.JBTW.01061003

Ochoa-Maganda, V. Y., Rangel-Castañeda, I. A., Suárez-Rico, D. O., Cortés-Zárate, R., Hernández-Hernández, J. M., Pérez-Rangel, A., Chiquete-Félix, N., León-Ávila, G., González-Pozos, S., Gaona-Bernal, J., and Castillo-Romero, A. (2020). Antigiardial activity of acetylsalicylic acid is associated with overexpression of HSP70 and membrane transporters. *Pharmaceuticals. 13*(12), 440. https://doi.org/10.3390/ph13120440

Olson, M. E., Hannigan, C. J., Gaviller, P. F., and Fulton, L. A. (2001). Short communications: The use of a Giardia vaccine as an immunotherapeutic agent in dogs. *Can. Vet. J. 42*(11), 865–868.

Oprea, T. I., and Overington, J. P. (2015). Computational and practical aspects of drug repositioning. *Assay Drug Dev Technol. 13*(6), 299–306. https://doi.org/10.1089/adt.2015.29011.tiodrrr

Palomo-Ligas, L., Gutiérrez-Gutiérrez, F., Ochoa-Maganda, V. Y., Cortés-Zárate, R., Charles-Niño, C. L., and Castillo-Romero, A. (2019). Identification of a novel potassium channel (GiK) as a potential drug target in Giardia lamblia: Computational descriptions of binding sites. *PeerJ*. 7(2), e6430. https://doi.org/10.7717/peerj.6430

Palomo-Ligas, L., Estrada-Camacho, J., Garza-Ontiveros, M., Vargas-Villanueva, J. R., Gutiérrez-Gutiérrez, F., Nery-Flores, S. D., Cañas Montoya, J. A., Ascacio-Valdés, J., Campos-Muzquiz, L. G., and Rodriguez-Herrera, R. (2022). Polyphenolic extract from Punica granatum peel causes cytoskeleton-related damage on Giardia lamblia trophozoites in vitro. PeerJ. *10*, e13350. https://doi.org/10.7717/peerj.13350

Paredes, J. L., Sparks, H., White, A. C., Martinez-Traverso, G., Ochoa, T., and Castellanos-González, A. (2017). Killing of cryptosporidium sporozoites by lactoferrin. *Am. J. Trop. Med. Hyg. 97*(3), 774–776. https://doi.org/10.4269/ajtmh.16-0804

Patra, J. K., Das, G., Fraceto, L. F., Campos, E. V. R., Rodriguez-Torres, M. D. P., Acosta-Raghunath, A., and Perumal, E. (2017). Metal oxide nanoparticles as antimicrobial agents: A promise for the future. *Int. J. Antimicrob. Agents. 49*(2), 137–152. https://doi.org/https://doi.org/10.1016/j.ijantimicag.2016.11.011

Patra, J. K., Das, G., Fraceto, L. F., Campos, E. V. R., Rodriguez-Torres, M. D. P., Acosta-Torres, L. S., Diaz-Torres, L. A., Grillo, R., Swamy, M. K., Sharma, S., Habtemariam, S., and Shin, H. S. (2018). Nano based drug delivery systems: Recent developments and future. *J. Nanobiotechnology*. 16(1). https://doi.org/10.1186/s12951-018-0392-8

Pérez-Villanueva, J., Romo-Mancillas, A., Hernández-Campos, A., Yépez-Mulia, L., Hernández-Luis, F., and Castillo, R. (2011). Antiprotozoal activity of proton-pump inhibitors. *Bioorganic Med. Chem. Lett. 21*(24), 7351–7354. https://doi.org/https://doi.org/10.1016/j.bmcl.2011.10.028

Pintong, A., Ruangsittichai, J., Ampawong, S., Thima, K., Sriwichai, P., Komalamisra, N., and Popruk, S. (2020). Efficacy of Ageratum conyzoides extracts against Giardia duodenalis trophozoites: an experimental study. *BMC Complement. Med. Ther. 20*(1), 63. https://doi.org/10.1186/s12906-020-2860-6

Pinzi, L., and Rastelli, G. (2019). Molecular docking: Shifting paradigms in drug discovery. *Int. J. Mol. Sci. 20*(18). https://doi.org/10.3390/ijms20184331

Piotrowicz, Z., Tabisz, Ł., Waligórska, M., Pankiewicz, R., and Łęska, B. (2021). Phenol-rich alternatives for Rosa x damascena Mill. Efficient phytochemical profiling using different extraction methods and colorimetric assays. *Sci Rep 11*. 23883 https://doi.org/10.1038/s41598-021-03337-1

Plancarte, A., and Nava, G. (2015). Purification and kinetic analysis of cytosolic and mitochondrial thioredoxin glutathione reductase extracted from Taenia solium cysticerci. *Exp. Parasitol.*, *149*, 65–73. https://doi.org/https://doi.org/10.1016/j.exppara.2014.12.009

Pogonka, T., Schelzke, K., Stange, J., Papadakis, K., Steinfelder, S., Liesenfeld, O., and Lucius, R. (2010). CD8+ cells protect mice against reinfection with the intestinal parasite Eimeria falciformis. *Microbes Infect. 12*(3), 218–226. https://doi.org/10.1016/j.micinf.2009.12.005

Popovic, O., Jojic, N., Milutinovic-Djuric, S., Djurdjevic, D., and Milovic, V. Propranolol for metronidazole-resistant giardiasis. *Ital J Gastroenterol. 23*(6), 359.

Popović OS, Milović V. Propranolol and metronidazole for the treatment of metronidazole-resistant giardiasis. *J Clin Gastroenterol. 12*(5), 604–605.

Prasad, R., Pandey, R., Barman, I. (2016). Engineering tailored nanoparticles with microbes: quo vadis. *WIREs Nanomed Nanobiotechnol.* 8, 316–330. DOI: 10.1002/wnan.1363

Prasad, R., Pandey, R., Varma, A., Barman, I. (2017). Polymer based nanoparticles for drug delivery systems and cancer therapeutics. In: H., Kharkwal, and S., Janaswamy, eds, *Natural Polymers for Drug Delivery*. CAB International, UK. 53–70.

Qaid, M. M., Al-Mufarrej, S. I., Azzam, M. M., Al-Garadi, M. A., Alqhtani, A. H., H Fazea, E., Suliman, G. M., and Alhidary, I. A. (2021). Effect of rumex nervosus leaf powder on the breast meat quality, carcass traits, and performance indices of eimeria tenella oocyst-infected broiler chickens. *Animals (Basel). 11*(6):1551. doi:10.3390/ani11061551.

Quintanilla-Licea, R., Mata-Cárdenas, B. D., Vargas-Villarreal, J., Bazaldúa-Rodríguez, A. F., Ángeles-Hernández, I. K., Garza-González, J. N., and Hernández-García, M. E. (2014). Antiprotozoal activity against entamoeba histolytica of plants used in northeast mexican traditional medicine. Bioactive compounds from lippia graveolens and ruta chalepensis. *Molecules. 19*(12), 21044–21065. https://doi.org/10.3390/molecules191221044

Raj, S., Sasidharan, S., Balaji, S. N., and Saudagar, P. (2020). An overview of biochemically characterized drug targets in metabolic pathways of Leishmania parasite. *Parasitol. Res. 119*(7), 2025–2037. https://doi.org/10.1007/s00436-020-06736-x

Rangel-Castañeda, I. A., Hernández-Hernández, J. M., Pérez-Rangel, A., González-Pozos, S., Carranza-Rosales, P., Charles-Niño, C. L., Tapia-Pastrana, G., Ramírez-Herrera, M. A., and Castillo-Romero, A. (2018). Amoebicidal activity of curcumin on Entamoeba histolytica trophozoites. *J. Pharm. Pharmacol. 70*(3), 426–433. https://doi.org/https://doi.org/10.1111/jphp.12867

Reaume, C., Moore, B., Hernández, P., Ruzzini, A., Chlebus, M., Wasserman, M., and Yee, J. (2013). Evaluation of drugs and stationary growth on the cell cycle of Giardia intestinalis. *Mol. Biochem. Parasitol. 187*(1), 72–76. https://doi.org/https://doi.org/10.1016/j.molbiopara.2012.11.005

Ren, P., Yang, X., Wang, T., Hou, Y., and Zhang, Z. (2022). Proteome-wide prediction and analysis of the Cryptosporidium parvum protein–protein interaction network through integrative methods. *Comput Structl Biotechnol J. 20*, 2322–2331. https://doi.org/https://doi.org/10.1016/j.csbj.2022.05.017

Reyes-Vivas, H., de la Mora-de la Mora, I., Castillo-Villanueva, A., Yépez-Mulia, L., Hernández-Alcántara, G., Figueroa-Salazar, R., García-Torres, I., Gómez-Manzo, S., Méndez, S. T., Vanoye-Carlo, A., Marcial-Quino, J., Torres-Arroyo, A., Oria-Hernández, J., Gutiérrez-Castrellón, P., Enríquez-Flores, S., and López-Velázquez, G. (2014). Giardial triosephosphate isomerase as possible target of the cytotoxic effect of omeprazole in Giardia lamblia. *Antimicrob. Agents Chemother. 58*(12), 7072–7082. https://doi.org/10.1128/AAC.02900-14

Riggs, M. W., Schaefer, D. A., Kapil, S. J., Barley-Maloney, L., and Perryman, L. E. (2002). Efficacy of monoclonal antibodies against defined antigens for passive immunotherapy of chronic gastrointestinal cryptosporidiosis. *Antimicrob. Agents Chemother. 46*(2), 275–282. https://doi.org/10.1128/AAC.46.2.275-282.2002

Roy, U., Panwar, A., Pandit, A., Das, S. K., and Joshi, B. (2016). Clinical and neuroradiological spectrum of metronidazole induced encephalopathy: Our experience and the review of literature. *J. Clin. Diagnostic Res.* 10(6), OE01–OE09. https://doi.org/10.7860/JCDR/2016/19032.8054

Saad, H. A., Soliman, M. I., Azzam, A. M., and Mostafa, B. (2015). Antiparasitic activity of silver and copper oxide nanoparticles against *Enatamoeba hystolitica* and *Cryptosporidium parvum. J. Egypt. Soc. Parasitol. 45*(3), 593–602. https://doi.org/10.12816/0017920

Sahoo, N., Labruyère, E., Bhattacharya, S., Sen, P., Guillén, N., and Bhattacharya, A. (2004). Calcium binding protein 1 of the protozoan parasite Entamoeba histolytica interacts with actin and is involved in cytoskeleton dynamics. *J. Cell Sci. 117*(16), 3625–3634. https://doi.org/10.1242/jcs.01198

Said, D. E., ElSamad, L. M., and Gohar, Y. M. (2012). Validity of silver, chitosan, and curcumin nanoparticles as anti-Giardia agents. *Parasitol. Res. 111*(2), 545–554. https://doi.org/10.1007/s00436-012-2866-1

Salgado-Martínez, A. I., Avila-Bonilla, R. G., Ramírez-Moreno, E., Castañón-Sánchez, C. A., López-Camarillo, C., and Marchat, L. A. (2021). Unraveling the relevance of the polyadenylation factor EhCFIm25 in Entamoeba histolytica through proteomic analysis. *FEBS Open Bio. 11*(10):2819–2835. doi:10.1002/2211-5463.13287

Sandholt, A. K. S., Xu, F., Söderlund, R., Lundén, A., Troell, K., Svärd, S. G., and Wattrang, E. (2021). Dual RNA-Seq transcriptome analysis of chicken macrophage-like cells (HD11) infected in vitro with Eimeria tenella. *Parasitology. 148*(6):712–725. doi:10.1017/S0031182021000111

Sandhu, H., Mahajan, R. C., and Ganguly, N. K. (2004). Flowcytometric assessment of the effect of drugs on Giardia lamblia trophozoites in vitro. *Mol. Cell. Biochem. 265*(1), 151–160. https://doi.org/10.1023/B:MCBI.0000044392.01704.5f

Santiago, Á., Razo-Hernández, R. S., and Pastor, N. (2020). The TATA-binding Protein DNA-binding domain of eukaryotic parasites is a potentially druggable target. *Chem. Biol. Drug Des. 95*(1), 130–149. https://doi.org/10.1111/cbdd.13630

Sarma, H., Forid, N., Prasad, R., Prasad, M. N. V., Ma., L. Q., Rinklebe, J. (2021). Enhancing phytoremediation of hazardous metal(loid)s using genome engineering CRISPR–Cas9 technology. *Journal of Hazardous Materials*. 414, 125493. https://doi.org/10.1016/j.jhazmat.2021.125493

Sarwono, A. E. Y., Mitsuhashi, S., Kabir, M. H. bin, Shigetomi, K., Okada, T., Ohsaka, F., Otsuguro, S., Maenaka, K., Igarashi, M., Kato, K., and Ubukata, M. (2019). Repurposing existing drugs: identification of irreversible IMPDH inhibitors by high-throughput screening. *J. Enzyme Inhib. Med. Chem. 34*(1), 171–178. https://doi.org/10.1080/14756366.2018.1540474

Serafin, M. B., and Hörner, R. (2018). Drug repositioning, a new alternative in infectious diseases. *Brazilian J. Infect. Dis.* 22(3), 252–256. https://doi.org/10.1016/j.bjid.2018.05.007

Serradell, M. C., Saura, A., Rupil, L. L., Gargantini, P. R., Faya, M. I., Furlan, P. J., and Lujan, H. D. (2016). Vaccination of domestic animals with a novel oral vaccine prevents Giardia infections, alleviates signs of giardiasis and reduces transmission to humans. *NPJ Vaccines*. 1: 16018. doi:10.1038/npjvaccines.2016.18

Shimasaki, C. (2020). Chapter 10 – Understanding Biotechnology Product Sectors. In: C. Shimasaki, ed., *Biotechnology Entrepreneurship (Second Edition)* (pp. 123–149). Academic Press. https://doi.org/https://doi.org/10.1016/B978-0-12-815585-1.00010-3

Shirley, D. A., Sharma, I., Warren, C. A., and Moonah, S. (2021). Drug repurposing of the alcohol abuse medication disulfiram as an anti-parasitic agent. *Front Cell Infect Microbiol. 11*: 633194. doi:10.3389/fcimb.2021.633194

Song, X., Li, Y., Chen, S., Jia, R., Huang, Y., Zou, Y., Li, L., Zhao, X., and Yin, Z. (2020). Anticoccidial effect of herbal powder "shi ying zi" in chickens infected with eimeria tenella. *Animals. 10*(9), 1484. https://doi.org/10.3390/ani10091484

Song, Y. Y., Zhang, Y., Yang, D., Ren, H. N., Sun, G. G., Jiang, P., Liu, R. D., Zhang, X., Cui, J., and Wang, Z. Q. (2018). The immune protection induced by a serine protease inhibitor from the foodborne parasite trichinella spiralis. *Front. Microbiol. 9*(July), 1544. https://doi.org/10.3389/fmicb.2018.01544

Stein, J. E., Radecki, S. V., and Lappin, M. R. (2003). Efficacy of Giardia vaccination in the treatment of giardiasis in cats. *J. Am. Vet. Med. Assoc. 222*(11), 1548–1551. https://doi.org/10.2460/javma.2003.222.1548

Stevanovic, S., Marjanović, D. S., Trailović, S. M., Zdravković, N., Perdih, A., and Nikolic, K. (2021). Potential modulating effect of the Ascaris suum nicotinic acetylcholine receptor (nAChR) by compounds GSK575594A, diazepam and flumazenil discovered by structure-based virtual screening approach. *Mol Biochem Parasitol. 242*: 111350. doi:10.1016/j.molbiopara.2021.111350

Sun, Y., Chen, D., Pan, Y., Qu, W., Hao, H., Wang, X., Liu, Z., and Xie, S. (2019). Nanoparticles for antiparasitic drug delivery. *Drug Deliv*. 26(1): 1206–1221. doi:10.1080/10717544.2019.1692968

Teichmann, K., Kuliberda, M., Schatzmayr, G., Pacher, T., Zitterl-Eglseer, K., Joachim, A., and Hadacek, F. (2016). In vitro inhibitory effects of plant-derived by-products against Cryptosporidium parvum. *Parasite (Paris, France)*. *23*, 41. https://doi.org/10.1051/parasite/2016050

Teixeira, J. E., and Huston, C. D. (2008). Participation of the serine-rich Entamoeba histolytica protein in amebic phagocytosis of apoptotic host cells. *Infect. Immun. 76*(3), 959–966. https://doi.org/10.1128/IAI.01455-07

Tejman-Yarden, N., Miyamoto, Y., Leitsch, D., Santini, J., Debnath, A., Gut, J., McKerrow, J. H., Reed, S. L., and Eckmann, L. (2013). A reprofiled drug, auranofin, is effective against metronidazole-resistant Giardia lamblia. *Antimicrob. Agents Chemother. 57*(5), 2029–2035. https://doi.org/10.1128/AAC.01675-12

Tessema, T. S., Dauber, E., and Petry, F. (2009). Adoptive transfer of protective immunity from Cryptosporidium parvum-infected interferon-γ and interleukin-12-deficient mice to naive recipients. *Vaccine. 27*(47), 6575–6581. https://doi.org/10.1016/j.vaccine.2009.08.036

Thakur R. K., and Patel S. P. (2021). *Mebendazole*. StatPearls Publishing. www.ncbi.nlm.nih.gov/books/NBK557705/

Trejo-Soto, P. J., Aguayo-Ortiz, R., Yépez-Mulia, L., Hernández-Campos, A., Medina-Franco, J. L., and Castillo, R. (2016). Insights into the structure and inhibition of Giardia intestinalis arginine deiminase: Homology modeling, docking, and molecular dynamics studies. *J. Biomol. Struct. Dyn. 34*(4), 732–748. https://doi.org/10.1080/07391102.2015.1051115

Tuvshintulga, B., Sivakumar, T., Yokoyama, N., and Igarashi, I. (2019). Development of unstable resistance to diminazene aceturate in Babesia bovis. *Int. J. Parasitol: Drug. 9*, 87–92. https://doi.org/https://doi.org/10.1016/j.ijpddr.2019.02.001

Tyagi, R., Elfawal, M. A., Wildman, S. A., Helander, J., Bulman, C. A., Sakanari, J., Rosa, B. A., Brindley, P. J., Janetka, J. W., Aroian, R. V, and Mitreva, M. (2019). Identification of small molecule enzyme inhibitors as broad-spectrum anthelmintics. *Sci. Rep. 9*(1), 9085. https://doi.org/10.1038/s41598-019-45548-7

Valenzuela-Salas, L. M., Blanco-Salazar, A., Perrusquía-Hernández, J. D., Nequiz-Avendaño, M., Mier-Maldonado, P. A., Ruiz-Ruiz, B., Campos-Gallegos, V., Arellano-García, M. E., García-Ramos, J. C., Pestryakov, A., Villarreal-Gómez, L. J., Toledano-Magaña, Y., and Bogdanchikova, N. (2021). New protein-coated silver nanoparticles: Characterization, antitumor and amoebicidal activity, antiproliferative selectivity, genotoxicity, and biocompatibility evaluation. *Pharmaceutics*. 13(1), 1–17. https://doi.org/10.3390/pharmaceutics13010065

Verdugo-González, L., Acosta-Cárdenas, A., Zapata-Zapata, A. D., Cadena-Chamorro, E. M., and Pulido-Muñoz, S. A. (2019). Antimicrobial potential of a hydrolyzed protein extract of the microalgae nannochloropsis sp. *DYNA (Colomb.)*. 86(211), 192–198. https://doi.org/10.15446/dyna.v86n211.78865

Vique-Sánchez, J. L., Jiménez-Pineda, A., and Benítez-Cardoza, C. G. (2021). Amoebicidal effect of 5,5′-[(4-nitrophenyl)methylene]bis-6-hydroxy-2-mercapto-3-methyl-4(3H)-pyrimidinone), a new drug against Entamoeba histolytica. *Arch. Pharm. 354*(2). https://doi.org/10.1002/ardp.202000263.e2000263.

Virlan, M. J. R., Miricescu, D., Radulescu, R., Sabliov, C. M., Totan, A., Calenic, B., and Greabu, M. (2016). Organic nanomaterials and their applications in the treatment of oral diseases. *Molecules. 21*(2). https://doi.org/10.3390/molecules21020207

Viswanath, A., Yarrarapu Siva, N. S., and Williams, M. (2021, August). *Trichuris Trichiura*. StatPearls publishing. www.ncbi.nlm.nih.gov/books/NBK507843/

Vivancos, V., González-Alvarez, I., Bermejo, M., and Gonzalez-Alvarez, M. (2018). Giardiasis: Characteristics, pathogenesis and new insights about treatment. *Curr. Top. Med. Chem. 18*(15), 1287–1303. https://doi.org/10.2174/1568026618666181002095314

Wang, J., Cardoso, R., Marreros, N., Müller, N., Lundström-stadelmann, B., Siffert, M., Vuitton, D. A., Boué, F., Lin, R., and Wen, H. Foxp3+ T regulatory cells as a potential target for immunotherapy against primary infection with echinococcus multilocularis eggs. *Infect Immun. 86*(10): e00542–18.

Weeks, J. C., Roberts, W. M., Leasure, C., Suzuki, B. M., Robinson, K. J., Currey, H., Wangchuk, P., Eichenberger, R. M., Saxton, A. D., Bird, T. D., Kraemer, B. C., Loukas, A., Hawdon, J. M., Caffrey, C. R., and Liachko, N. F. (2018). Sertraline, paroxetine, and chlorpromazine are rapidly acting anthelmintic drugs capable of clinical repurposing. *Sci. Rep.*, *8*(1), 975. https://doi.org/10.1038/s41598-017-18457-w

Williams, A. R., Soelberg, J., and Jäger, A. K. (2016). Anthelmintic properties of traditional African and Caribbean medicinal plants: Identification of extracts with potent activity against Ascaris suum in vitro. *Parasite (Paris, France)*. *23*, 24. https://doi.org/10.1051/parasite/2016024

World Health Organization. (2020). *The top 10 causes of death.* www.who.int/news-room/fact-sheets/detail/the-top-10-causes-of-death

Wu, J.-H., Tung, S.-Y., Ho, C.-C., Su, L.-H., Gan, S.-W., Liao, J.-Y., Cho, C.-C., Lin, B.-C., Chiu, P.-W., Pan, Y.-J., Kao, Y.-Y., Liu, Y.-C., and Sun, C.-H. (2021). A myeloid leukemia factor homolog involved in encystation-induced protein metabolism in Giardia lamblia. *Biochim Biophys Acta Gen Subj. 1865*(6), 129859. https://doi.org/https://doi.org/10.1016/j.bbagen.2021.129859

Yarahmadi, M., Fakhar, M., Ebrahimzadeh, M. A., Chabra, A., and Rahimi-Esboei, B. (2016). The anti-giardial effectiveness of fungal and commercial chitosan against Giardia intestinalis cysts in vitro. *J. Parasit. Dis. 40*(1), 75–80. https://doi.org/10.1007/s12639-014-0449-z

Yason, J. A., Koh, K. A. R. P., and Tan, K. S. W. (2018). Viability screen of LOPAC (1280) reveals phosphorylation inhibitor auranofin as a potent inhibitor of blastocystis subtype 1, 4, and 7 Isolates. *Antimicrob. Agents Chemother. 62*(8), e00208–18. https://doi.org/10.1128/AAC.00208-18

Yong, T., Chen, M., Li, Y., Song, X., Huang, Y., Chen, Y., Jia, R., Zou, Y., Li, L., Yin, L., He, C., Lv, C., Liang, X., Ye, G., and Yin, Z. (2020). Anticoccidial effect of Fructus Meliae toosendan extract against Eimeria tenella. *Pharm. Biol. 58*(1), 636–645. https://doi.org/10.1080/13880209.2020.1784234

Younis, M. S., Abououf, E. A. E. R., Ali, A. E. S., Abd Elhady, S. M., and Wassef, R. M. (2020). In vitro effect of silver nanoparticles on blastocystis hominis. *Int. J. Nanomedicine. 15*, 8167–8173. https://doi.org/10.2147/IJN.S272532

Yousofi Darani, H., Yousefi, M., Safari, M., and Jafari, R. (2016). Parasites and immunotherapy: with or against? *J. Parasit. Dis. 40*(2), 217–226. https://doi.org/10.1007/s12639-014-0533-4

Zea, A. H., Rodriguez, P. C., Culotta, K. S., Hernandez, C. P., DeSalvo, J., Ochoa, J. B., Park, H.-J., Zabaleta, J., and Ochoa, A. C. (2004). l-Arginine modulates CD3ζ expression and T cell function in activated human T lymphocytes. *Cell. Immunol. 232*(1), 21–31. https://doi.org/https://doi.org/10.1016/j.cellimm.2005.01.004

Zhang, X. F., Liu, Z. G., Shen, W., and Gurunathan, S. (2016). Silver nanoparticles: Synthesis, characterization, properties, applications, and therapeutic approaches. *Int. J. Mol. Sci.* 17(9). https://doi.org/10.3390/ijms17091534

Zielińska, A., Alves, H., Marques, V., Durazzo, A., Lucarini, M., Alves, T. F., Morsink, M., Willemen, N., Eder, P., Chaud, M. v., Severino, P., Santini, A., and Souto, E. B. (2020). Properties, extraction methods, and delivery systems for curcumin as a natural source of beneficial health effects. *Medicina. 56*(7):336. https://doi.org/10.3390/medicina56070336

Index

H

I

J

K

L

M

N

V

W

X

Z